# Philosophy of Engineering and Technology

Volume 48

The *Philosophy of Engineering and Technology* book series provides the multifaceted and rapidly growing discipline of the philosophy of technology with a central overarching and integrative platform. It publishes on all topics in the philosophy of technology and is open to all research communities across the world.

Specifcally, it publishes edited volumes and monographs in:

- the phenomenology, anthropology and socio-politics of technology and engineering
- the emergent felds of the ontology and epistemology of technology and design
- engineering ethics and the ethics of specifc technologies ranging from nuclear technologies to artifcial intelligence
- written from philosophical and practitioners' perspectives and authored by philosophers and practitioners

The series also welcomes proposals that bring these felds together or advance philosophy of engineering and technology in other integrative ways.
Proposals should include:

- A short synopsis of the work or the introduction chapter
- The proposed Table of Contents
- The CV of the lead author(s) or editor(s)
- If available: one sample chapter

We aim to make a frst decision within 1 month of submission. In case of a positive frst decision the work will be provisionally contracted; the fnal decision about publication will depend upon the result of anonymous peer review of the complete manuscript. We aim to have the complete work peer-reviewed within 3 months of submission.

The series discourages the submission of manuscripts that contain reprints of previous published material and/or manuscripts that are below 150 pages/75,000 words

For inquiries and submission of proposals authors can contact the editor-in-chief Pieter Vermaas via: p.e.vermaas@tudelft.nl, or contact one of the associate editors.

Christelle Didier • Aurélien Béranger
Antoine Bouzin • Hugo Paris • Jérémie Supiot
Editors

# Engineering and Value Change

Springer

*Editors*
Christelle Didier
CIREL
University of Lille
Lille, France

Antoine Bouzin
Centre Émile Durkheim
Bordeaux, France

Jérémie Supiot
Catholic University of Lyon
Lyon, France

LUMSA
Roma, Italia

Aurélien Béranger
Costech
Université de Technologie de Compiègne
Compiègne, France

Hugo Paris
ADEF
Aix-Marseille University
Marseille, France

INSA Lyon
Villeurbanne, France

ISSN 1879-7202 ISSN 1879-7210 (electronic)
Philosophy of Engineering and Technology
ISBN 978-3-031-83548-3 ISBN 978-3-031-83549-0 (eBook)
https://doi.org/10.1007/978-3-031-83549-0



This Springer imprint is published by the registered company Springer Nature Switzerland AG
The registered company address is: Gewerbestrasse 11, 6330 Cham, Switzerland

# Contents

# Chapter 1
# Values Change, So Does Philosophy of Technology and Engineering

**Christelle Didier, Aurélien Béranger, Antoine Bouzin, Hugo Paris, and Jérémie Supiot**

**Abstract** The aim of this first chapter is to introduce the current Philosophy of Engineering and Technology volume *Engineering and Changing Values*, and explain its origin, originality, and purpose. While growing out of a meeting of the forum on Philosophy, Engineering, and Technology (fPET) held in Delft in 2023, its final form is the fruit of a singular editorial, selection and support effort. This introductory chapter aims to highlight its singularity, which is based as much on the theme addressed as on the geographical, cultural, and disciplinary diversities that characterize the group of contributors, and the editorial team, which is in the image of the forum from which it comes. This chapter also describes the general philosophy of the book and the four thematic sections that make it up: ethics, justice, epistemology, and theoretical approaches to value change in design. This introduction

C. Didier (✉)
CIREL, University of Lille, Lille, France
e-mail: christelle.didier@univ-lille.fr

A. Béranger
Costech, Université de Technologie de Compiègne, Compiègne, France
e-mail: aurelien.beranger@utc.fr

A. Bouzin
Centre Émile Durkheim, Bordeaux, France
e-mail: antoine.bouzin@u-bordeaux.fr

H. Paris
ADEF, Aix-Marseille University, Marseille, France
e-mail: hugo.paris@insa-lyon.fr

INSA Lyon, Villeurbanne, France

J. Supiot
Catholic University of Lyon, Lyon, France

LUMSA, Roma, Italia
e-mail: jeremie.supiot@philosophie.ucly.fr



C. Didier et al. (eds.), *Engineering and Value Change*, Philosophy of Engineering and Technology 48, https://doi.org/10.1007/978-3-031-83549-0_1

also provides an opportunity to thank the many people who helped make this project a reality.

**Keywords** fPET · Value change · Engineering · Technology · EPSI

## 1.1 Introduction

The chapters making up this book are based on selected presentations at the forum on Philosophy, Engineering, and Technology (fPET) held at Delft University of Technology in April 2023. Organized every 2 years since 2010, fPET aims at bringing together engineers and philosophers concerned with the challenges of engineering in a world undergoing major transformations due to climate change and pandemics (to name but a few), but also transformed by technologies themselves. The fPET conferences are an opportunity for scholars and practitioners from a wide range of disciplinary, cultural and geographical backgrounds to meet, present and discuss technology and engineering research at the interface of philosophy and engineering, in a spirit of openness towards various approaches. The 2023 forum, themed "Value Change and Engineering", explored the interrelationship between technology and the evolution of values in society.

The question of the normative dimension of technologies and their links with the transformations of society, although still of obvious topicality, leads us back to fundamental problems in the philosophy of technology. A historical perspective on this question allows us to grasp the interrelationship between technology and the evolution of values in society. For an extended period, considerations of value were absent from technical endeavors, particularly those of engineers, as the responsibility for the consequences of technology was attributed solely to users. Without going into the genesis of this idea in detail, it is noteworthy that in the *Gorgias*, Plato posited that it is the utilization of techniques, rather than the techniques themselves, that should be held accountable. In his view, techniques were regarded as neutral.

Closer to our times, workers whose living conditions were threatened by industrialization may have found themselves breaking machines they saw as destroying their social world. The Luddite movement reached its peak in England in the early nineteenth century (Hobsbawm, 1952). But then again, for prominent socialist thinkers of the era, such as Karl Marx (1818–1883) and Pierre-Joseph Proudhon (1809–1865), the issue was not with the technologies themselves. Technologies were no longer regarded as mere neutral instruments; rather, they were viewed as "intrinsically positive" (Gordon, 2009), providing freedom from the necessities of nature (Proudhon, 1972), with the real issue lying in the capitalist organization of production relations.

In contrast to this optimistic outlook, which can be traced back to the Enlightenment, the "classical" philosophies of technology emerged in the mid-twentieth century. Darkly marked by the two World Wars, these approaches were resolutely critical of the ideal of social progress that was to flow from technological development. These approaches share the common feature of amalgamating all modern technologies into an abstract whole, which are perceived as an external, autonomous sphere that poses a significant risk of superseding the social world (Brey, 2010). The postulate of the neutrality of techniques is radically challenged by the heterogeneous authors associated with these approaches, including Martin Heidegger (1889–1976), Herbert Marcuse (1898–1979), Jacques Ellul (1912–1994), Ivan Illich (1926–2002), Manuel Ortega y Gasset (1883–1965), Hans Jonas (1903–1993), and Lewis Mumford (1895–1990), to name but a few. Despite the diversity of their perspectives, a common thread runs through their critiques of technology. This thread is characterized by two key elements: an essentialist approach, which views technology as an inherently problematic entity, and a deterministic perspective, which posits that a given object will inevitably produce a specific set of effects. This determinism is particularly evident in the classes of techniques that these authors develop. Some techniques are inherently "authoritarian" or "democratic" (Mumford, 1964), while others are "convivial" (Illich, 1973). Each category assumes the fixed inscription of values in their intrinsic materiality.

The pioneering work of Langdon Winner (1944) in the 1980s marks a major shift in thinking about determinations between the technical and the social. In direct opposition to Ellul's all-encompassing approach (Winner, 1977), he refocuses analysis at the level of singular technical devices. He suggests that their politics be subjected to direct scrutiny, specifically the "arrangements of power and authority" that they conceal (Winner, 1980). Winner posits that the politics of artifacts can be analyzed in two ways. Firstly, certain technologies are contingent upon, or are inherently compatible with, specific modes of social and political organization. Rather than reiterating the deterministic argument in the manner of Mumford, he suggests a certain openness in the way a device will reconfigure social relations (Winner, 1986). The second method for examining the political characteristics of artifacts is to examine how technologies reinforce or alter the existing power dynamics within a society. This approach focuses on social and political matters at the level of design, thereby shifting the question of responsibility to those who conceive and implement the artifacts.

The 1970s and 1980s witnessed a proliferation of theoretical approaches to the study of technological issues. In the wake of Science and Technology Studies, philosophy turned to an approach that was more constructivist, empirical (Kroes & Meijers, 2000; Achterhuis, 2001), and oriented towards the analysis of technologies themselves and engineering activity (Mitcham, 1994). This abundant theoretical renewal, part of a broader context of politicization of technologies in society (Jarrige, 2014), resituates part of the responsibility in the stage of elaboration and design of technologies.

With the recognition that technology is not value-neutral, the challenge becomes to prevent the negative social consequences of technology. This resulted in the establishment of governmental agencies in numerous Western countries whose objective was to gain a deeper understanding of the emerging technological risks and to provide more effective guidance to inform public action. The field of Technology Assessments (TA) emerged in this context (Grunwald, 2009). Initially, these assessments were conducted far from the very places where technologies were conceived and designed. Additionally, the initial TA activities, frequently assigned to experts, were characterized by a techno-scientific approach. The objective was to facilitate the development of technologies and their social acceptability, rather than to direct their trajectory. In a second phase of development, new approaches to technology assessment emerged, including Constructive TA, Deliberative TA, Political TA, and Real-Time TA. These approaches sought to integrate evaluation into the early stages of technical development (Schuurbiers et al., 2013). The contemporary objective of TA can be summarized as follows: to support political decision-making with the aim to developing technologies that best meet society's needs, by means of informing both institutions and citizens.

In the field of engineering, various approaches and methods have been explored to investigate the value issues that are intrinsic to the design process, particularly at the nascent stage of technology development.

As a variation of the multiple Value Sensitive Design (VSD) methods developed by Bayta Friedman (Friedman et al., 2002), the "design for values" perspective has been developed at Delft University of Technology to integrate value considerations into the fundamental stages of technology development (van den Hoven et al., 2015). Nevertheless, this approach still lacks consideration of a crucial aspect:

> Different design approaches, tools and guidelines have been developed to take values systematically into account during the development of new technologies. But here is a fundamental challenge for any such approach: although values are relatively stable, they do sometimes change over time. What types of value change exist, and how can engineers and designers deal with them? (van der Weij et al., 2023a)

The 2023 fPET meeting wished to address this matter of the changing nature of values in society in relation to technologies, and how it affects engineering. This book provides a rich overview of this crucial subject.

## 1.2 A Few Words About the Contributors

Several publication opportunities were offered to presenters following the Forum. A book on the theme of pedagogical issues in engineering education, based on the German concept of *Bildung*, was already in preparation and announced as a volume in the POET (*Philosophy of Engineering and Technology*) series, with our late colleague and friend Steen Hyldgaard Christensen (1950–2024). The paper presented by Steen and Anders Buchs, co-editor of the forthcoming volume, naturally found its place in it. The research work developed at the University of Delft on the notion of "Value Change" led to the publication of a white paper before fPET and a synthesis afterward, also a special issue (van der Weij et al., 2023a, 2023b, Melnyk et al., 2024). An invitation went out to all who to contribute to the volume, and 40 participants expressed interest in doing that. The volume resulting from the selection of the abstracts by our editorial team, followed by a double-blind evaluation and the revision of most chapters comprises 19 contributions that the editorial team chose to divide into four thematic sections.

A third of the contributors are women, which is a decent number the low feminization of engineering on the one hand, and the long exclusion of women from the field of philosophy on the other. Some 15 countries are represented, reflecting the diversity of the fPET community. Half of the contributors' places of initial training, doctoral and postdoctoral research or current professional practice are in Europe (France, Netherlands, Italy, Germany, Belgium, Ireland, Bulgaria, and Poland), and half outside (USA, Brazil, Australia, Mexico, Argentina, South Africa). Contributors include young researchers (doctoral and postdoctoral students) as well as more advanced ones, teachers, assistants, professors and even a professor emerita in the person of Diane Michelfelder, also co-founder of fPET. There are as many contributors who began their higher education in engineering and the experimental sciences (physics, biology) or (industrial) design as there are contributors whose primary training was in philosophy. Other fields of initial academic training are varied: English studies, music, art history, architecture, psychology, sociology, education sciences, law, economics, public administration, public policy… Many authors also have dual disciplinary backgrounds.

The editorial team is made up of four French PhD students, one of whom has become a doctor since the Forum, and a senior researcher, also French. The team reflects the disciplinary diversity of the volume's contributors. Thus, Aurélien Béranger and Antoine Bouzin are engineering graduates, but now respectively PhD students in Information and Communication Sciences and Sociology. Hugo Paris, an architect and engineering graduate, is a recent doctor in Education and Training Sciences (Paris, 2024). Jérémie Supiot is a PhD student in philosophy specializing in epistemology, who has made technologies his object of study. Finally, Christelle Didier, a sociologist by training (after receiving an undergraduate degree in engineering) and a specialist in engineering ethics, is now teaching in a department of educational sciences at the University of Lille. The members of the editorial team are the founding members of a very active research network set up in 2020 aiming at developing a multidisciplinary approach on engineering in the French-speaking world (Algeria, Belgium, Morocco, Quebec, Switzerland, Tunisia). This network is called EPSI for (in french) *Études Pluridisciplinaires Sur l'Ingénierie.*

This distribution of contributors and editorial team nicely reflects the evolution of the fPET community since it began in 2007 as a Workshop (WPE) in Delft. The challenge of this Workshop was to bring together engineers and philosophers in a single conference. Over the years, the core of the workshop, now a forum, made up of engineers, philosophers, and engineer-philosophers, has opened up to other disciplinary approaches. The biographies of the contributors to this volume clearly demonstrate this richness.

## 1.3 Structure of the Book

### *1.3.1 Ethical Issues, Value Change, and Engineering*

The first part of the volume concerns the field of engineering ethics, and shows how difficult it is to find a common ground for engineering ethics: is it an external framework from which engineers should decide? A set of shared values? The search for universal principles? Should engineers decide what is valuable and what is not, or should they be looking for what is valued by their fellow citizens and institutions? Is being (explicitly or implicitly) valued by someone equivalent to being valuable to someone? How do engineers connect ethics, laws, and design? Beyond the meta-ethical questions concerning the status to be given to the values and principles that guide our actions (does moral knowledge exist? If so, does it concern universal principles, or is it always situated in a context? Are values in the world or in the mind?) we can at least observe a concern for ethical issues in engineering. This concern appears in many ways and at different scales (individuals, disciplines, institutions…), so that the terms "value" and "ethics" cover a variety of meanings in both philosophy and engineering, as can be seen in the chapters gathered in this initial section.

In Chap. 2, Diane P. Michelfelder and Sharon A. Jones study the ethical issues of engineering in the context of value changes, through the analysis of the ethical principles that underpin the rhetoric of professional engineering organizations in the United States. They suggest that commitments to sustainability could be strengthened by integrating the ethos of care into engineering codes of ethic. Erhardt Graef in Chap. 3 questions the limits of such codes given the culture of disengagement of engineers in the USA, and suggests to educate engineers to Civic Professionalism. Here, when it comes to ethics, the object is more the values shared within a specific profession, probably closer to something like an ethos, in the sense of Aristotle's virtue ethics. It is questioned not in terms of ethical principles that should apply to engineering, but from an educational point of view: what values should we convey through the education of future engineers? In the fourth chapter, Sarah Carter and Heike Felzmann explore how people value data privacy through their behaviors and discuss the implications of their findings for designing privacy technologies and for regulators. Here the approach is not normative but descriptive: Carter and Felzman do not care about what is valuable but actually valued by people. Antoine Bouzin in Chap. 5 explains how engineers themselves, particularly environmental engineers, question the values generally attributed to technologies (efficiency, reliability, utility) and try to find compromises (Boltanski & Thévenot, 2006) between industry expectations and the ecological values they defend. Here again we change scale, looking specifically at engineers concerned about the environmental impact of the technologies they develop, to understand how this translates into concrete negotiations to ensure that their values are heard.

Depending on the authors, the term value and the concern for ethics refer to their political and social dimensions (Bouzin, Chap. 5), the normative dimension of ethics (Michelfelder and Jones, Chap. 2), education (Graef, Chap. 3), behavior and design (Carter and Felzman, Chap. 4). It varies so much that research in engineering ethics needs to spend a great deal of time defining its objects, bringing many disciplines into dialogue (some of which are not even represented here, like psychology or anthropology).

### *1.3.2 Questions of Justice in Engineering and Technology*

The following chapters are dedicated to an in-depth examination of a specific—and highly polymorphic—moral value: justice. The chapters gathered together here address a number of common issues, including the social effects of technological development, the diversity of data-related social practices (collection, processing, management, use), environmental problems, and questions of ethics and governance. The proximity of the research topics studied in this section gives rise to research that is both distinctive and complementary in nature. The conceptual repertoires do not entirely overlap between the different chapters. In order to address the complex issue of justice, they call upon the notions of "algorithmic fairness" (Richter, Chap. 6), "ethical, legal, and socio-economic (ELSE) challenges" (Bogucki, Chap. 7), "distributive justice and fairness" (Kreiken and McCarthy, Chap. 8), "global justice" (Bögner, Chap. 9), "energy justice" (Van Uffelen, Chap. 10), and "epistemic injustice" (Quaresmini et al., Chap. 11). This section also highlights the diversity of scientific disciplines and theoretical backgrounds used by the various authors, who draw on a spectrum that includes philosophy, history, law, economics, ethics, and administrative sciences. Secondly, the diversity of the analyzed fields offers a truly comparative overview of several areas of activity, including mobility, biodiversity, agriculture, and mining.

In Chap. 6, Florian Richter analyzes the justice issues involved in the development of smart cities, and in particular the various conceptions of justice that are mobilized in discussions of what is referred to in the literature as "smart cities", "smart government", and "cities 5.0". He defends the idea that making these conceptions more explicit would make it possible to evaluate the technologies implemented in context, and to better decide which conception is the most appropriate for each case. Finally, he sketches the condition for a Model of "Technologically Mediated Justice". In Chap. 7, Arthur Bogucki analyzes in the ELSE-related literature (both academic and more practical works) the arguments and trends about data governance in the field of precision agriculture, a field which, like smart cities, is very dependent on AI and data storage. He then assesses the European Commission's Ethics Guidelines for Trustworthy AI on the basis of several concrete case studies in three countries using the ALTAI (Assessment List for Trustworthy AI) Framework. In Chap. 8, we carry on in the field of agricultural and biological engineering. Bob Kreiken and Adam McCarthy examine the negotiations of the United Nations Convention on Biological Diversity (CBD), and in particular the issue of the digitization of genetic resources. The CBD provides a site for better understanding the relationships between science and technology policies, innovation, values, and distributive justice. The authors highlight conflicts between the utilitarian aims of open science and claims to sovereignty over digital sequence information. In Chap. 9, Frieder Bögner remains on an international scale and deals with the subject of rare-earth extraction in the context of ecological transition. In particular, the author analyzes the techno-solutionist tendencies observed in the collective decision-making process and its tendency to dismiss non- or low-technological options. Also interested in the subject of transition, Nynke van Uffelen examines the role of normative assumptions in technological development in Chap. 10. Anchored in a field linked to energy storage, the author highlights the notion of "normative uncertainty" and points to the European-centrism of the dominant conceptions of justice. Finally, in Chap. 11, Camilla Quaresmini, Valentina Breschi, Samira Maghool, Viola Schiaffonati & Mara Tanelli study the phenomenon of the acceptance of sustainable mobility technologies and the spread of new habits. In particular, the authors stress the importance of credibility, mutual influences, homophily, and epistemic justice in the process of adopting new behaviors.

### *1.3.3 Epistemic Values*

The examination of evolving values is not solely a concern within the domain of ethics. It is also a matter of epistemology. Firstly, epistemology is concerned with the understanding of what values are and they evolve. Secondly, the knowledge that individuals possess depends on epistemic values that can also undergo change. Usually, philosophers explore those questions separately: the first is about values, the other is about facts. However, this distinction is not always clear-cut in engineering, where concepts like *transparency* can simultaneously be regarded as both an epistemic value and an ethical one. To understand engineering value change, it is essential to examine how epistemology can contribute to engineering ethics.

In Chap. 12, Jérémie Supiot, Philippe Estival, and Mathieu Guillermin take precisely this point of departure as a foundation for their analysis, and discuss the modes of moral evaluation of values that engineers could or should implement, highlighting the importance of public debate. In Chap. 13, Vivek Kant and Michael Poznic focus on the epistemic activities of engineers, drawing attention to an underestimated facet based on a discussion of the literature in philosophy of technology. They present understanding as a type of knowledge constituted by the integration of existing knowledge to form new general knowledge. Thus, it is by having to deal with complex systems, involving diverse scientific and technical knowledge of which they cannot master each and every subtlety, that engineers would develop the global understanding they need. The case study presented by Hans Voordijk, Faridaddin Vahdatikhaki, and Sergei Miller in Chap. 14 provides an illustrative example of a more specific technological mediation. They reflect on the reasons why roller operators for road pavement do not always utilize intelligent systems during asphalt compaction work, employing the concept of technological and composite intentionality. This chapter discusses how it would be possible to better articulate the benefits of these intelligent systems with the operators' experience and tacit knowledge. In Chap. 15, Hugo Paris, Nicolas Freud, and Caroline Ladage examine how some scientists and engineers develop a technocratic understanding of how to address contemporary socio-ecological issues, drawing upon Michel Foucault's works on neoliberalism and pastoral power. This leads them to put forth a perspectivist epistemological framework as a way to better grasp the diversity of knowledge. The potential benefits for engineering students of a better understanding of the plurality of narratives on ecological transition are illustrated by an example from a course. This reflection resonates with the ideas presented in Chap. 12, suggesting ways in which engineers can more effectively contribute to public debate by adopting a more comprehensive perspective.

In Chap. 16, David Spurett studies the ambiguous role that technologies may play by manipulating their users and lead them to adopt specific values and behaviors. In Chap. 17, Alexandra Karakas suggests another epistemic contribution of engineers through a reflection on the maintenance of scientific instruments. Through the example of the Hubble Space Telescope, she shows how the anticipated functioning of an instrument is more contentious than it may appear, and brings into tension diverse visions and values. In the following section, devoted to theoretical approaches to value change in design, Chaps. 18 and 20 build on this reflection by questioning the very notion of function.

### *1.3.4 Theoretical Approaches to Value Change in Design*

The fourth and last part of this book puts forth a series of theoretical elaborations with the aim of shedding new light on the challenges of value change in design. The three contributions included in this section adopt a philosophical-analytical approach, taking an artifact-based perspective to discuss its meaning, evolving functions, and transformation processes. While contributing to theoretical discussions in engineering and design philosophy, they also provide practitioners with useful tools for situational analysis.

In Chap. 18, Clint Hurshman examines the relationship between the concepts of function in artifacts and in biology. He suggests a shift in focus from ontological considerations to the development of a "pragmatic ontology", in which the distinction between the artificial and the natural is based not on the intrinsic nature of the entities in question, but on the intentions involved in the process of assigning functions to them. The role of design in this conceptual framework sheds new light on the intricacies of changing values in artifacts. In Chap. 19, Alok Srivastava and Gunter Bombaerts put forth a proposal to develop a framework and analytical tools to comprehend the evolution of values across the life cycle of a technological project or object. This approach shifts the conventional retrospective perspective on such matters. This theoretical contribution analyze the processual dimension of value change in technology. In Chap. 20, Ryan Mitchell Wittingslow proposes an analogy between Austinian language theory and the technical functions of objects. This analogy clarifies how the technical object makes sense in a given context, particularly from the point of view of values, and the impact on its use. This promising application of speech act theory allows us to identify discrepancies between the functions initially conceived by designers and the actual performance of the artifact in use. These developments present novel perspectives on the engineer's responsibility and the affordances of objects for their users.

Last but not least, the conclusive chapter reviews the different contributions of the book on value change and highlights some of the remaining questions. These questions open up promising new perspectives for philosophy of engineering and technology. In particular, there seems to be a need to forge links with other approaches, such as environmental philosophy to help us clarify the notion of value in the context of the anthropocene.

**Acknowledgments** We, co-editors of this volume, are particularly grateful to the organizing committee of the 2023 fPET meeting and to the fPET steering committee for the trust placed in a young team to produce this volume. We are also grateful to the contributors who were willing to take part in this adventure for the quality of their proposals, the quality of their exchanges, the respect of deadlines, and the rewriting work carried out, thanks to the many reviewers asked to accompany us. We would also like to thank our colleagues and friends, including many members of our EPSI network: Catherine Adam, Arnaud Billon, Sébastien Chailleux, Nathan Coutable, Hadrien Coutant, Fabrizio Delage, Neelke Doorn, Marie-Pierre Escudié, Yann Ferguson, Vincent Gerbaud, Sophie Gerber, Sarah Ghaffari, Xavier Guchet, Manon Guillet, Sacha Hodencq, Pablo Jensen, Sylvain Lavelle, Elise Jacquier Lefevre, Fabrizio Li Vigni, Diana Adela Martin, Diane Michelfelder (also for her help all along the process), Natasha McCarty, Thierry Ménissier, Stéphanie Merle, Glenn Miller, Johan Milleret, Alexandre Monnin, Jullie Nolland, Patrick Obertelli, Antoine Peltier, Laetitia Piet, David Pontille, Adrien Poujon, Pascal Ragouet, Zoë Robaey, Catherine Roby, Nino Rodrigues, Camille Roelens, Alberto Romele, Emmanuel Rozières, Romain Sauzet, Ibo van de Poel, Pieter Vermaas, and Dominique Vinck.

We would like to end this introductory chapter with a word of tribute to our colleague and friend Steen Hylgaard Christensen, who passed away in June 2024. Steen was an important person for the academic world in many ways:

- For the community of researchers interested in engineering culture and education, from an international and interdisciplinary perspective, through his original work and, above all, his ability to involve a wide range of researchers in publishing projects that were always ambitious (ten volumes since 2007, with three different publishers).
- For the POET collection, for which he has been editor-in-chief of several high-quality volumes over the years. His latest volume, co-directed by his faithful friend and colleague Anders Buch, on the concept of *Bildung* in relation to engineering education, was under review when he passed away. Anders saw it through to publication.
- For the fPET community, to whom he was close: he had taken part in several conferences, most notably in 2023, where he presented a paper with Anders Buch on a subject close to his heart: the role of the educational philosophy of Bildung in engineering education.
- For our young editorial team. Thanks to Christelle, who has worked extensively with Steen (she has contributed to most of the volumes he published and co-edited four of them), our small team has been able to rely on an efficient and structured working method.

## References


Achterhuis, H. (Ed.). (2001). *American philosophy of technology. The empirical turn*. Indiana University Press.

Boltanski, L., & Thévenot, L. (2006). *On justification: Economies of worth, translated by Catherine Porte*. Princeton University Press.

Brey, P. (2010). Philosophy of technology after the empirical turn. *Techné: Research in Philosophy and Technology, 14*(1), 36–48. https://doi.org/10.5840/techne20101416

Christensen, S. H., & Buch, A. (2025). *Framing conversations between engineering, social sciences, and the humanities through the concept of Bildung*. Springer (POET).

Friedman, B., Kahn, P. H., & Borning, A. (2002). Value sensitive design: Theory and methods, University of Washington, Technical report.

Gordon, U. (2009). Anarchism and the politics of technology, In: Working USA, 12(3), 489–503. https://doi.org/10.1111/j.1743-4580.2009.00250.x.

Grunwald, A. (2009). Technology assessment: Concepts and methods. In A. Meijers (Ed.), *Handbook of the philosophy of science* (Philosophy of technology and engineering sciences) (Vol. 9, pp. 1103–1146). Elsevier.

Hobsbawm, E. (1952). The machine-breakers, *Past & Present, 1*, 57–70.

Illich, I. (1973). *Tools for conviviality*. Harper & Row.

Jarrige, F. (2014). *Technocritiques: du refus des machines à la contestation des technosciences*. La Découverte.

Kroes, P., & Meijers, A. (Eds.). (2000). *The empirical turn in the philosophy of technology*. JAI.

Melnyk, A., Edmonds, B., Ghorbani, A., & van de Poel, I. (2024). Editorial: Modelling values in social, technical, and ecological systems. *Journal of Artificial Societies and Social Simulation, 27*(1). 10.18564/jasss.5361

Mitcham, C. (1994). *Thinking through technology: The path between engineering and philosophy*. University of Chicago Press.

Mumford, L. (1964). Authoritarian and democratic technics. *Technology and Culture, 5*(1), 1–8. https://doi.org/10.2307/3101118

Paris, H. (2024). *Comment accompagner les enseignants d'école d'ingénieur pour former leurs étudiants à affronter les enjeux écologiques ?* Doctoral Dissertation in educational sciences, Aix Marseille University.

Proudhon, P. J. (1972). *The philosophy of poverty*. Arno Press.

Schuurbiers, D., N. Doorn, N., van de Poel, I. E. Gorman, M. (2013). Mandates and methods for early engagement. In Doorn et al., Early engagement and new technologies: opening up the laboratory, Philosophy of Engineering and Technology, vol 16. Springer, Dordrecht, 3–14. https://doi.org/10.1007/978-94-007-7844-3_1

van den Hoven, J., Vermaas, P. E., & van de Poel, I. (2015). Handbook of Ethics, Values, and Technological Design: Sources, Theory, Values and Application Domains. Springer Reference. Dordrecht Heidelberg Heidelberg New York London: Springer. https://doi.org/10.1007/978-94-007-6970-0

van der Weij, F., Steinert, S., van de Poel, I., Alleblas, J., Malnyk, A., de Wildt, T. (2023a). *Value change and technological design*, white paper.

van der Weij, F., Steinert, S., van de Poel, I., Alleblas, J., Malnyk, A., & de Wildt, T. (2023b). Value change and technological design. *IEEE Technology and Society Magazine, 42*(3), 25–32.

Winner, L. (1977). *Autonomous technology: Technics-out-of-control as a theme in political thought*. MIT Press.

Winner, L. (1980). Do artifact have politics? *Daedalus, 109*(1), 121–136.

Winner, L. (1986). *The whale and the reactor: A search for limits in an age of high technology*. University of Chicago Press.

**Christelle Didier** BS Electrochemistry Eng., MA in Education, PhD in Sociology. 1993–2023 Assistant Professor in Engineering Ethics, Lille Catholic University, France. Today, Education Department, Lille University. Author of Ethique industrielle (1998); Penser l'éthique des ingénieurs (2008), Les ingénieurs et l'éthique (2008); several POET Volumes with Steen Hyldgaard Christensen (2015, 2015, 2019, 2022). Co-founder of collectif EPSI (*Études Pluridisciplinaires sur l'Ingénierie*). Research areas: Engineering cultures, identities and ethics, including historical and comparative perspectives.

**Aurélien Béranger** BA Humanities and technology, Computer Science Eng., PhD student in Communication sciences and Philosophy of technology at the Costech Lab of the UTC (Université de Technologie de Compiègne). The focus of his thesis is on the low-tech movement in France. Member of Collectif EPSI (*Études Pluridisciplinaires sur l'Ingénierie*—Multi-disciplinary engineering studies). His main research interests include low-tech, technology-oriented social movements, appropriate technologies, engineering, open source, makers, and degrowth studies.

**Antoine Bouzin** Hydraulics and fluid mechanics Eng., Master of Sociology, PhD Student in Sociology and Political Science (Centre Émile Durkheim, France). Co-founder of Collectif EPSI (*Études Pluridisciplinaires sur l'Ingénierie*—Multi-disciplinary engineering studies) and Collectif Biomasse & SHS (Biomass studies from human and social sciences). Research areas: Engineering studies, social movements, public action and policies, environment, expertise.

**Hugo Paris** MA Architecture (ENSA Lyon, 2017) & MSc Civil Engineering and Urbanism (INSA Lyon, 2019), PhD Educational sciences (Aix-Marseille University, 2024). Co-founder of collectif EPSI (*Études Pluridisciplinaires sur l'Ingénierie*) and Associate Editor for the European Journal of Engineering Education. 2024–2025: Post Doc researcher at INSA Lyon, UNESCO Chair: Training engineers for the socio-ecological transition. Research areas: Superior education for sustainability, teacher continuous training, curriculum studies, anthropological approach to didactics (TAD).

**Jérémie Supiot** PhD student in philosophy at the CONFLUENCE Sciences and Humanites research unit (EA 1598) of UCLy (Lyon Catholic University) and LUMSA University at Rome. His thesis focuses on ethical empowerment at the interface between society and the university. It is supervised by Mathieu Guillermin (UCLy, Lyon) and Stefano Biancu (LUMSA, ROME) and is part of the Activity 4 of the NHNAI project. His main research topics are epistemology and engineering ethics. He co-organizes the ETIUS seminars at the Université Catholique de Lyon. Co-founder of collectif EPSI (*Études Pluridisciplinaires sur l'Ingénierie*), which organizes the Cultures d'Ingénieurs et Formation seminars.

# Part I
# Ethics

# Chapter 2
# Aligning the Ethics of Care with Commitments to Sustainability in US Professional Engineering Codes

**Diane P. Michelfelder and Sharon A. Jones**

**Abstract** In recent years, interest in re-envisioning engineering activities through the lens of an ethics of care has grown. At the same time, several US engineering societies revised the canons for their codes of ethics in ways that raise their commitments to sustainability. The intent of this chapter is to bring these parallel trajectories together and demonstrate the promise their intersection holds for enhancing the codes' commitments to sustainability even further. Following a brief introduction, we look at the progress made with respect to these two trajectories before turning to the advantages of infusing engineering codes of ethics with the ethics of care. With this in mind, we offer some concrete suggestions for how the National Society of Professional Engineers (NSPE) code might be revised. We then acknowledge and respond to possible objections before concluding with some directions for future research.

**Keywords** Engineering codes of ethics · Sustainability · Ethics of care · Intragenerational justice · Intergenerational justice

D. P. Michelfelder (✉)
Macalester College, St. Paul, MN, USA
e-mail: michelfelder@macalester.edu

S. A. Jones
University of Washington Bothell, Bothell, WA, USA
e-mail: sjones23@uw.edu



C. Didier et al. (eds.), *Engineering and Value Change*, Philosophy of Engineering and Technology 48, https://doi.org/10.1007/978-3-031-83549-0_2

## 2.1 Introduction

How might the ethical codes adopted by professional engineering societies in the United States benefit from being integrated with and grounded on an ethics of care? In taking steps toward answering this question, in particular by appealing to work in feminist care ethics by Carol Gilligan (1982), Nel Noddings (1984, 2002), Joan Tronto (1993), and Daniel Engster (2007), Elisa Warford (2018) suggested that incorporating care into engineering codes of ethics would emphasize the importance of being aware of the context within which engineering decisions are made.

Since the publication of Warford's paper, interest in re-envisioning engineering by looking at it through the lens of the ethics of care continues to trend upwards (see for instance Van Grunsven et al., 2023, Baas et al., 2022; Fore et al., 2020; Nair & Bulleit, 2020; Russell & Vinsel, 2019). In parallel, across the past 5 years, several US engineering societies revised the canons for their codes of ethics (ASME, 2021; ASCE, 2020; IEEE, 2020), raising their commitments to sustainability. In this chapter we bring these separate trajectories together to explore the promise for their intersection. Specifically, we expand on Warford (2018) to imagine how enhancing engineering codes of ethics from a care ethics perspective might elevate their commitment to sustainability, both with respect to codes that have already made positive shifts in this direction and to ones that haven't.

Following this introduction, the next two sections provide relevant background. First, we look at the evolution of US professional engineering societies with regards to their commitments to sustainability. Second, we offer an overview of scholarly work connecting the ethics of care to engineering activities. Third, we highlight several ways in which sustainability could be advanced by integrating engineering codes of ethics with the ethics of care. Fourth, we give concrete suggestions for how the language of the canons of the National Society of Professional Engineers (NSPE) code could be changed in accordance with what we have been saying. Fifth, we acknowledge and respond to possible objections to our proposed merger. We conclude the chapter by identifying directions for future research.

## 2.2 Sustainability and US Engineering Codes of Ethics

In reviewing the evolution of the Institute of Electrical and Electronics Engineers (IEEE) code of ethics, Tang and Nieusma (2017) suggested that professional engineers must continue to question the boundaries of their responsibilities as they improve what are accepted as ethical standards. An engineer's responsibility for sustainability, a macro ethics issue, is one of these boundaries that continues to be questioned. Herkert has defined sustainable development as "*achieving objectives in three realms: ecological (sustainable scale), economic (efficient allocation), and social (just distribution)*" (Herkert, 1998, p. 333). Although primarily focused on the environmental aspects of sustainability, Vesilind (2002) argued that the engineer's responsibility to the public also includes future generations (intergenerational equity). In 2013, we proposed, somewhat controversially, that engineering to ensure the public's safety, health, and welfare (for all subpopulations) cannot be achieved without engineering for sustainability (Michelfelder & Jones, 2013). In addition, we argued that sustainability requires achieving a fair and equitable distribution of environmental, economic, and social outcomes for both current and future generations (Michelfelder & Jones, 2013). We later suggested that sustainability remains an elusive target and rests on "*creating a culture of sustainability*" (p. 1) which requires elevating sustainability within professional engineering codes (Jones et al., 2017).

In this section, we describe how the codes of ethics adopted by US engineering societies have evolved since 2013 with regards to elevating sustainability. We also describe how sustainability definitions compare across the most recent engineering societies in terms of both inter- and intragenerational equity and the full spectrum of environmental, economic, and social outcomes. We also note the placement of sustainability in the codes and how it is defined by each engineering society. While we do not address how sustainability should be prioritized within the codes, this is an area for future research.

We focused on engineering societies as opposed to engineering licensing boards given the regulatory responsibilities that boards have (Burgess, 2019). Professional societies exist to advance a community of professionals in similar disciplines. For example, the NSPE states that their purpose is to "*1) To do everything that can be done to advance the interests of the engineering profession; and 2) To act whenever the interests of the engineering profession are jeopardized*" (Steinman, 1935, p. 6). As with most engineering societies, it provides members with continuing education courses, networking opportunities, job boards, government advocacy, professional news, recognition opportunities, and so on. Besides engineering societies, there are 60 engineering and surveying licensing boards across the US that require a combination of education, experience, and examination to designate an individual as a licensed professional engineer (NCEES, 2024). While licensure is an important mechanism used by states to safeguard the health, safety, and welfare of the public, only a small fraction of practicing engineers is licensed: typically, in areas where failure of the engineered system results in larger-scale impacts beyond an individual e.g., infrastructure such as bridges, waterways, etc. as opposed to a personal vehicle or an electronic device.

### 2.2.1 Current State of Sustainability Within US Engineering Codes of Ethics

Table 2.1 shows the relevant excerpts from several engineering codes of ethics that directly address sustainability. The first two entries are the professional codes of ethics adopted by the World Federation of Engineering Organizations (WFEO) and

**Table 2.1** Latest versions of several engineering codes of ethics

| Code | Last revision | How sustainability is addressed (code excerpts) |
|---|---|---|
| World Federation of Engineering Organizations (WFEO) | 2001 | *Principles*<br>Create and implement engineering solutions for a sustainable future. Be mindful of the economic, societal, and environmental consequences of actions or projects |
| National Society of Professional Engineers (NSPE) | 2019 | *Professional obligations*<br>Engineers are encouraged to adhere to the principles of sustainable development in order to protect the environment for future generations |
| Institute of Electrical and Electronics Engineers (IEEE) | 2020 | *Principles*<br>1. To hold paramount, the safety, health, and welfare of the public, to strive to comply with ethical design and sustainable development practices |
| American Society of Mechanical Engineers (ASME) | 2021 | *Fundamental canons*<br>Engineers shall consider environmental impact and sustainable development in the performance of their professional duties |
| American Society of Civil Engineers (ASCE) | 2020 | *Preamble*<br>Create safe, resilient, and sustainable infrastructure<br>*Stakeholders*<br>2. Natural and built environment:<br>a. Adhere to the principles of sustainable development<br>b. Consider and balance societal, environmental, and economic impacts, along with opportunities for improvement, in their work<br>c. Mitigate adverse societal, environmental, and economic effects<br>d. Use resources wisely while minimizing resource depletion |
| American Institute of Chemical Engineers (AIChE) | 2015 | N/A |
| American Academy of Environmental Engineers & Scientists (AAEES) | 2021 | *Canon #2*<br>They will support and contribute to sustainable development practices in the built environment, considering and balancing societal, environmental, and economic impacts locally and globally |

the NSPE that apply to all engineers (Davis, 2021). The remaining entries are organizational codes of ethics that apply to members of those societies arranged by membership totals (Davis, 2021). We intentionally omitted references to the environment (built and/or natural) and focused solely on how sustainability is captured in these codes. We did this not to minimize environmental responsibility, but to emphasize that sustainable design is now commonly understood as encompassing much more than environmental impact. Except for the WFEO, this is a US-centric analysis, and we are cognizant that engineering codes differ across countries in terms of micro versus macro ethical issues with the US preferencing a micro ethical focus (Polmear et al., 2019).

While we won't address all code changes, the American Society of Civil Engineers (ASCE) provides an example of significant change that is relevant for this discussion. In 2017 the ASCE code of ethics' fundamental canon no. 1 stated that "*engineers shall hold paramount the safety, health and welfare of the public and shall strive to comply with the principles of sustainable development in the performance of their professional duties*." For some time, ASCE was the only US professional society that included sustainable development (or sustainability) in the paramountcy clause. With the revised 2020 code, ASCE abandoned using canons and moved to a values-based framework where the engineer's obligations are organized as a hierarchy of stakeholders so that the engineer is clear on which stakeholder takes priority (ASCE, 2020).

The 2020 code begins with a preamble that highlights sustainability as follows, "*Engineers govern their professional careers on the following fundamental principles: create safe, resilient, and sustainable infrastructure; …*" The natural and built environment is second in the hierarchy of stakeholders, and behind society, though ahead of several other stakeholders. This stand-alone section stipulates that engineers must, among other responsibilities, "*a) adhere to the principles of sustainable development; b) consider and balance societal, environmental, and economic impacts, along with opportunities for improvement, in their work; c) mitigate adverse societal, environmental, and economic effects; and d) use resources wisely while minimizing resource depletion*."

The 2020 ASCE code elevated a responsibility to create "*sustainable infrastructure*" and shifted the language from "*strive to comply*" to "*adhere to*." This treatment of sustainability now aligns with the 2001 WFEO code; however, by placing sustainability within the stakeholder category for natural and built environment, the ASCE continues to separate and treat sustainability as a lower priority than society.

In 2023, most of the other US engineering societies now commit to a macro ethics responsibility for sustainability in their codes, however not to the degree of the ASCE and the WFEO. NSPE *encourages* engineers to adhere, IEEE says they *should strive* to comply, ASME asks engineers to *consider*, and AAEES wants engineers to *support and contribute*. A noticeable exception is the American Institute of Chemical Engineers (AIChE, 2015) which still does not refer to sustainability (or sustainable development) within its code. The terms chosen by these other societies indicate that while a responsibility for sustainability is important, it is a lower priority than an engineer's other responsibilities.

### 2.2.2 How the Professional Societies Define Sustainability and/or Sustainable Development

The most widely known definition of sustainable development comes from the Brundtland Report (World Commission on Environment and Development, 1987) that states, "*Sustainable development is development that meets the needs of the present without compromising the ability of future generations to meet their own needs*." While *Our Common Future* was written as part of a call to action to prevent environmental harms at the global scale, the evolution of sustainable development is illustrated best by the 17 interconnected sustainable development goals (SDGs) that range from the elimination of poverty and hunger to issues such as inequality, health and well-being, infrastructure, communities, peace, etc. (United Nations, 2016).

Given the broad and interconnected nature of the SDGs, we examined how professional societies define sustainability as part of codes, policies, or other official statements. Our findings (Table 2.2) illustrate a scarcity of guidance to engineers about what it means to engineer sustainably, along with an emphasis on environmental and natural resources aspects of sustainability, and little clarity on intragenerational equity.

The ASCE code captures all three elements of sustainability which again aligns with the WFEO code. The American Academy for Environmental Engineers and Scientists (AAEES) delineates that sustainable development practices include attention to societal, environmental, and economic impacts. NSPE's code includes a footnote that defines sustainable development in terms of protecting the environment and natural resources, while the American Society of Mechanical Engineers (ASME) and the IEEE do not delineate what sustainable development refers to; the ASME's phrasing suggests that they also limit sustainable development to the environment.

While the ASCE code implies intergenerational equity, it does not directly address intragenerational equity (the distribution of benefits and harms) as part of the obligations for engineers. Intragenerational equity also remains absent from the latest code revisions for all engineering societies, while the NSPE's code is the only one that specifically refers to future generations (intergenerational equity).

NSPE's position statement (05-100) highlights that sustainable communities are "*critical to public health, safety, and welfare*" which is the closest that any of the organizations comes to inferring a cause and effect between sustainability and society.

Several engineering societies also address sustainability in separate policy/position statements and other forms of professional development.

**Table 2.2** Engineering society sustainability definitions

| Organization | Date | Definition |
|---|---|---|
| NSPE | 2019 | *Footnote*: Sustainable development is the challenge of meeting human needs for natural resources, industrial products, energy, food, transportation, shelter, and effective waste management while conserving and protecting environmental quality and the natural resource base essential for future development. (National Society of Professional Engineers, 2019) |
| NSPE | 2023 | *Position Statement No. 05-100:* Because sustainable communities—communities that meet the needs of the present without compromising the ability of future generations to meet their own needs—are critical to public health, safety and welfare, it is the position of NSPE that licensed professional engineers must be engaged in creating, maintaining and renewing sustainable communities. (National Society of Professional Engineers, 2023). |
| ASCE | 2023 | *Policy Statement 418*: The American Society of Civil Engineers (ASCE) defines sustainability as a set of environmental, social, and economic conditions (aka "The Triple Bottom Line") in which all of society has the capacity and opportunity to maintain and improve its quality of life indefinitely without degrading the quantity, quality, or the availability of environmental, social, and economic resources... Sustainable development is the application of these resources to enhance the safety, welfare, and quality of life for all of society. (American Society of Civil Engineers, 2023) |
| AIChE | Undated | *AIChE Institute for Sustainability*: Sustainable development describes those efforts to guide economic growth, especially in less-developed countries, in an environmentally sound manner, with an emphasis on natural resource conservation. (American Institute of Chemical Engineers, n.d.) |
| AIChE | 2008 | *Sustainability Index:* strategic commitment, sustainability innovation, environmental performance, safety performance, product stewardship, social responsibility, and value-chain management (American Institute of Chemical Engineers, 2008) |
| AAEES | Undated | *Sustainability Specialty:* Environmental Sustainability specialty covers the range of topics about which a specialist in environmental sustainability would be knowledgeable including general sustainability issues. Sustainability science focuses on the understanding of the science underlying basic concepts of environmental sustainability, including some economic and social aspects of sustainability (American Academy of Environmental Engineers and Scientists, n.d.) |

## 2.3 The Ethics of Care and Engineering Codes of Ethics

Eliza Warford (2018)'s work incorporating the ethics of care into US engineering codes of ethics is just one example, as mentioned in Sect. 2.1, of how interest among engineers in incorporating the ethics of care into different engineering activities has been trending upwards in recent years. We open this section with a brief description of the theoretical features of the ethics of care in general, and follow with an overview of the trend just mentioned before looking specifically at Warford's proposal.

### 2.3.1 Key Theoretical Dimensions of the Ethics of Care

Someone hearing about the ethics of care for the first time might be most struck not by what it involves but by what is conspicuously missing from it. Unlike two of the three major ethical theories in the Western normative ethics tradition, consequentialism and Kantian ethics (deontologism), decision-making under the ethics of care is not characterized by a universal, rational, principle-based procedure. The starting point for deliberation under the ethics of care, by contrast, is the specific lived context within which the moral agent is located. Such an agent would begin their deliberations by taking stock of the moral dilemma at hand and think about what they might do to address the needs of a particular other that come to the fore so that others might flourish. Because of the emphasis on context, every moral dilemma "must be approached and addressed on its own merits and not as an instantiation of a particular moral 'problem'" (Michelfelder & Jones, 2016, p. 178). And, while reason is certainly an important element in the tool-box of a care-centered moral agent, it is not the only one: empathy, compassion, concern for the vulnerability of others, and trust can also have an important role to play.

This points to another significant difference between "conventional" moral theory and the ethics of care. While the former sees the moral agent as a stand-alone, autonomous individual, care ethics reflects a relational ontology. It is "propelled by the fundamental assumption that human beings are first and foremost social creatures who, as interconnected, relational beings, are 'primed' to pay attention to the needs of others and be responsive to them" (Michelfelder, forthcoming). Aristotelian virtue ethics, the other main ethical theory within the Western tradition, diverges from the ethics of care in that it stresses individuals rather than the relations between them: it is an individual person who is the bearer of virtues (Held, 2006, p. 52). It is though similar to the ethics of care in that ethical activity needs to be sustained, improved, and cultivated over time.

### 2.3.2 The Ethics of Care, Engineering Education, and Engineering Practice

Credit goes to Pantazidou and Nair (1999) for being the first to call attention to the positive role that the ethics of care, in particular as articulated by Joan Tronto (1993), could play in engineering education and practice. From their perspective, a "*natural analogy*" exists between care and engineering: both "*respond to a need*

**Table 2.3** Tronto's four stages of care

| | |
|---|---|
| Attentiveness | Perceiving a need to care |
| Responsibility | Accepting the need to care |
| Competence | Providing care to another |
| Responsiveness | Getting/responding to feedback |

*and are oriented towards action*" (Pantazidou & Nair, 1999, p. 139). A key feature of this work is how Tronto's four stages of care (1993), plus the requirement to integrate them into a cohesive whole, could be mapped onto specific steps that an engineering student could take in working through a richly complex problem, the kind of problem they would encounter as a professional engineer. Being motivated by a concern for others as the starting point for engineering design and keeping these needs first and foremost in the design process has the potential, as they see it, for more environmentally and socially responsible outcomes (Table 2.3).

As interest in centering the ethics of care within engineering education and practice has grown, Pantazidou and Nair (1999) remain a touchstone for further reflection. Campbell et al. (2012) also believe in including the ethics of care in the curriculum to "*re-humanize*" engineering education by offering students the ethical resources they need for addressing complex problems involving multiple stakeholders, and to be better prepared for dealing with issues such as climate change and social justice (p. 6). Michelfelder and Jones (2016) argued that an engineering student's ability to frame an engineering problem with both its technical and socio-ethical dimensions can be improved by leveraging the deep commitment that many students have to sustainability, and connecting it to an ethics of care framework. Unfortunately, this premise may not hold up based on a 2014 survey showing that fewer than half the students who begin as engineering majors in the US believe that climate change is caused by humans (Shealy et al., 2019).

More recently, Nair and Bulleit (2020) suggestively proposed joining the ethics of care with pragmatism to create a pragmatic-care ethical framework that would be particularly well suited to support good design choices related to the climate and other complex adaptive systems. Van Grunsven et al. (2023) also make a case for how it would be promising to shift the orientation of engineering ethics education toward an ethics of care. Their approach stands out from others in identifying two specific forms of caring—generalized care and universalized care—that would work within engineering practice even when, as is almost always the case, engineers have limited direct contact with those who will be using the product from their designs. We will come back to Van Grunsven et al. (2023) in Sect. 2.7.

### 2.3.3 *Revising Engineering Codes of Ethics to Reflect an Ethic of Care*

Warford (2018) expanded on the growing interest in incorporating the ethics of care into engineering activities by opening the door to see how it could be specifically incorporated into US engineering codes of ethics. A limitation of engineering codes of ethics, as she sees it, is that they are built around general moral rules and principles, which makes it difficult for engineers to use them as a guide for action in real-life situations. Because care ethics are "*dependent on context and particulars rather than universal moral principles*," she suggests that such an approach can be more helpful for the engineer in their everyday practice, particularly given the fact that many engineering activities are intended to support humans in their own activities of care (Warford, 2018, p. 4).

In thinking about how the ethics of care might best be worked into the codes, Warford (2018) took her cue from what Engster (2007) calls the "*caring virtues*" of attentiveness, responsiveness, and respectfulness along with Tronto's categories of competence and responsibility. Using the 2007 NSPE code as a general model, she proposed (p. 10) several changes including these revisions to the section on Professional Obligations:

> Current language: "*Engineers shall at all times strive to serve the public interest.*"
>
> Proposed revision: Add a sub-rule: "*Engineers shall make design choices that are attentive and responsive to public needs and interests.*"
>
> Current language: "*Engineers shall be guided in all their relations by the highest standards of honesty and integrity.*"
>
> Proposed revision: "*Engineers shall be guided in all their relations by the highest standards of honesty, integrity, and respect for others.*"

Before we turn to discuss our own suggestions for how the more recent 2019 NSPE code could be revised, we first take up the question of just what might be gained from infusing an ethics of care into engineering codes of ethics to bolster an engineer's commitment to the breadth of sustainability.

## 2.4 Rationale

The fundamental presuppositions of care ethics, such as the belief in the interdependency of human beings on one another and on the environment, and the importance of caring for whom and for what is most vulnerable, sync up well with the idea that engineering codes of ethics need to more explicitly acknowledge that engineers should work toward sustainability as part of their responsibility to others. These are some of the ways in which this integration might be beneficial.

### 2.4.1 Motivating a Commitment to Responsible Action

Certainly, the move away from "*should strive to adhere*" language to simply "*adhere*" as we saw in the revision of the ASCE code in Sect. 2.2.1, is applaudable. But while stronger language is needed within the other codes, the question of how sufficient this language would be to motivate engineers to design for sustainability is a worrisome one. Likely many engineers would say they care about sustainability, but they might not be motivated, even with stronger language in the codes, to act on what they care about. Integrating the ethics of care into the engineering codes could help to close the gap between someone simply voicing their support for sustainability and taking responsible, competent action to support sustainability in their professional practice. Schmidt (2014) called for a new paradigm for engineering ethics centered on virtue ethics to undo the conventional separation between engineering practice and engineering ethics, something that Schmidt captures in the phrase "*your practice is your ethics*" (p. 1008). Integrating the ethics of care can work in the same way as an accelerant and motivation for action.

### 2.4.2 Putting Innovation More in Line with People's Needs

Another reason for integrating the ethics of care into the codes in the interests of sustainability rests on the importance of recognizing and responding to the needs of others. When innovation is pursued primarily for its own sake, as happens when something new is introduced into the world simply because it can be done rather than as a response intended to address human needs, attention to its environmental impacts risk being back-burnered. One has only to look at the immense amount of electricity and water used when training large language and other AI models (Li et al., 2023, Kumar & Davenport, 2023, Marr, 2023, van Wynsberghe, 2021) to find a contemporary case in point; the rapidity of their use is outpacing needed environmental mitigation. Similarly to how an enormous number of actors are incentivized to make small individual choices that together make the world more vulnerable by contributing to global warming (Bostrom, 2019), the world becomes more environmentally vulnerable as individual demand for and use of generative AI escalates. While it is unrealistic to expect that incorporating the ethics of care into the codes would create firm guardrails against innovating for the sake of innovation, it could help to counter this sustainability blind spot.

### 2.4.3 Responding to Engineering's Impacts on Intragenerational Equity

As described earlier, the codes are not explicit about sustainability in terms of intragenerational equity issues such as environmental justice. Robert Bullard's work starting in the 1970s contributed to an environmental justice movement that continues to push against the concept that the environment is neutral in terms of who is harmed and who benefits from significant decisions (Bullard, 1990; Funes, 2023). The Environmental Protection Agency (EPA) (n.d.) defines environmental justice as "*… the fair treatment and meaningful involvement of all people regardless of race, color, national origin, or income, with respect to the development, implementation, and enforcement of environmental laws, regulations, and policies.*" In addition, Mohai (2020) explains that a generation from one socioeconomic group can harm another from a different socioeconomic group via legacy decisions (e.g., redlining) that continue to contribute to current environmental injustices.

Thomas et al. (2022) suggested that a community's relationship with an environmental issue needs to be examined in terms of differentiated impacts and harms, equal and meaningful access to information and participation in decision making, equal access to environmental amenities, and access to effective remedies for environmental harms. Avelino et al. (2024) in a recent paper also emphasized the need to democratize decision-making as a key strategy towards achieving sustainable and just cities and one that is often neglected by those with formal authority. In addition, while life cycle assessment, a proven engineering method for sustainable design, embraces the need to consider the social impacts of an engineered product, these methods remain the least developed (Valdivia, 2011). Like the innovation example above, while it is unrealistic to expect that incorporating the ethics of care into the codes alone would provide engineers with methods to incorporate all the facets of environmental justice into their designs, it could serve to elevate these considerations.

### 2.4.4 Operationalizing Engineering for Sustainability

Nair and Bulleit (2020) considered how each of the elements of care identified by Tronto (1993) could be applied to engineering in general. These applications become more relevant when thinking about engineering for sustainability. Even using the most limited definitions in the codes, to engineer for sustainability means broadening one's scope to consider and balance the environmental, economic, and social aspects while being attentive to who and what are affected. Being responsive aligns with sustainable engineering since engineering for the long term depends on using ongoing feedback and adjustments "*…along with opportunities for improvement, in their work…*" (ASCE, 2020).

A commitment to sustainability also asks the engineer to go beyond the technical design to consider the broader contextual issues which may require different levels of competence. This breadth of competencies is captured by the Accreditation Board for Engineering and Technology recent criteria changes (ABET, 2024) requiring engineering graduates to have "*…an ability to recognize ethical and professional responsibilities in engineering situations and make informed judgments…*" Similarly, WFEO's (2001) "*create and implement engineering solutions for a sustainable future*" embraces a responsibility to care beyond the technology, including care for those affected now and into the future in both the human and the natural spheres. In summary, context is at the core of sustainable engineering and is partially reflected by several of the recent changes to the engineering codes, though more is needed.

## 2.5 Using the Ethics of Care to Strengthen the Sustainability Commitments in the NSPE Code of Ethics

We applied the ethics of care to the six canons within the NSPE code of ethics to demonstrate how to strengthen a code's commitment to sustainability, and an engineer's commitment to sustainability in turn. We chose the NSPE code because it applies to all engineers in the US and many of the discipline-specific codes are based in part on its traditional structure of canons, rules of practice that expand on these canons, and professional obligations that further expand on the rules of practice. Certainly, NSPE and other organizations may choose to abandon a canonical framework and use a values-based framework like ASCE, however, we hope that this example serves to show how such an approach may apply to the codes as they currently stand. Showing which framework is preferable or what other changes may be needed to update the codes is outside the scope of this paper.

Below is a starting point for the suggested changes (indicated in italics) that we hope will spark further discussion:

1. Hold paramount the safety, health, and welfare of the public, *and adhere to the principles for sustainability with respect for both intergenerational and intragenerational equity.*
2. Perform services only in areas of their competence *defined as technical competence, the competence to determine economic, societal, and environmental consequences along with how they are distributed among varying populations, and the competence to subsequently make informed judgments.*
3. Issue public statements only in an objective, *inclusive,* and truthful manner *with respect for others.*
4. Act for each employer or client *or affected community member* as faithful agents or trustees *while communicating the broad sustainability context for the engineering services, along with the distributional impacts for others.*

5. Avoid deceptive acts *while* committing *to a responsive approach that provides ongoing opportunities for improvement based on feedback and reflection.*
6. Conduct themselves honorably, responsibly, ethically, and lawfully so as to enhance the honor, reputation, and usefulness of the profession.

In making these suggestions, we follow both Warford (2018) and Riley and Lanbrinindou (2015) in seeing them as complementary to values that the NSPE code already reflects.

## 2.6 Objections and Rejoinders

The benefits described in Sect. 2.4 do not of course create an open and shut case for our suggested integration. Thinking specifically about this integration from the side of the ethics of care, what might be some reasons to be skeptical?

### *2.6.1 The Open-Endedness of the Concept of Care*

One very basic concern has to do with just what the activity of caring consists of. As an ethical concept, care is very open-ended (see for example Puech, 2016). No consensus exists about its meaning and the definition of care given in Fischer's and Tronto's seminal work (1991, p. 40) as involving "*everything we do to maintain, continue, and repair our 'world' so that we can live in it as well as possible*" is highly elastic. Randall (2019) sums up criticisms of this definition by saying that if this is what care means, then "*caring becomes indistinguishable from any given kind of practice*" (p. 532). Does this imply that there are strong limitations on how productively the ethics of care can be infused into engineering codes of ethics?

One way of responding to this question would be to say that when it comes to giving care, not just anything can count. Tronto's (1993) identification of *responsiveness* as a stage in the ethics of care giving those extending care a chance to adjust when their actions are found to be unsatisfactory, either by the cared-for or another entity (Tronto, 2013, p. 35), provides a helpful limit. To say that care is an open-ended concept does not imply that it is too contested a concept to be used. In fact, embedding the ethics of care into the codes will require imposing boundaries on what "*care*" means within the context of sustainable engineering. The previous section provides such an example for the NSPE.

### 2.6.2 The "Fit" Between Care and the Complexity of Engineering Problems

Another difficulty has to do with *attentiveness:* the first step in problem-solving from an ethics of care perspective. Attending to individual and collective needs is difficult. Identifying the needs that should be addressed in any given situation is not always easy, and what one ought to pay attention to might not be the first thing that meets the eye. Given that in an engineering context there are also competing needs and/or time/resource constraints, it can be extremely complicated to prioritize those needs. Some might also wonder whether changing the language of the codes to more intentionally reflect an ethics of care might have unintended consequences. Could the difficulty just mentioned about prioritization prompt engineers working together on a particular project to move toward mitigating adverse effects in the name of sustainability rather than promoting other goals?

Shifts in how engineers are trained could help to ameliorate these reasonable concerns. In exploring the use of case studies in ethics engineering education, Martin et al. (2021) found evidence to suggest they could be even more valuable if they were "*immersive*" in character, giving students an opportunity to reflect on the ethical dimensions of actual, contemporary situations. Such case studies could also afford opportunities for students to develop their skills of attentiveness and to go beyond reading engineering problems as only technical in nature. This thought aligns with the vision outlined by Martin et al. (2023) that one of the ways for a technological university to be responsible is to emphasize intra- and inter-generational responsibility in the training of future engineers.

### 2.6.3 The "Fit" Between Care and Sustainable Development

As mentioned in 2.4.2 one advantage of intentionally integrating the ethics of care into professional engineering codes is that it could serve as a brake against innovation that does not prioritize meeting human needs. But might it serve as too much of a brake? The idea that care includes "*everything we do to maintain, continue, and repair our 'world' so that we can live it as well as possible*" raises the question of whether the ethics of care supports overly precautionary rather than innovative engineering activities, and so might work to stifle innovation that could support sustainability and other goals.

Behind this question is the understandable belief that to maintain and continue something is at odds with making it better, and that the aim of repair is to return something as close as possible to the condition it was in when new. But this fails to consider that in an engineering context, a system where careful attention is given to its maintenance will have fewer negative environmental impacts than one whose maintenance is neglected (Vinsel, quoted in Brock, 2021); innovation that supports maintenance can also support sustainability. Understanding repair as a social-cultural practice rather than a purely technical activity could also shed light on how the boundary between repairing and innovating is more porous than we generally assume.

## 2.7 Directions for Future Research

Our demonstration of how the ethics of care might be integrated into engineering codes of ethics to enhance commitments to sustainability prompts several questions meriting further exploration.

How could this demonstration be extended beyond the canons of the NSPE code to other codes and other aspects of the codes such as the rules of practice and professional obligations? Since undergraduate engineering programs in the US are required to address ethics within the curriculum, code changes that are framed within an ethics of care and include the breadth of sustainability should influence what and how engineering students are taught. However, influencing such changes to the codes of ethics for the main professional engineering societies has not been very effective to date as can be seen by the prolonged and incomplete attempts to include the breadth of sustainability into the codes. There are at least four main actors that could influence such transformative change: ABET via accreditation requirements for undergraduate programs, NCEES via licensing exams, clients demanding greater competency with the breadth of sustainability and justice, and engineering companies similarly demanding such competency. Understanding how to mobilize these actors towards the type of changes described in this paper is an important research question to explore.

Looming especially large over the revisions to the NSPE canons proposed here is the related question of how undergraduate engineering education would need to change. The challenges here are far from insignificant. While sustainability has been identified as displaying the characteristics of a "*wicked problem*" (Blok et al., 2015), a study of eight university engineering programs by Schuelke-Leech (2020) showed that the problems students confronted in their curricula were almost entirely "*well-structured*". They were not asked to engage with wicked ones. Cech (2014, especially p. 48) has forcibly argued that US engineering culture encourages engineers in their work to cognitively separate the technical from the social in such a way that the competency to be able to approach the social aspects of a problem is downplayed while technical competencies are uplifted. It is also worth asking how such changes might promote not only the commitment of engineers to sustainability, but also work to mitigate gender and racial imbalance within many of the engineering disciplines.

Van Grunsven et al. (2023) suggest another area where more research is needed when they refer to generalized care and universalized care for "*their other*"; do either or both address the care needed to achieve intragenerational justice? They acknowledge that an engineer's care for "*their other*" is marked by distance in such a way that the engineer prioritizes those in closer proximity such as colleagues, direct clients, and technologies, while operating in an activity that is perceived as neutral (Van Grunsven et al., 2023). This is problematic in at least two aspects. Given the USA's racialized past, the engineer is likely to be even more distant from those who are susceptible to environmental injustices given the lack of diversity within the profession. In addition, environmental justice frameworks emphasize the need for those most vulnerable to participate in the decisions that may affect them. Given that generalized care is likely determined by stakeholders who do not represent marginalized populations, and universalized care may hide intragenerational justice by aggregating populations, do engineers need to practice a different type of care for "*their other*" to ensure that environmental justice is integral to sustainable engineering?

Along with these possibilities, this chapter provides only a partial perspective since we relied on the US context for engineering codes to illustrate how adopting an ethics of care framework complements and elevates an engineer's commitment to sustainability. This claim should be examined beyond the US context to identify where the rationale holds and where it fails, and to explore how engineers in other countries are responding to the sustainability mandate. While a recent survey of engineering educators in the US, non-US Anglo countries, and western European countries found that in general, non-US countries included sustainability in engineering curricula more frequently than the US (Polmear et al., 2019), it did not consider countries beyond the western hemisphere (other than Australia and New Zealand), nor those with lesser developed economies where context may be quite different.

In Warford (2018), sustainability is mentioned only once, as an example of how difficult it is to change engineering codes of ethics. Specifically, Warford notes the "*late and tepid 'encouragement' for engineers to consider sustainability in their designs*." And it is certainly the case that efforts to change codes of ethics can face strong political and cultural headwinds. But when it comes to sustainability, the changes that we mentioned in Sect. 2.2.1 are a hopeful sign that these headwinds are abating. Coupling this with the increasing interest among engineers and ethicists in the ethics of care leads us to believe that our proposal is well-timed. In the interview mentioned earlier in this chapter (Care Ethics Research Consortium, 2017), Joan Tronto was asked: "*How might care ethics contribute to society as a whole, do you think?*" Tronto replied that it might contribute by "*reorienting us towards the things that really matter*." It is in the hope of continuing this reorientation of the professional codes of ethics toward sustainability that we offer the considerations found in this chapter.

**Acknowledgments** Our gratitude goes out to those who attended the fPET 2023 session where the paper on which this chapter is based was presented for the lively discussion and insightful comments and suggestions. We also want to thank the two anonymous reviewers of the first draft of this chapter for their attentive reading and excellent feedback.

## References

ABET. (2024). Accreditation Criteria & Supporting Documents. https://www.abet.org/accreditation/accreditation-criteria/.

American Academy of Environmental Engineers and Scientists. (n.d.). https://www.aaees.org/.

American Institute of Chemical Engineers. (n.d.). Institute for Sustainability. https://www.aiche.org/ifs.

American Institute of Chemical Engineers. (2008). Sustainability Index. https://www.aiche.org/ifs/resources/sustainability-index.

American Institute of Chemical Engineers (AIChE). (2015). AIChE code of ethics. https://www.aiche.org/about/governance/policies/code-ethics.

American Society of Civil Engineers (ASCE). (2020). ASCE code of ethics. https://www.asce.org/career-growth/ethics/code-of-ethics.

American Society of Civil Engineers (ASCE). (2023). Policy statement 418—the role of the civil engineer in sustainable development. https://www.asce.org/advocacy/policy-statements/ps418%2D%2D-the-role-of-the-civil-engineer-in-sustainable-development.

American Society of Mechanical Engineers (ASME). (2021). ASME code of ethics. https://www.asme.org/about-asme/governance/ethics-in-engineering.

Avelino, F., van Steenbergen, E., Schipper, K., Steger, T., Henfrey, T., Cook, I., Bach, M., Crowley, D., Kotsila, P., & Crudi, F. (2024). Mapping the diversity and transformative potential of approaches to sustainable just cities. *Urban Transformation, 6*. https://doi.org/10.1186/s42854-023-00058-8

Baas, L., Metselaar, S., & Klaassen, P. (2022). Circles of care for safety: A care ethics approach to safe-by-design. *NanoEthics, 16*(2), 167–179.

Blok, V., Gremmen, B., & Wesselink, R. (2015). Dealing with the wicked problem of sustainability: The role of individual virtuous competence. *Business and Professional Ethics Journal, 34*(3), 297–327.

Bostrom, N. (2019). The vulnerable world hypothesis. *Global Policy, 10*(4), 455–476.

Brock, D. C. (2021). Decoding the innovation delusion, nurturing the maintenance mindset → why we celebrate innovation while ignoring maintenance and care. *IEEE Spectrum*. 5 February. https://spectrum.ieee.org/decoding-the-innovation-delusion-nurturing-the-maintenance-mindset

Bullard, R. D. (1990). *Dumping in Dixie: Race, class, and environmental quality* (3rd ed.). Routledge. https://doi.org/10.4324/9780429495274

Burgess, R. (2019). Professional codes of ethics: Some observations and lessons learned. *Journal of Forestry, 117*(4), 406–411.

Campbell, R. C., Yasoharu, K., & Wilson, D. M. (2012). Care ethics in engineering education: Undergraduate student perceptions of responsibility. *Frontiers in Education Conference Proceedings*, 1–6.

Care Ethics Research Consortium. (2017). Interview with Joan Tronto. https://care-ethics.com/2017/05/31/joan-tronto-2/.

Cech, E. A. (2014). Culture of disengagement in engineering education? *Science, Technology, & Human Values, 39*(1), 42–72. https://doi.org/10.1177/0162243913504305

Davis, M. (2021). Professional codes of ethics. In D. Michelfelder & N. Doorn (Eds.), *The Routledge handbook of the philosophy of engineering* (pp. 580–591). Routledge.

Engster, D. (2007). *The heart of justice: Care ethics and political theory*. Oxford University Press.

Environmental Protection Agency. (n.d.). Environmental Justice. Accessed November 26, 2023. https://www.epa.gov/environmentaljustice.

Fischer, B., & Tronto, J. (1991). Toward a feminist theory of caring. In I. E. Abel & M. Nelson (Eds.), *Circles of care: Work and identity in women's lives* (pp. 36–54). State University of New York Press.

Fore, GA, Hess, J. L, & Katz, A. (2020). Ethics in engineering or engineering in ethics? ASEE annual conference and exhibition (held virtually). https://doi.org/10.18260/1-2%2D%2D34590.

Funes, Y. (2023). The father of environmental justice exposes the geography of inequity. *Scientific American*.

Gilligan, C. (1982). *A different voice*. Harvard University Press.

Held, V. (2006). *The ethics of care: Personal, political, and global*. Oxford University Press.

Herkert, J. R. (1998). Sustainable development, engineering and multinational corporations: Ethical and public policy implications. *Science and Engineering Ethics, 4*(3), 333–346. https://doi.org/10.1007/s11948-998-0025-y

Institute of Electrical and Electronics Engineers (IEEE). (2020). IEEE code of ethics. https://www.ieee.org/content/dam/ieee-org/ieee/web/org/about/corporate/ieee-code-of-ethics.pdf.

Jones, S., Michelfelder, D., & Nair, I. (2017). Engineering managers and sustainable systems: The need for and challenges of using an ethical framework for transformative leadership. *Journal of Cleaner Production, 140*(1), 1–8.

Kumar, A., & Davenport, T. (2023). How to make generative AI greener. *Harvard Business Review*. July 20. https://hbr.org/2023/07/how-to-make-generative-ai-greener

Li, P., Yang, J., Islam, M.A., & Ren, S. (2023). Making AI less "Thirsty": Uncovering and addressing the secret water footprint of AI models. arXiv2304.03271.

Marr, B. (2023). Green intelligence: Why data and AI must become more sustainable. *Forbes*. 22 March. https://www.forbes.com/sites/bernardmarr/2023/03/22/green-intelligence-why-data-and-ai-must-become-more-sustainable

Martin, D. A., Conlon, E., & Bowe, B. (2021). Using case studies in engineering ethics education: The case for immersive scenarios through stakeholder engagement and real life data. *Australasian Journal of Engineering Education, 26*(1), 47–63.

Martin, D. A., Bombaerts, G., Horst, M., Papageorgiou, K., & Viscusi, G. (2023). Pedagogical orientations and evolving responsibilities of technological universities: A literature review of the history of engineering education. *Science and Engineering Ethics, 29*(6). https://doi.org/10.1007/s11948-023-00460-2

Michelfelder, D. (Forthcoming). Moving from the margins: Multistability, engineering ethics, and the ethics of care. In G. Wellner, R. Rosenberger, & L. Friedman (Eds.), *Postphenomenology and feminist theory: Intersectionality, gender, and technology*. Lexington Books.

Michelfelder, D., & Jones, S. A. (2013). Sustaining engineering codes of ethics for the 21st century. *Science and Engineering Ethics, 19*(1), 237–258.

Michelfelder, D., & Jones, S. A. (2016). From caring about sustainability to developing care-ful engineers. In W. A. Filho & S. Nesbit (Eds.), *New developments in engineering education for sustainable development* (pp. 173–184). Springer.

Mohai, P. (2020). Weighing the evidence: Measuring environmental justice. In M. Mascarenhas (Ed.), *Lessons in environmental justice: From civil rights to black lives matter and idle no more* (pp. 52–76). Sage.

Nair, I., & Bulleit, W. M. (2020). Pragmatism and care in engineering ethics. *Science and Engineering Ethics, 26*(1), 65–87.

National Council of Examiners for Engineering and Surveying (NCEES). (2024). https://ncees.org/licensure.

National Society of Professional Engineers (NSPE). (2019). NSPE code of ethics. https://www.nspe.org/sites/default/files/resources/pdfs/Ethics/CodeofEthics/NSPECodeofEthicsforEngineers.pdf.

National Society of Professional Engineers (NSPE). (2023). NSPE Position Statement No. 05-100-Sustainability. https://www.nspe.org/sites/default/files/sites/default/files/resources/PSdownloadables/Sustainability.pdf.

Noddings, N. (1984). *Caring: A feminine approach to ethics and moral education*. University of California Press.

Noddings, N. (2002). *Starting at home: Caring and social policy*. University of California Press.

Pantazidou, M., & Nair, I. (1999). Ethic of care: Guiding principles for engineering teaching & practice. *Journal of Engineering Education, 88*(2), 205–212.

Polmear, M., Bielefeldt, A., Knight, D., Canney, N., & Swan, C. (2019). Analysis of macroethics teaching practices and perceptions in engineering: A cultural comparison. *European Journal of Engineering Education, 44*(6), 866–881.

Puech, M. (2016). *The ethics of ordinary technology*. Routledge.

Randall, T. (2019). Care ethics and obligations to future generations. *Hypatia, 34*(3), 527–545.

Riley, D. M., & Lambrinidou, Y. (2015). Canons against cannons?: Social justice and the engineering ethics imaginary. *ASEE Annual Conference and Exhibition*, 26.322.1–26.322.19.

Russell, A., & Vinsel, L. (2019). Make maintainers and an ethic of care. In M. Wisnioski, E. H. Hintz, & M. S. Kleine (Eds.), *Does America need more innovators?* (pp. 249–269). MIT Press.

Schmidt, J. A. (2014). Changing the paradigm for engineering ethics. *Science and Engineering Ethics, 20*(4), 985–1010.

Shealy, T., Klotz, L., Godwin, A., Hazari, Z., Potvin, G., Barclay, N., & Cribbs, J. (2019). High school experiences and climate change beliefs of first year college students in the United States. *Environmental Education Research, 25*(6), 925–935.

Schuelke-Leech, B.-A. (2020). The place of wicked problems in engineering problem solving: A proposed taxonomy. *IEEE International Symposium on Technology and Society (ISTAS)*, 361–372.

Steinman, D.E. (1935). What the National Society of Professional Engineers can accomplish. The American engineer, January, 6, 14, 19.

Tang, X., & Nieusma, D. (2017). Contextualizing the code: Ethical support and professional interests in the creation and institutionalization of the 1974 IEEE code of ethics. *Engineering Studies, 9*(3), 166–194.

Thomas, A. S., Escobedo, F. J., Sloggy, M. R., & Sanchez, J. J. (2022). A burning issue: Reviewing the socio-demographic and environmental justice aspects of the wildfire literature. *PLoS One, 17*(7). https://doi.org/10.1371/journal.pone.0271019

Tronto, J. (1993). *Moral boundaries: A political argument for an ethic of care*. Routledge.

Tronto, J. (2013). *Caring democracy: Markets, equality, and justice*. New York University Press.

United Nations. (2016). Sustainable development goals. https://www.un.org/sustainabledevelopment/sustainable-development-goals/.

Van Grunsven, J. B., Marin, L., Stone, T., Doorn, N., & Roeser, S. (2023). How engineers can care from a distance: Promoting moral sensitivity in engineering ethics education. In I. G. Miller, H. M. Jeronimo, & Q. Zhu (Eds.), *Thinking through science and technology: Philosophy, religion, and politics in an engineered world* (pp. 141–163). Rowman & Littlefield International.

Valdivia, Sonia (2011). UNEP/SETAC life cycle initiative. https://wedocs.unep.org/20.500.11822/32747.

Van Wynsberghe, A. (2021). Sustainable AI: AI for sustainability and the sustainability of AI. *AI and Ethics, 1*, 213–218.

Vesilind, P. A. (2002). Vestal virgins and engineering ethics. *Ethics and the Environment, 7*(1), 92–101.

Warford, E. (2018). Toward a more caring code of engineering ethics. ASEE annual conference and exhibition, Salt Lake City, Utah.

World Commission on Environment and Development. (1987). *Our common future*. Oxford University Press.

World Federation of Engineering Organizations. (2001). WFEO model code of ethics. https://www.wfeo.org/wp-content/uploads/code_of_ethics/WFEO_MODEL_CODE_OF_ETHICS.pdf.

**Diane P. Michelfelder** AB Philosophy Bryn Mawr College. PhD Philosophy, The University of Texas at Austin. Professor of Philosophy, 2005–2023; Professor Emerita, 2023–, Macalester College. 2 Springer-POET Volumes: Philosophy and Engineering: Reflections on Practice, Principles and Process (with Natasha McCarthy and David E. Goldberg, 2014) and Philosophy and Engineering: Exploring Boundaries, Expanding Connections (with Byron Newberry and Qin Zhu, 2016). Co-editor with Neelke Doorn of The Routledge Handbook of The Philosophy of Engineering, 2020. Editor, Test-Driving the Future: Autonomous Vehicles and the Ethics of Technological Change. Co-founder of the Forum for Philosophy, Engineering and Technology (fPET). Research areas: Philosophy of technology and engineering, engineering ethics, ethics of care.

**Sharon A. Jones** BS Civil Engineering Columbia University, PhD Engineering and Public Policy Carnegie Mellon University). Professor, University of Washington Bothell. Research areas: Sustainability, systems analysis, climate adaptation, engineering ethics.

# Chapter 3
# Using Civic Professionalism to Frame Ethical and Social Responsibility in Engineering

**Erhardt Graeff**

**Abstract** Most common approaches to ethical and social responsibility in engineering are insufficient to addressing the growing need to ensure engineers and technologists serve the common good. In particular, professional codes of ethics, grand challenges and social entrepreneurship, and corporate adoption of self-policed ethical principles are often toothless in shaping individual and corporate behavior and tend to reinscribe irresponsible technocratic ideologies at the heart of engineering culture. Erin Cech argues there is a "culture of disengagement" in engineering that depoliticizes engineering, separates and differentially values technical and social aspects of engineering work, and embraces the problematic values and worldview of meritocracy. Looking beyond STEM (science, technology, engineering, and mathematics) and STEM education to civic education and democratic theory, I argue civic professionalism, based on the work of Harry Boyte and Albert Dzur, offers a framing of professional identity and practice to engineers which articulates a positive ethics of virtue and resists technocratic forms of professionalism. It proactively engages in the broader sociopolitical questions connected to engineering work and embraces a democratic epistemology and way of working. Educating engineers to become civic professionals will require cultivating reflexivity and civic skills and virtues, and the creation of experiential learning opportunities that engage authentically with sociopolitical complexity.

**Keywords** Professionalism · Engineering ethics · Civic-mindedness · Engineering education · Civic education

---

E. Graeff (✉)
Olin College of Engineering, Needham, MA, USA
e-mail: egraeff@olin.edu



C. Didier et al. (eds.), *Engineering and Value Change*, Philosophy of Engineering and Technology 48, https://doi.org/10.1007/978-3-031-83549-0_3

## 3.1 Introduction

In recent years, professional societies, philanthropists, scholars, and corporations within engineering and computer science have been grasping for better principles and frameworks for the ethical and social responsibility that engineers, computer scientists, and technologists should embody. Professional societies, such as the National Society of Professional Engineers (NSPE), the Institute of Electrical and Electronics Engineers (IEEE), and the Association of Computing Machinery (ACM), have revised their codes of ethics (National Society of Professional Engineers, 2019; Association for Computing Machinery, 2018). Philanthropists and certain scholars and educators have suggested social entrepreneurship should be the model for responsible innovation (Spicer et al., 2019). Corporations have publicized their ethical commitments and formed consortia to govern ethical approaches to research and development in areas like artificial intelligence (Jobin et al., 2019; Attard-Frost et al., 2023).

But engineering's ethical and social responsibility problem is more deeply rooted, often capturing superficial approaches. In 2014, Erin Cech identified a "culture of disengagement" in engineering that weakens the commitments to public welfare of engineering students in the United States during their undergraduate years. She argued that ideologies of (1) depoliticization, (2) technical/social dualism, and (3) meritocracy undermine engineers' social responsibility by positioning technical expertise as supremely relevant and perceiving existing social, economic, and political structures as fair and just. These ideologies overlap with the "myths" and cultural practices of meritocracy, openness, and entrepreneurialism that Alice Marwick argues are at the heart of Silicon Valley's political economy (2017). Such myths dominate the contemporary professional culture of technologists and have convinced many around the world to put faith in Silicon Valley's models of change (Marwick, 2017). Lilly Irani calls this "entrepreneurial citizenship," where hackathons and other fixtures of tech entrepreneurship subsume political desire into speculative value production and reinscribe elitist and technocratic values that afflict countries like India as a result of colonial and international development agendas (Irani, 2015, 2019).

This is not only a professional crisis but a societal one. Technocracy and meritocracy undermine democracy—ethical approaches and professional identities based on them fail, in the end, to serve the common good (Sandel, 2020). To counter these social and professional issues, and resist design and entrepreneurship standing in for civic pedagogy (Irani, 2019), engineering education and epistemologies of engineering must repoliticize the profession and its work, as Cech and Sherick argue (2015).

I believe *civic professionalism* answers this call. Based on Harry Boyte's concept of "citizen professionalism" and Albert Dzur's concept of "democratic professionalism" (Boyte, 2008, 2009; Boyte & Fretz, 2010; Dzur, 2004, 2010, 2018; Graeff & Wood, 2021), civic professionalism offers a virtue ethics approach to professional identity and practices that embrace epistemological humility and pluralism and center the common good by actually centering the public in engineering work.

## 3.2 Reinforcing a Culture of Disengagement

Public consciousness is catching up with internal and academic critiques of engineering culture and practice that go back decades, as documented by Matthew Wisnioski (Wisnioski, 2016; Hendren, 2022) and seen in the complicated history of efforts by Computing Professionals for Social Responsibility (Finn & DuPont, 2020). Over the years, engineering has tried on various approaches and frameworks for promoting ethical and social responsibility with limited success.

### *3.2.1 Codes of Ethics*

Professionalism embodied by a code of ethics has long been a site of concern about safety and upstanding conduct relative to one's employer. Some professional codes of ethics, particularly the U.S.-based NSPE (National Society of Professional Engineers, 2019) and the ACM (Association for Computing Machinery, 2018), have been recently revised with language explicitly calling professionals to serve the public interest or public good and to only practice within the limits of one's competence—an aspect of epistemic humility. Professional codes of ethics can enable engineers to refuse work based on a defined professional standard rather than personal whim. However, this is widely acknowledged as a meager moral baseline.

There is little evidence that these codes actually help create a more ethical and socially responsible culture by themselves (Rosenberg, 1998; Giorgini et al., 2015; McNamara et al., 2018). Codes primarily codify negative rules, are inherently conservative, assume saintly professionals, and "neglect the messy, complex realities of both people and the world" (Sadowski, 2014). As Deborah Johnson (2020) has noted, codes of ethics can be useful in socializing young engineers under a common ethical standard; however, they struggle to help the many engineers working in international contexts or on global systems (Harris, 2004). Moreover, they continue to miss contemporary demands of ethics for sustainability and social justice (Michelfelder & Jones, 2013).

### *3.2.2 Grand Challenges and Social Entrepreneurship*

In 1998, the U.S.'s National Academy of Engineering offered the Grand Challenges, which implied that the primary role of engineers in changing the world was to focus on solving a few particularly difficult technical problems (National Academy of

Engineering, 2017). Schools like my own Olin College of Engineering and other early partners offered Grand Challenges Scholars Programs to honor students committed to this kind of work. This coincided with social entrepreneurship's emergence as a model for changemaking (Spicer et al., 2019).

Growing since the late 1970s, the movement for social entrepreneurship and enterprise argued for market-based approaches to complex problems applying novel "innovations" and turning citizens into customers (Spicer et al., 2019). One trenchant critique of the social entrepreneurship movement describes it as field encroachment—a neoliberal social movement displacing the work of the state and of nonprofits; downplaying the reality of systemic problems, politics, and power; and entrenching a neoliberal model of action and responsibility despite a lack of any findings that this model is more effective at making change (Spicer et al., 2019). Social entrepreneurship appeals to dominant narratives historically defining engineering in service of industry and economy (Cheville, 2014), and faculty who commonly tell their students engineering is primarily about "problem-solving" and "making things" (Pawley, 2009).

Grand challenges and social entrepreneurship can appear to be exciting paths to engineering the world for the better. But they also reinscribe an engineer's default moral purpose of designing efficient technology for a client and their hubris that individual achievement can offer generalizable solutions to complex problems. Design and entrepreneurship tends to invoke a hero narrative that obscures collective politics, the existence of privilege, and the roles of coercion and non-entrepreneurial labor as part of how change often is made (Irani, 2019, pp. 78–79).

### 3.2.3 *Corporate Ethical Principles*

If Silicon Valley's ethos predominates the engineering and technology professions, then its approach to ethics is similarly predominate and anemic. An ethnography of "ethics owners" at Big Tech companies concluded that "the commitment to ethics is in tension with—and at risk of being absorbed within—broader and longer-standing industry commitments to meritocracy, technological solutionism, and market fundamentalism" (Metcalf et al., 2019). Among multinational technology firms, a race for supremacy in artificial intelligence (AI) has exploded. Technology critics and scholars have alerted the public to the dangers posed by AI and machine learning's applications in war, surveillance, criminal justice, welfare services, search engine results, and myriad other contexts, which has led to an explosion of AI ethics principles defined by scholars, governments, and corporations (Jobin et al., 2019; Attard-Frost et al., 2023).

In order to retain the best computer scientists and placate technology workers to want to believe they work for firms that won't misuse AI, individual companies and corporate consortia have adopted ethical frameworks promising they will be treated carefully. Unfortunately, it turns out these AI ethics principles are largely "toothless" platitudes (Ességuier & Rodrigues, 2020; Green, 2021; Munn, 2022). They suffer from many of the same limitations of professional codes of ethics. It's unclear where accountability lies, as corporate ethics statements are self-policed. Corporate ethical frameworks and industry lobbying that defines the threat of AI in terms of extreme existential threats rather than the many known, immediate concerns serve their economic interests and attempt to dodge accountability to a common good defined and policed by governments (Kak & West, 2023).

In his experience as a graduate student researcher at the MIT Media Lab working with the Microsoft, Google/DeepMind, Facebook, IBM, and Amazon co-founded Partnership on AI to Benefit People and Society, Rodrigo Ochigame found that the corporate stakeholders explicitly tried to promote ethical principles in lieu of regulation and watered down those principles according to their corporate interests rather than in the service of what was ethical according to research and what served the public interest (Ochigame, 2019). Many scholars have similarly concluded that corporate ethical principles around AI amount to "ethics washing" to avoid regulation (Wagner, 2018; Bietti, 2021; Green, 2021). Ironically, socio-technical scholars and other critics seem to have found themselves in the position of having called for more ethics in technology and corporations having responded by twisting ethics into something unrecognizable that ultimately undermines the moral responsibility of engineers and technologists (Greene et al., 2019).

### 3.2.4 Need for Reflexivity and Broadened Participation

Engineering's culture is a magnet for frameworks like the Grand Challenges, social entrepreneurship, and toothless corporate ethical principles. Each of these approaches to responsibility, along with other popular suggestions for how to make engineering more ethically and socially responsible, miss what Cech describes in her critique of the Grand Challenges as the need for "reflexivity" and "broadened participation" (Cech, 2012). Reflexivity being "a critical examination of engineers' role in the past, present and future of societies—in both problem definition and problem solution" (Cech, 2012, p. 92). And broadened participation meaning, "The voices of other professionals and the broader constituencies impacted by those solutions should be included at various stages in the process" (Cech, 2012, p. 93). A successful approach to ethical and social responsibility should ensure that reflexivity and broadened participation are at the heart of the values and practices put into actions by professionals trained and governed by such a framework.

The dominant logics of engineering culture in the USA, and globally in forms such as Irani's entrepreneurial citizenship, follow Cech's ideological pillars of a culture of disengagement:

1. The ideology of depoliticization, which "frames any 'non-technical' concerns such as public welfare as irrelevant to 'real' engineering work" (Cech, 2014, p. 45)
2. The technical/social dualism, which "devalues 'social' competencies such as those related to public welfare" (Cech, 2014, p. 45)

3. The meritocratic ideology, which "frames existing social structures as fair and just" (Cech, 2014, p. 45)

Unfortunately, these ideological pillars are socialized before engineers and technologists even join the workplace. Over four years of undergraduate engineering education at several schools—including Olin College of Engineering—students exhibited declines in survey measures of beliefs in the importance of professional and ethical responsibilities, understanding the consequences of technology, understanding how people use machines, and social consciousness (Cech, 2014). This is the background to engineering's struggle to adopt good frameworks for ethical and social responsibility. Engineering culture and its definitions of ethical and social responsibility must be shaped by reflexive and inclusive approaches that resist depoliticization and meritocratic and dualistic assumptions about technical expertise and public value.

## 3.3 Civic Professionalism for Engineers

For nearly 18 years I have been studying the design and use of technology for civic engagement, which led me to start investigating civic education and democratic theory, asking about how people become good and effective civic actors. It is from this literature, looking outside of science, technology, engineering, and mathematics (STEM) and STEM education, that I propose civic professionalism as a virtuous framework for ethical and social responsibility in engineering.

Civic professionalism is an umbrella term for the concepts "citizen professionalism" and "public work" from Harry Boyte (Boyte, 2008; 2009; Boyte & Fretz, 2010) and "democratic professionalism" from Albert Dzur (Dzur, 2004, 2010, 2018). Civic professionalism is both a professional identity, anchored by civic attitudes and related values, and a set of normative professional practices that rely on civic knowledge, skills, and habits, which augment specialized technical competencies. In an earlier paper, I summarized this as: (1) "civic professionals do not distinguish between their civic and professional responsibilities"; "they are motivated by the common good and seek to produce common goods," and (2) "civic professionalism is defined by its commitment to democratic participation" (Graeff & Wood, 2021).

But if we look at the source texts, we see the contrast between the outside expert and the civic professional. Citizen professionalism explicitly pushes back on technocratic models of professional expertise. In contrast to typical outside experts, Boyte (2008) argues citizen professionals see their role as co-creators and facilitators of problem-solving. They acknowledge that they too are citizens alongside many other stakeholders and should share power over decisions. They acknowledge the limits of their knowledge and expertise in different contexts and embrace local knowledge. Their work serves the common good by solving technical problems while also building and strengthening relationships. Boyte calls this "public work" to emphasize the idea that people are making meaningful contributions to society through their core efforts rather than on the side as volunteers or ad hoc community service.

Similar to Boyte's description, Dzur (2018) describes democratic professionalism as resisting the technocratic urge to flatten complex problems into challenges well-suited to professional methods beyond the lay public's reach, and instead create space for deliberation and collective action regarding social and political issues beyond the borders of their own professional domains. Dzur specifically offers democratic professionalism as a middle ground between a "social trustee" model of professional and its radical critique, which would seek to deprofessionalize expertise and recover all such power for the public. Dzur argues that there is value to the expertise of professionals but that value is a social compact with laypeople. Both have a stake in professional decisions and should share oversight of professional decisions and ethics democratically.

Philosophy of technology scholars have attempted to more clearly articulate the long-held but ambiguously defined responsibility of engineers to the common good. Deborah Johnson has acknowledged the ways the social practice of engineering in complex socio-technical systems inherently involves navigating political interests to achieve accountability (Johnson 2017). Pirtle and Szajnfarber have called for engineers to engage with public policy and more explicitly fold the goals of democratic society into their identities (Pirtle and Szajnfarber 2017).

Civic professionalism unites identity and practices with essential ingredients of epistemic humility and epistemological pluralism (defined here as knowing the limits of your expertise and perspective and valuing the expertise of different forms of knowledge), political engagement, and the common good as a primary goal of work. It responds to Cech's concerns about engineering by pushing back on technocratic professionalism, embracing politics by explicitly negotiating power and competing values, and by engaging in the broader sociopolitical questions connected to the context of their work rather than assuming the status quo. Compared to social entrepreneurship and the way it encroaches and depoliticizes engineering work (Spicer et al., 2019), civic professionalism is better aligned with the questions of politics and power embodied by community organizations. In this way, it also responds to calls to make sustainability and social justice paramount in engineering (Michelfelder & Jones, 2013).

In his article "The Good Engineer," Harris offers a compelling argument that virtue ethics, rather than negative or preventive ethics (such as codes/rules), should be the basis for engineering ethics and responsibility (2008); an argument later echoed by Schmidt (2014). Civic professionalism fits Harris's (2008) five components of virtue by:

1. Embodying a wide-ranging "character trait that includes emotions, perceptions, attitudes and sensibilities" (p. 158).
2. Connecting professional satisfaction to professional virtue by working toward meaningful contributions to communities.

3. Investing in “practical wisdom gained over time” learning from local contexts specific to each project (p. 158).
4. Comprising both technical and nontechnical virtues such as integrity and “a sensitivity to the social implications of technology” (p. 158).
5. The possibility of sketching a portrait of the civic engineer, as Boyte (2008) does when he tells the story of Laurie McGinnis’s journey from traditional civil engineer to civic professional in the ways she pays attention to a wide variety of stakeholders, facilitates public deliberation, and navigates social context, policy, and technical requirements.

## 3.4 Applications

In a first year course on Democracy and Media, I include a module on civic professionalism where we read Boyte and Dzur and I ask students to write a reflective essay about whom they see as examples of civic professionals in engineering and whether every engineer can and should be a civic professional. Reflexivity, as Cech calls for, is key. But I agree with Caroline Kreber, who argues that forging civic professional identities (what she calls “civic-mindedness”) must be through “transformative” undergraduate educational experiences, which are action-oriented and community-engaged, forcing students to challenge and revise their assumptions, values, and identities (Kreber, 2016).

In a capstone course organized around community-engaged design projects, my students write a personal reflection that asks them to consider their starting point at the beginning of the term; their outsider status in the communities they are working; how their knowledge, skills, and abilities have been stretched; and their professional identity and the cultures within their professional communities. In response to one student’s personal reflection essay, a past student team found and discussed the Spicer, Kay, and Ganz article (2019) critiquing social entrepreneurship that I cite in this paper.

Other engineering education scholars, in particular Athena Lin and Justin Hess, are working on models of “civic responsibility in engineering education” (Lin & Hess, 2021). Darshan Karwat’s work on “activist engineering” also emphasizes the political role engineers can take on in their practice and teaching (Karwat et al., 2015; Karwat, 2020).

An emerging practice I have been studying that embodies civic professionalism is “design refusal” (Graeff, 2020; Chowdhary et al., 2020). In 2019, student teams I advised refused to design projects for partners they were working with, defying traditional expectations of engineers to solve the problems they are presented with and bracket out the social and political concerns from the technical. In one case, undergraduate students in a public interest technology clinic were asked by an anti-trafficking organization to build a web scraper that would automatically capture data about sex solicitation online. The organization planned to analyze the data and pass it on to local law enforcement.

This was a challenging problem to engage with for multiple reasons. Initially, we were concerned about exposure to the sexual and possibly traumatic content that the students would be viewing in order to build a tool that could automatically scrape pictures and text from websites where mostly young girls were advertised for sex. I reached out to a friend I knew who had worked on sex trafficking as a socio-technical problem earlier in their career. They pointed me to several resources including a 2013 white paper on "How to Responsibly Create Technological Interventions to Address the Domestic Sex Trafficking of Minors," co-signed by faculty members from Olin's sister schools Babson College and Wellesley College (Bain et al., 2013).

These mentors helped the team think through the ethical challenges presented by building a tool that would be designed to eventually hand data about sex workers over to law enforcement without any assurance that the people identified on these websites would be treated with care and dignity. Ultimately, the team refused to build the tool, sending a detailed letter to their partner explaining why, and delivering a presentation to the Olin community describing their process and exhorting the audience to "design no harm."

Silicon Valley technologists have also practiced refusal in recent campaigns like "NoTechForIce" and "TechWontBuildIt." These movements have connected technical work to political consciousness and efforts to shift power within corporate engineering contexts. Refusal only succeeds through collective action. It requires civic and political engagement among engineers. They must develop and use civic skills, relying on formal or informal civic learning and reflection to integrate those skills and freshly clarified values with their professional identities and practices.

## 3.5 Conclusion

I strongly believe engineering needs a civic professionalism movement that explicitly connects technical work to the common good and civic responsibility, rather than the broader and more ambiguous ethical and social responsibility frames captured by historical definitions of engineering (Cheville, 2014) or codes of ethics (Sadowski, 2014). This is necessary not only to counteract depoliticization and disengagement in engineering and meritocracy's threat to democracy, but to prepare engineers to deal with "value change" (van der Weij et al., 2023). Changes in the form of value emergence, relevance, reprioritization, meaning shifts, or specification all demand that engineers be engaged with the sociopolitical contexts of their work, be able to facilitate deliberative processes, and be reflexive and broadly participatory as they work. Achieving this goal requires an emphasis on civic education, not just ethical education, in training engineers—one that creates many opportunities for learning experiences capable of transforming the identities and practices of young engineers.

## References

Association for Computing Machinery. (2018). ACM code of ethics and professional conduct. *Association for Computing Machinery*. https://www.acm.org/code-of-ethics.

Attard-Frost, B., De los Ríos, A., & Walters, D. R. (2023). The ethics of AI business practices: A review of 47 AI ethics guidelines. *AI and Ethics, 3*(2), 389–406. https://doi.org/10.1007/s43681-022-00156-6

Bain, C., Blair, D., Boyd, Danah, Bryan, N., Clark, D., Finklehor, D., Foot, K., Korytova, S., Latonero, M., Leary, M., Malone, P., Mitchell, K., Musto, J., McIntyre, S., Poucki, S., Shavers, A., Thakor, M., Tidball, S., & Wolak, J. (2013, April 8). *How to responsibly create technological interventions to address the domestic sex trafficking of minors*. http://www.danah.org/papers/TechnologistsCSEC.pdf.

Bietti, E. (2021). From ethics washing to ethics bashing: A moral philosophy view on tech ethics. *Journal of Social Computing, 2*(3), 266–283. https://doi.org/10.23919/JSC.2021.0031

Boyte, H. C. (2008). *The citizen solution: How you can make A difference*. Minnesota Historical Society Press.

Boyte, H. C. (2009). Civic agency and the cult of the expert. *Kettering Foundation*. https://eric.ed.gov/?id=ED510128

Boyte, H. C., & Fretz, E. (2010). Civic professionalism. *Journal of Higher Education Outreach and Engagement, 14*(2), 67–90.

Cech, E. (2012). Great problems of grand challenges: Problematizing Engineering's understandings of its role in society. *International Journal of Engineering, Social Justice, and Peace, 1*(2), 85–94.

Cech, E. A. (2014). Culture of disengagement in engineering education? *Science, Technology, & Human Values, 39*(1), 42–72. https://doi.org/10.1177/0162243913504305

Cech, E. A., & Sherick, H. M. (2015). Depoliticization and the structure of engineering education. In S. H. Christensen, C. Didier, A. Jamison, M. Meganck, C. Mitcham, & B. Newberry (Eds.), *International perspectives on engineering education: Engineering education and practice in context* (Vol. 1, pp. 203–216). Springer. https://doi.org/10.1007/978-3-319-16169-3_10

Cheville, A. (2014, June). Defining engineering education. In *2014 ASEE annual conference & exposition*. https://doi.org/10.18260/1-2-20248

Chowdhary, S., Daitzman, S., Eisenbud, R., Pan, E., & Graeff, E. (2020). Care and liberation in creating a student-led public interest technology clinic. In *2020 IEEE international symposium on technology and society (ISTAS)* (pp. 164–175). https://doi.org/10.1109/ISTAS50296.2020.9462188

Dzur, A. W. (2004). Democratic professionalism: Sharing authority in civic life. *The Good Society, 13*(1), 6–14. https://doi.org/10.1353/gso.2004.0026

Dzur, A. W. (2010). *Democratic professionalism: Citizen participation and the reconstruction of professional ethics, identity, and practice*. Penn State University Press.

Dzur, A. W. (2018). *Democracy inside: Participatory innovation in unlikely places*. Oxford University Press.

Finn, M., & DuPont, Q. (2020). From closed world discourse to digital utopianism: The changing face of responsible computing at computer professionals for social responsibility (1981–1992). *Internet Histories, 4*(1), 6–31. https://doi.org/10.1080/24701475.2020.1725851

Giorgini, V., Mecca, J. T., Gibson, C., Medeiros, K., Mumford, M. D., Connelly, S., & Devenport, L. D. (2015). Researcher perceptions of ethical guidelines and codes of conduct. *Accountability in Research, 22*(3), 123–138. https://doi.org/10.1080/08989621.2014.955607

Graeff, E. (2020). The responsibility to not design and the need for citizen professionalism. *Tech Otherwise*. https://doi.org/10.21428/93b2c832.c8387014

Graeff, E., & Wood, A. (2021). Undergraduate engineering as civic professionalism. *The Good Society, 30*(1), 76–95.

Green, B. (2021). The contestation of tech ethics: A sociotechnical approach to technology ethics in practice. *Journal of Social Computing, 2*(3), 209–225. https://doi.org/10.23919/JSC.2021.0018

Greene, D., Hoffmann, A. L., & Stark, L. (2019). Better, nicer, clearer, fairer: A critical assessment of the movement for ethical artificial intelligence and machine learning. *Hawaii international conference on system sciences 2019 (HICSS-52)*. https://aisel.aisnet.org/hicss-52/dsm/critical_and_ethical_studies/2.

Harris, C. E. (2004). Internationalizing professional codes in engineering. *Science and Engineering Ethics, 10*(3), 503–521. https://doi.org/10.1007/s11948-004-0008-6

Harris, C. E. (2008). The good engineer: Giving virtue its due in engineering ethics. *Science and Engineering Ethics, 14*(2), 153. https://doi.org/10.1007/s11948-008-9068-3

Hendren, S. (2022). *Sketch Model Episode 2*(2) [Mp3]. Retrieved April 17, 2023, from https://www.olin.edu/sketchmodel.

Irani, L. (2015). Hackathons and the making of entrepreneurial citizenship. *Science, Technology & Human Values, 40*(5), 799–824. https://doi.org/10.1177/0162243915578486

Irani, L. (2019). *Chasing innovation: Making entrepreneurial citizens in modern India*. Princeton University Press.

Jobin, A., Ienca, M., & Vayena, E. (2019). The global landscape of AI ethics guidelines. *Nature Machine Intelligence, 1*, 389–399. https://doi.org/10.1038/s42256-019-0088-2

Johnson, D. G. (2017). Rethinking the social responsibilities of engineers as a form of accountability. In D. P. Michelfelder, B. Newberry, & Q. Zhu (Eds.), *Philosophy and engineering: Exploring boundaries, expanding connections* (pp. 85–98). Springer. https://doi.org/10.1007/978-3-319-45193-0_7

Johnson, D. G. (2020). *Engineering ethics: Contemporary and enduring debates*. Yale University Press.

Kak, A., & West, S. M. (2023, November 9). The AI debate is happening in a Cocoon. *The Atlantic*. https://www.theatlantic.com/ideas/archive/2023/11/focus-problems-artificial-intelligence-causing-today/675941/.

Karwat, D. M. A. (2020). Self-reflection for activist engineering. *Science and Engineering Ethics, 26*(3), 1329–1352. https://doi.org/10.1007/s11948-019-00150-y

Karwat, D. M. A., Eagle, W. E., Wooldridge, M. S., & Princen, T. E. (2015). Activist engineering: Changing engineering practice by deploying praxis. *Science and Engineering Ethics, 21*(1), 227–239. https://doi.org/10.1007/s11948-014-9525-0

Kreber, C. (2016). *Educating for civic-mindedness*. Routledge.

Lin, A., & Hess, J. L. (2021, July). Developing a framework for civic responsibility in engineering education. In *2021 ASEE virtual annual conference content access, virtual conference*. https://doi.org/10.18260/1-2-36933

Marwick, A. (2017). Silicon Valley and the social media industry. In J. Burgess, A. Marwick, & T. Poell (Eds.), *The SAGE handbook of social media* (pp. 314–329). SAGE Publications Ltd.

McNamara, A., Smith, J., & Murphy-Hill, E. (2018). Does ACM's code of ethics change ethical decision making in software development? *Proceedings of the 2018 26th ACM joint meeting on European software engineering conference and symposium on the foundations of software engineering*. (pp. 729–733). https://doi.org/10.1145/3236024.3264833.

Metcalf, J., Moss, E., & boyd, danah. (2019). Owning ethics: Corporate logics, Silicon Valley, and the institutionalization of ethics. *Social Research, 86*, 449–476. https://doi.org/10.1353/sor.2019.0022

Michelfelder, D., & Jones, S. A. (2013). Sustaining engineering codes of ethics for the twenty-first century. *Science and Engineering Ethics, 19*(1), 237–258. https://doi.org/10.1007/s11948-011-9310-2

Munn, L. (2022). The uselessness of AI ethics. *AI and Ethics*. https://doi.org/10.1007/s43681-022-00209-w

National Academy of Engineering. (2017). NAE grand challenges for engineering. *National Academy of Engineering*. http://www.engineeringchallenges.org/File.aspx?id=11574&v=34765dff

National Society of Professional Engineers. (2019). Code of ethics for engineers. *National Society of Professional Engineers*. https://www.nspe.org/resources/ethics/code-ethics.

Ochigame, R. (2019, December 20). The invention of "Ethical AI": How big tech manipulates academia to avoid regulation. *The Intercept*. https://theintercept.com/2019/12/20/mit-ethical-ai-artificial-intelligence/.

Pawley, A. L. (2009). Universalized narratives: Patterns in how faculty members define "Engineering". *Journal of Engineering Education, 98*(4), 309–319. https://doi.org/10.1002/j.2168-9830.2009.tb01029.x

Pirtle, Z., & Szajnfarber, Z. (2017). On ideals for engineering in democratic societies. In D. P. Michelfelder, B. Newberry, & Q. Zhu (Eds.), *Philosophy and engineering: Exploring boundaries, expanding connections* (pp. 99–112). Springer. https://doi.org/10.1007/978-3-319-45193-0_8

Rességuier, A., & Rodrigues, R. (2020). AI ethics should not remain toothless! A call to bring back the teeth of ethics. *Big Data & Society, 7*(2), 2053951720942541. https://doi.org/10.1177/2053951720942541

Rosenberg, R. S. (1998). Beyond the code of ethics: The responsibility of professional societies. In *Proceedings of the ethics and social impact component on shaping policy in the information age* (pp. 18–25). https://doi.org/10.1145/276755.276768

Sadowski, J. (2014). Leaning on the ethical crutch: A critique of codes of ethics. *IEEE Technology and Society Magazine, 33*(4), 44–72. https://doi.org/10.1109/MTS.2014.2363985

Sandel, M. J. (2020). *The tyranny of merit: What's become of the common good?* (1st ed.). Farrar, Straus and Giroux.

Schmidt, J. A. (2014). Changing the paradigm for engineering ethics. *Science and Engineering Ethics, 20*(4), 985–1010. https://doi.org/10.1007/s11948-013-9491-y

Spicer, J., Kay, T., & Ganz, M. (2019). Social entrepreneurship as field encroachment: How a neoliberal social movement constructed a new field. *Socio-Economic Review, 17*(1), 195–227. https://doi.org/10.1093/ser/mwz014

van der Weij, F., Steinert, S., van de Poel, I., Alleblas, J., Melnyk, A., & de Wildt, T. (2023). Value change and technological design. *IEEE Technology and Society Magazine, 42*(3), 25–32. https://doi.org/10.1109/MTS.2023.3302406

Wagner, B. (2018). Ethics as an escape from regulation. From "ethics-washing" to ethics-shopping? In E. Bayamlioğlu, I. Baraliuc, L. Janssens, & M. Hildebrandt (Eds.), *Being profiled: Cogitas Ergo Sum. 10 years of 'Profiling the European Citizen* (pp. 84–88). Amsterdam University Press. https://doi.org/10.25969/mediarep/13281

Wisnioski, M. (2016). *Engineers for change: Competing visions of technology in 1960s America (reprint edition)*. The MIT Press.

**Erhardt Graeff** BS Information Technology (RIT 2006), BS International Studies (RIT 2007), MPhil Sociology (Cambridge 2008), PhD Media Arts and Sciences (MIT 2018). 2018–Today, Associate Professor of Social and Computer Science, Olin College of Engineering (Needham, MA), United States. Research areas: Design and use of technology for civic engagement, civic learning, and social justice, and the ethical responsibility of technologists as stewards of democracy.

# Chapter 4
# How Do We Value Data Privacy? Insights and Design Implications

**Sarah E. Carter and Heike Felzmann**

**Abstract** While many of us claim to value privacy, we often share data online or on our smartphones. The privacy literature is rich with discussions around how our own cognitive biases and heuristics are exploited to encourage such incongruous actions. However, further research is required to understand how we value data privacy—in particular, how personal values influence our decisions and why we may act against our values. Here, we discuss recent findings from a semi-structured interview investigation exploring how we value data privacy. Interview questions probed the relationship between privacy and values when selecting smartphone applications and privacy decisions more generally. To elucidate often-unconscious values, questions aimed to obtain statements pertaining to self-governing commitments representing values. Reflexive thematic analysis was used to interrogate instances where participants felt unable to act according to their values or where values appeared to conflict. From the interviews, we identified six major value themes that were motivationally related to privacy decisions, shaped by an individual's understanding of the value and the context. Tensions between values were also linked to context with individual variation. However, values *Use*, *Community*, and *Control* were frequent and in tension with each other. We discuss the implications of these findings for smartphone app developers and stores, for designing privacy technologies, and for regulators. We also point to areas of future research into how we value privacy, particularly engagement with traditionally underrepresented groups.

**Keywords** Values · Data privacy

S. E. Carter (✉)
Delft Univeristy of Technology, Delft, The Netherlands
e-mail: S.E.Carter@tudelft.nl

H. Felzmann
University of Galway, Galway, Ireland
e-mail: heike.felzmann@universityofgalway.ie



C. Didier et al. (eds.), *Engineering and Value Change*, Philosophy of Engineering and Technology 48, https://doi.org/10.1007/978-3-031-83549-0_4

## 4.1 Introduction

### *4.1.1 Dark Patterns and Privacy Self-Management*

While most of us claim to value our data privacy, we nevertheless constantly share our data online. The privacy literature is rich with discussions of this "privacy paradox" phenomenon (Spiekermann et al., 2001). Indeed, service providers that collect data utilize design strategies to coax users into data sharing, employing nudges, choice architectures, and cognitive biases and heuristics, leading to incongruous-seeming privacy behavior (Solove, 2021). These "dark patterns" or "deceptive design patterns" are ubiquitously used on privacy notices and have noticeable, measurable effects on data privacy decision-making (Brignull, n.d.; Gray et al., 2018; Mathur et al., 2019; Utz et al., 2019). In addition, existing structures—such as the attention economy, where service providers compete for our attention (Davenport & Beck, 2002; Goldhaber, 1997); surveillance capitalism, where data is utilized to predict and shape our future behavior (Zuboff, 2019); and platform economies controlling large swathes of the world market—make it all the more difficult to manage our privacy in keeping with our values.

In addition, too many privacy notices can cause privacy notice fatigue, where we simply "click through" notices rather than reading them (Ben-Shahar & Schneider, 2010; Schaub et al., 2015). In the EU, this issue has been further exacerbated due to mandated notices under the ePrivacy Directive and the General Data Protection Regulation (GDPR) (General Data Protection Regulation (GDPR), 2016; EPrivacy Directive, 2009). A study of online social networking services estimated this "click through" rate could be as high as 74% (Obar & Oeldorf-Hirsch, 2020). Eventually, this notice fatigue can create "apathetic users"—those who decide to consent every time to a service's data collection practices because they no longer care about their data privacy (Carter, 2022).

### *4.1.2 Personal Values in Privacy Decisions*

To address these challenges, one target has been the redesign and improvement of privacy notices to be more effective. Nearly every aspect of notice design—from timing, channel, modality, and control—has been extensively researched regarding nudging and the design of choice architectures (Schaub et al., 2015). However, more research is required to fully understand how we value privacy—in particular, how our personal values influence our data privacy decisions, and why we may not always act in accordance with them. Carter (2022, 2023) defines value-centered privacy decisions as the decisions that result in ends that accurately reflect one's personal values. Individual privacy decisions can therefore be evaluated to the extent to which they result in ends that accurately reflect the individual's personal values. Some privacy decisions represent users' values fully, whereas in other cases users may be nudged towards privacy choices that do not reflect their values.

To provide a differentiated conceptualization of value-centered privacy decision-making, Carter (2022, 2023) applied the four-dimensional theory of self-governance (4DT) to data privacy decisions (Killmister, 2017). This systematic theory of autonomy understands values as clusters of commitments pertaining to how to be or act that share a similar end state or object (Killmister, 2017). For example, consider an agent who values *personal security*.[1] Her value of *personal security* involves commitments not to share pictures of herself online and to not sharing her location with smartphone apps, both oriented towards the desirable end state of maintaining her *personal security*. Values can be in resolvable tension with each other (an agent deliberates on their values, forms an intention of what they best ought to do, and acts accordingly) or in double binds, here understood as values held in equal esteem with no means of acting in accordance with both. If the agent who holds *personal security* also holds *efficiency* in equal esteem regarding her privacy decision-making, she would not be able to act in accordance with both of her values given the current flurry of privacy notices.

Carter (2021, 2022, 2023) argues that designing for value-centered privacy decisions in privacy technologies, such as machine-learning personalized privacy assistants (PPAs) (Das et al., 2018; Liu et al., 2016; Story et al., 2020; Warberg et al., 2019), could be a means of promoting value-centered privacy decisions. Personalized privacy assistants are constructed to help individuals with managing the overwhelming number of privacy notices they face. PPAs utilize user privacy preferences to trigger personalized privacy notifications to help users manage their data privacy. PPAs can be contrasted with privacy-enhancing technologies (PETs), a collection of techniques and design choices inspired by privacy-by-design (Cavoukian, 2009), such as differential privacy and anonymization techniques (Garrido et al., 2022; Heurix et al., 2015).

Building off PPAs, Carter (2021, 2022, 2023) used the 4DT-based understanding of value-centered privacy decision to propose a value-centered privacy assistant (VcPA) for smartphone users. In particular, the VcPA aims to help users make privacy decisions based on their values and privacy preferences, rather than privacy preferences alone as is done in current PPAs. Theoretically, this approach will better promote autonomous, value-centered privacy decisions than existing systems (Carter, 2022). To accomplish this, the VcPA incorporates insights from human-computer interaction (HCI) and user experience (UX) research—in particular, utilizing friction, or small obstacles during a user's interaction with technology. Researchers have used friction to induce a shift from "fast" to "slow" user thinking and thereby facilitate more autonomous decision-making (Cox et al., 2016; Peters et al., 2018; Sandhaus, 2023; Terpstra et al., 2019). As described by Kahneman (2011), "Fast" System 1 thinking is automatic, mindless, and ripe with heuristics and biases, while "Slow" System 2 thinking is more conscious, deliberate, mindful, and value aware. Using these insights along with 4DT, the VcPA was designed to help users choose smartphone apps on the app store whose privacy practices are

[1] Values, where mentioned, will be italicized for clarity.

consistent with their personal values. The VcPA consists of three features: selective notices (notices based on a user's values and privacy preferences, instilled in a VcPA profile); exploratory notices (to confirm one's value profile is still relevant); and suggesting alternative apps (that are more value-consistent).[2]

In this work, we discuss findings from a qualitative, semi-structured interview investigation that explores how we value data privacy. In the interviews, questions probed the relationship between privacy and values when selecting a smartphone application as well as online privacy decisions made more generally.

## 4.2 Methodology

### *4.2.1 Participant Recruitment*

Participants were recruited from a concurrent user study of a prototype VcPA (Carter et al., 2023). This consisted of a pool of 273 English-speaking adult smartphone users who were recruited in the researchers' networks using snowball sampling. From these, we aimed to recruit 15–20 participants, considering both saturation (no new content) and informational power (sufficient richness of data) (Braun & Clarke, 2022; Glaser & Strauss, 1999; Malterud et al., 2016; Sandelowski, 1995). Interviews were conducted July-October 2022.

### *4.2.2 Interview Design*

Before the interview, participants provided their demographic details through a survey, received an information sheet, and completed a consent form. The interviews were 20 min to an hour in length, conducted in English over Microsoft Teams. They were semi-structured to allow participants sufficient space to explore their personal privacy and app decision-making process, while also covering defined issues of interest. Interview questions were designed to elicit affective responses to ascertain participants' values and value tensions, aiming to probe participants' commitments as a representation of values instead of asking directly about values (Fig. 4.1). This was based on the 4DT-informed understanding of value-centered privacy decisions defined in Carter (2022, 2023).

[2] For more on how these features can be operationalized, please see Carter et al. (2023), where a prototype VcPA system was developed and tested with users.

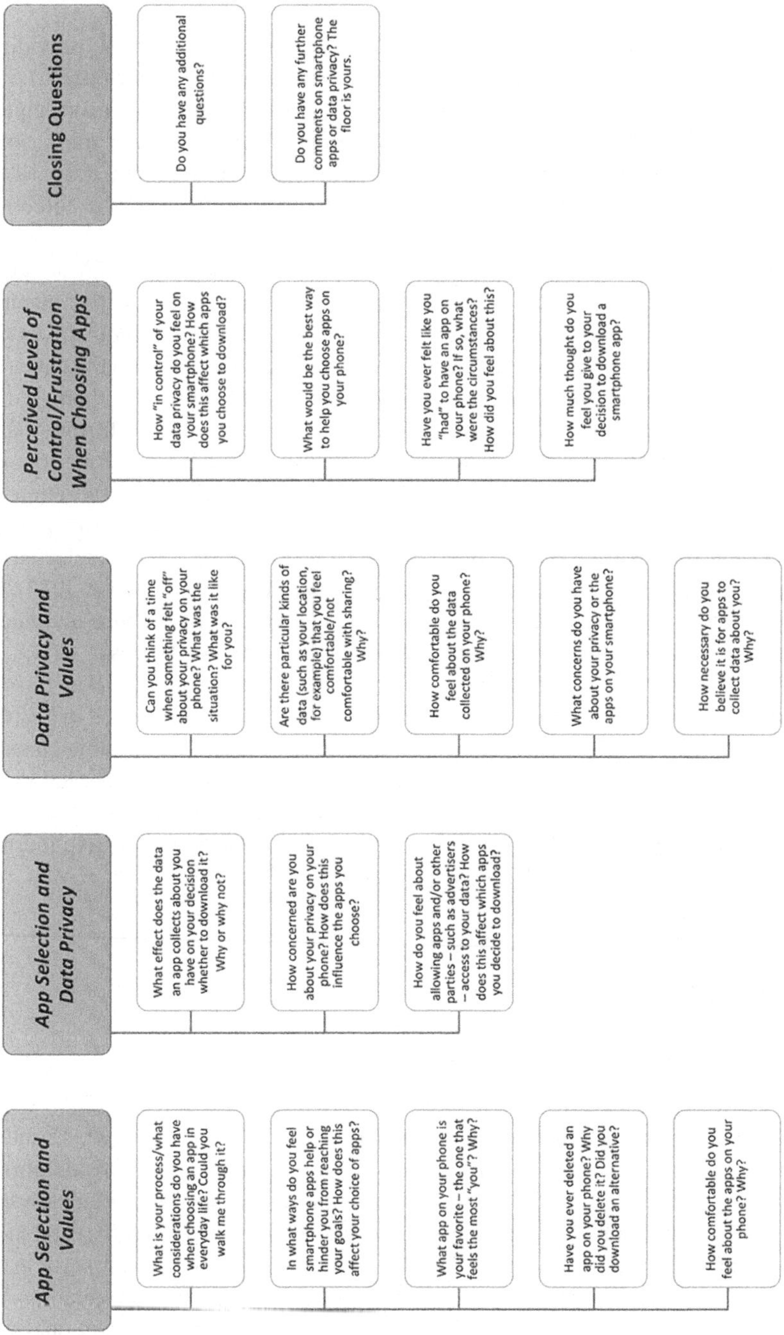


**Fig. 4.1** Interview Question Bank

### 4.2.3 Transcription and Data Analysis

Transcription and data analysis was conducted by author SEC with HF providing input, guidance, and feedback throughout the process. Otter.AI (https://otter.ai) was used to generate the initial transcripts, which SEC checked against the recording for accuracy and identifying information was removed. Interviews were transcribed clean verbatim—fillers words such as "um" and laughter were not transcribed unless deemed necessary for understanding the text (e.g., capturing affective responses).

Analysis was done using a process of Reflexive Thematic Analysis (TA) to identify values, tensions, and feedback in a predominately latent and inductive manner using a critical realist paradigm (Braun & Clarke, 2022). Reflexive TA involves six phases, which are iterative and nonlinear: (1) familiarizing oneself with the data; (2) coding; (3) generating initial themes; (4) developing and reviewing themes; (5) refining, defining, and naming themes; and (6) writing up results. To further interrogate the meaning, researchers conducting Reflexive TA keep a reflexive journal throughout the entire process. TA was selected because it allowed exploration of relevant values by grouping codes of commitment statements or affective responses into a theme for each identified value. Specifically Reflexive TA was selected to capture the researcher's role in knowledge production (Braun and Clarke, 2022). In this case, reflexivity was crucial due to the personal, yet abstract nature of values, and the role of the researcher's positionality in the data analysis process.

SEC completed this process twice, with interview transcripts shuffled for the second pass to ensure an evenly coded dataset. The interviews were coded in NVivo, with initial theme development occurring while coding to explore possible clustering of codes. During coding, codes were also reviewed, refined, and merged to be appropriately broad and representative. Themes were similarly checked to ensure coherence and were merged or split when deemed appropriate. A reflexive journal was kept by SEC throughout the process.

## 4.3 Results

### 4.3.1 Participant Demographics

In total, 46 participants agreed to be contacted for an interview. Of these, 18 interviews were conducted based on saturation and informational power considerations (Braun & Clarke, 2022; Glaser & Strauss, 1999; Malterud et al., 2016; Sandelowski, 1995). Participants' stated gender was evenly split between women (nine) and men (nine). Nearly all were adults (ages 25–64), with only one young adult participant (18–24) (Table 4.1). The majority also had (or were in the process of obtaining) a doctoral or master's degree (15) and three had a bachelor's degree. Nationalities were grouped by continent, the majority with European nationalities (nine),

**Table 4.1** Participant demographics

| | |
|---|---|
| Age | 1 Young Adults (18-24)<br>17 Adults (25-64)<br>0 Older Adults (65+)<br>0 Prefer not to say |
| Gender | **9** Women<br>**9** Men<br>**0** Other or prefer not to say |
| Education (have or in the process of obtaining) | **15** Doctoral or Master's Degree<br>**3** Bachelor's Degree<br>**0** Secondary Degree<br>**0** Prefer Not to Say |
| Nationality (by continent) | **9** Europe<br>**6** Asia<br>**2** North America<br>**1** Other or prefer not to say |
| Total | **18** |

followed by Asia (six), and all others (three). Fluent and native English speakers were evenly split, and all participants currently owned a smartphone.

### *4.3.2 Values Relevant to Privacy Decision-Making*

From the interviews, we identified six major value themes—*Community*, *Control*, *Use*, *Growth, Pleasure,* and *Safety*—that were motivationally related to privacy decisions, driven by both an individual's understanding of the value and the context in question (Table 4.2). However, three values—*Control*, *Community*, and *Use*—were prevalent in the data and spanned contexts, suggesting that these values could be especially relevant to privacy decision-making. These values were identified considering both the depth of discussion about these values and how distributed these discussions were in the data. For the sake of brevity, we focus on these three main values and only briefly present the other three values.

The value of *Community* was focused on social and world connection; it was "other facing." There were five sub-values: *accessibility*, *authenticity*, *benevolence and universalism*, *conformity*, and *connection*. While this general value was quite prevalent, some of the sub-values were observed with only one (*accessibility*) or a handful (e.g., *benevolence and universalism*) of participants. Even more prevalent sub-values, such as *connection*, could be understood in different manners. In addition, *Community* values were largely, but not exclusively, related to apps and services that had a social component. This suggests that *Community*'s relevance was quite context-dependent, and the applicability of its sub-values could vary by individual.

The first sub-value, *accessibility*, focused on data and technological implications on the most vulnerable. It was only expressed by one participant.

**Table 4.2** Summary of value themes and sub-values, ordered in terms of prevalence

| Values | Core concept | Sub-values | Example: |
|---|---|---|---|
| 1. *Use* | Convenience and function | 1. *Utility and function*<br>2. *Time and convenience* | **P06:** I choose [apps] for functionality [...] I have one game that I regularly play. And most of the others they like, use some digital linkage with my gym [and] my university. So [...] **my phone is very functional** |
| 2. *Control* | Freedom to make and the power that restricts choice | 1. *Power and choice*<br>2. *Tolerance*<br>3. *Dignity and respect* | **P15:** It's like **a war**, right. I mean, who's going to win? I mean, I'm sure [**the companies] are going to win** one way or another way, but I'm not going to give them **an easy win** |
| 3. *Community* | "Other-facing," social and world connection | 1. *Connection*<br>2. *Conformity*<br>3. *Authenticity*<br>4. *Benevolence and universalism*<br>5. *Accessibility* | **P06:** I am in contact with **[friends] on WhatsApp**, which I very much appreciate [...] we keep very, like **very connected** even though we haven't seen each other in two years [...] |
| 4. *Safety* | Safe and secure in personal life and world at large | 1. *Non-maleficence*<br>2. *Security*<br>3. *Trust and trustworthiness* | **P26:** As I go like, "only the absolute **necessary [cookies]**" [...] simply because people can track you [...] you're not doing anything illegal [...] but the fact is that identity fraud is a big problem [...] people's **identities have been stolen** |
| 5. *Pleasure* | Enjoyment, hedonism | | **P07:** There's been like, a couple of **games** that I've downloaded to like use on the airplane and things like that. And [...] they're **just not fun,** so I'll delete them |
| 6. *Growth* | Improving oneself and meeting goals | 1. *Learning and staying informed*<br>2. *Self-improvement* | **P22:** I think [the app] shows me probably **things that I value**, like how much exercise [I have had], it's kind of a nice reminder [...] and I guess this feedback is important in **helping me improve** [and] live a better lifestyle |

*Authenticity* focused on the meaningfulness and value of our online and offline interactions and was mostly consistent across participants. For most of the participants, the real world was viewed as more meaningful than the online world that constituted just one "large abstraction" (P08). However, one participant found that apps helped them engage with the authentic real world, such as using Google Maps "for exploring" (P13).

*Benevolence and universalism* centered on choosing actions that benefited others and the world. While this understanding was consistent across the few participants who found this value relevant, there was no consensus on what the "good" being promoted should be.

*Conformity* centered around fitting in with what others were doing when it came to apps or sharing data—that is, not becoming a "stone age type person" (P08). The sub-value spanned multiple contexts and had similar overall effects on app and data sharing behavior when present. For example, participants frequently stated that they would delete apps if an insufficient number of peers were using them, but some participants gave weight to specific groups (such as bloggers and influencers) that others did not.

Closely related to *conformity* was *connection*, which focused on forming and maintaining interpersonal relationships. Particularly prevalent was the use of WhatsApp and other social media services for connecting with loved ones, such as keeping "very connected even though [they] haven't seen [their friends] in two years" (P06). While largely associated with social media and messaging services, the value of sharing and supporting each other was also present in other contexts—for example, apps for translating and communicating with others in another language.

*Control* focused on the freedom to make choices and the power that restricts choice when it comes to our digital interactions. This includes online interactions, apps, data privacy choices, and their results. There were three sub-values: *dignity and respect*; *power and choice*; and *tolerance*, with *power and choice* in particular spanning contexts and quite prevalent in the data.

Firstly, *dignity and respect* encompassed the belief that humans are not data points and are deserving of respect and consideration. This was especially relevant when it came to data sharing and its effects, such as targeted advertisements, and not being "sneaky" when asking for consent to data collection (P15).

Secondly, *power and choice*, one of the most prevalent sub-values throughout the interviews, focused on one's perceived agency over their data privacy and app decisions as well as the power structures that may manipulate or coerce their choices. In this case, the power to manipulate was understood as an external other encouraging a choice against the user's intentions, such as a privacy notice designed to get someone to share their data (e.g., dark patterns) and targeted advertising to "sway [their] decision [...] [to] what [these] shops want [them] to do" (P22). Others expressed concerns around how data could be used to shape other aspects of their lives, with P09 describing how companies "have formed such an effectively predictive algorithm of your behaviors, that they know what you want before you know you want it, or they know what you're going to discuss before you think of discussing it". Understandably, there was a sense of futility among most of the participants, who viewed companies as simply too big to protect your data from. However, some participants viewed the futility in the face of powerful companies as a call-to-arms—comparing managing their data privacy to "a war," where you are "sure [the companies] are going to win," but unwilling to "give them an easy win" (P15).

Lastly, *tolerance* focused on accepting that others' data sharing choices and technological interactions may be different from one's own. Unlike other values and sub-values, this understanding was relatively universal among participants and valued in data sharing. A few participants, however, were more critical of others' data privacy choices, wishing others to take more responsibility to manage their data privacy.

The value *Use* was oriented towards the function, usability, or technical specifications of an app or service. There were two sub-values for *Use*: *time and convenience* and *utility and function*. It was one of the most prevalent values in the data that motivated privacy decisions. It spanned contexts and individuals; it was similarly understood; and it tended to motivate similar decisions. The degree of *Use* (especially *utility and function*) was also intimately associated with context-specific sub-values, such as defining a social media service's *utility and function* based on how well it promotes *connection*.

Firstly, *time and convenience* was consistently understood as promoting ease and speed in all aspects of life. Half of participants mentioned the value of apps and related data sharing for making everyday life more convenient and efficient, such as being able to book an appointment at a "nail bar" using an app (P26) or "double jobbin" by listening to audiobooks or podcasts while exercising (P08). Participants also mentioned the role of time and convenience on their data sharing behaviors, with three participants describing choosing to take "the path of least resistance" when encountering a privacy notice (P14).

Secondly, *utility and function* focused on the usefulness, necessity, purpose, and technical specifications of an app or service. This sub-value was mentioned by nearly every participant and in many different contexts and situations. Apps (and some online services) were consistently viewed as modern-day utilities whose necessity was intimately linked to the kind of app or service, including the degree to which context-related (sub-)values were being promoted by their use. For example, WhatsApp and social media were frequently viewed as necessary utilities for news gathering (*staying informed*) and communicating (*connection*). Lacking functionality or utility in relation to these sub-values motivated users to disengage with online services that were ostensibly dedicated to *connection*, such as deleting Facebook (P14: "90% of the things on your profile [are] competitions"). In addition, email apps and apps for online cloud services, such as Google Drive, were also viewed as necessary utilities for doing their jobs. The *utility and function* of the app was also considered in terms of its technical capabilities and requirements, including size, battery life, and interoperability with other devices.

The three other, less prevalent values were *Growth*, *Pleasure*, and *Safety*, with *Safety* being the most prevalent of the three. *Growth* focused on improving oneself and meeting one's goals. It had two sub-values: *learning and staying informed* and *self-improvement*. This value was highly context-dependent, constrained to apps and services related to education, news, social media, and health. The value *Pleasure* focused on the importance of enjoyment, hedonism, and engagement with apps and online services. It had no sub-values, nor did it relate too closely with the other values and was mostly relevant to apps for entertainment purposes. *Safety* focused on being safe and secure in one's personal life and the world at large. It had three sub-values—*non-maleficence*, *security*, and *trust and trustworthiness*. While relatively prevalent in the data, *Safety* sub-values tended to be more relevant in specific contexts, such as when engaging with health apps or when sharing sensitive data.

### 4.3.3 Values-in-Tension

In total, 14 tensions between the six values were coded in the interview data (Table 4.2). Like the values themselves, the tensions between values were also linked to context and individual understandings of the values. Even though value relevance to privacy or app decisions mostly varied based on the individual and the context, a handful of values such as *Use*, *Community*, and *Control* were quite prevalent in the data. Participants also frequently described ways of resolving tensions, with tensions existing on a spectrum from easily resolvable tensions to tensions that were difficult to resolve, often due to structural hindrances. However, resolving tensions between the more prevalent values was generally perceived as difficult, if not impossible. For the sake of brevity, we will focus on the tensions between the major values (*Control* vs. *Community*; *Control* vs. *Use*; *Community* vs. *Use*) before briefly introducing the other value tensions for completeness.

*Control* vs. *Community* could be best summarized as a "love/hate" (P22) relationship with *Community*-promoting apps and services and the limited ability to say "no" to them (*Control*). This was especially prevalent with social media or messaging services, such as WhatsApp. WhatsApp was seen as valuable to connect (subvalue of *Community*) with others, but participants also felt they had little choice in sharing data with it (*power and choice*), "feel[ing] like [they] need to be on WhatsApp [because] not many people meet up and chat like they used to" (P08). To manage this tension between *Community* and *Control*, participants described several strategies that varied based on the individual. Limiting one's social media presence (e.g., limiting social media accounts or only "friending" close friends or family) was the most frequent strategy to manage this tension. Regardless of strategy, however, participants did not find them very satisfactory. Attempts to resolve the tension between data privacy concerns (*power and choice*) and connecting with friends (*connection*) by moving to another platform was difficult, if not impossible, because their networks would not move with them.

*Control* vs. *Use* usually captured the feeling participants had that they would rather not share their data or have this app (*power and choice*), but that it was necessary to do so because of the utility provided (*utility and function*). This sentiment captures the recognition of the necessity of these apps and their required information sharing to exist in the modern world while still wishing for more control. Apps have become necessary to access critical, everyday services, such as transit (e.g., planes, taxi, bikes) or bank apps. Limited functionality or availability of alternative services also pressured participants into downloading apps. As one participant put it, "it's difficult to […] keep following your values when you have to download apps that are necessary for [wherever] you're traveling" (P13). Nearly every participant resolved the *Control* vs. *Use* tension with a strategy of data privacy pragmatism (Hoofnagle & Urban, 2014), understood as a cost/benefit analysis, a balancing of *utility* of the app or services against one's *control* over their data. A smaller number of participants utilized certain self-imposed rules, rather than pragmatic strategies, to manage the *Control* vs. *Use* tension. For example, if an app requires sharing one's home address, "then it doesn't matter how good the app is. That's very sensitive information" (P09). Regardless of the approach, however, participants did not feel that the tension was resolvable and suggested improvements, such as clarity on privacy notices of what data is required, to automatically "select their preferences, which are already to that standard" (P13), and making it easy "to not accept things [and] if you don't accept it, you can [still] use the app" (P22).

The *Community* vs. *Use* tension arose primarily between the sub-values *connection* and *time and convenience* where social media apps were described by participants as taking too much time from other aspects of their life, "a continuous distraction" (P02). Conversely, participants also mentioned how personalized social media feeds can cause different realities that make connecting with others difficult. One participant also expressed the *Community* vs. *Use* tension in terms of sub-values *authenticity* vs. *time and convenience*, where social media and messaging services were viewed as convenient but less authentic modes of connecting with others. To manage the tension between *Community* and *Use,* participants described several distinct strategies, such as deleting apps that were distracting or exiling them to a separate device. These strategies, usually satisfactory to some extent, focused primarily on resolving the tension between *connection* and *time and convenience.* The conflict between *authenticity* vs. *time and convenience* that results from personalized social media feeds causes different realities that make connecting with others difficult. This conflict also did not seem to be resolvable, making it a pressing value tension.

In addition to the tensions between *Control*, *Use*, and *Community*, two other tensions were prevalent in the data: *Pleasure* vs. *Use* and *Community* vs. *Safety*. Like *Community* vs. *Use, Pleasure* vs. *Use* also involved concerns around excessive time spent engaging with technology. In this case, however, the value of such interactions was not to connect with others, but for pleasure. Six participants expressed this tension in a variety of contexts, with social media (e.g., Tik-Tok), YouTube, game apps (e.g., Sudoku), and gaming platforms (e.g., PS2). *Community* vs. *Safety* primarily focused on the tension between destabilizing society (*security*) and the value of connecting with others on social media (*connection*), expressing concerns that "echo chambers" (term used by: P01, P07, P30) had "a detrimental effect on the world in things like politics and misinformation" (P14). Participants were also concerned about the mental health harms of consuming social media, whether it be the constant stream of "bad news" (P07) or getting into online arguments (sub-value tensions between *connection* and *non-maleficence*). To manage these tensions between *Safety* and *Community*, participants proposed mostly unsatisfactory management strategies (such as trying to be careful what one shares on social media) and more frequently pointed to ways social media could be improved, such as returning to a "true social media platform" (P07) as previously described in *Community* vs. *Use.*

In addition to these, there were six minor tensions which were mentioned in a more individualized manner, sometimes by as few as one participant (Table 4.2, previous page).[3] These were: *Control (power and choice)* vs. *Growth (self-improvement)*; *Growth (learning and staying informed)* vs. *Community (conformity)*; *Growth (self-improvement)* vs. *Use (time and convenience)*; *Pleasure* vs. *Control (power and choice)*; *Safety (non-maleficence)* vs. *Growth (self-improvement)*; and *Safety (non-maleficence/security)* vs. *Use (time and convenience/utility and function)*.

## 4.4 Discussion

The value themes described here suggest a complex, motivational relationship between a user's values and their privacy decision-making. Many of the values tended to gravitate towards specific contexts or situations. However, values of *Control* (especially *power and choice*), *Use* (*utility and function*), and *Community* (*connection*) were quite prevalent, and *Control* and *Use* were context-spanning.

These values also existed in frequent tension with each other. In addition, the ability to resolve tensions existed on a spectrum, with those that were the least resolvable caused by existing power structures, such as the attention economy. The tensions between *Control*, *Use*, and *Community* were especially difficult to resolve, resembling double binds. Our work suggests that to support value-centered privacy decision-making, each of these binds needs to be tackled in turn.

Firstly, *Control* vs. *Use* seems to result from lack of relevant controls or alternative courses of action for participants. While some participants were satisfied with a resolution strategy, such as data privacy pragmatism, most strategies were ultimately deemed unsatisfactory. To resolve this tension, then, we would need to provide greater privacy controls or alternative avenues of action that satisfy both *Control* and *Use*. In addition, we must also address how power structures, such as surveillance capitalism and platform economies, lead to a sense of futility of resistance, resulting in "apathetic users" (Carter, 2022, 2023).

Secondly, *Community* vs. *Control* appears to be an issue of data portability and interoperability of social media networks and platforms. Participants perhaps wished not to share their data with a social media service but felt unable to do so if they also highly valued *Community*. To resolve this tension, cross talk between platforms and messaging services or the ability to move one's individual network from one service to another would need to be facilitated.

Thirdly, the *Community* vs. *Use* tension was embodied by a desire to return to a "true social media platform" where the emphasis was less on time-wasting (*time and convenience*) addictive content and more on the personal news of loved ones

[3] Besides these six, there were three low prevalence tensions within one value but between sub-values (intra-value tensions): *authenticity* vs. *connectivity* and *conformity* vs. *benevolence and universalism* within *Community*; and *time and convenience* vs. *utility and function* within *Use*. These results suggest that, while sub-values within a value can come into tension with each other, they are less relevant to data privacy than tensions between values.

(*connectio*n and *authenticity*) (P07). This would involve moving away from the attention economy, which fuels societal polarization using subversive tactics targeting our reactive selves over our higher cognitive capacities (Davenport & Beck, 2002; Goldhaber, 1997).

### 4.4.1 Implications for PET and PPA Design

These findings could guide the design of privacy technologies, such as the PETs and privacy assistants. Below, some implications for the design features of one privacy assistant, the VcPA, and PETs are outlined.

To address the *Control* vs. *Use* tension, the "suggest alternatives" feature of the VcPA could allow users to find another app with similar *utility and function* while promoting their *power and choice*. This would require that VcPA user profiles are built using the major values of *Control*, *Community*, and *Use*. In addition, depending on what kind of app (the context) the user is looking at, our results show that other (less prevalent) values will likely also be relevant. However, by helping resolve this first tension between *Use and Control* via the "suggest alternatives" feature, the list of alternative apps might help identify alternatives that match the user's value set.

PETs, which use privacy-by-design as their default (Cavoukian, 2009), may disallow users from resolving the tensions between *Control*, *Use*, and *Community* if they deprioritize privacy preservation. In one sense, using privacy as default decreases rather than increases user *Control*. While there are valid reasons from a broader societal point of view for implementing privacy-as-default, this will require careful balancing of broader norms and values with an approach that respects individual value priorities. The VcPA, which captures personal values, could perhaps complement PETs and other technologies that are designed using theories that better capture relevant social norms and values on data privacy, such as Contextual Integrity (Nissenbaum, 2004), group privacy (Mittelstadt, 2017), or Privacy-as-Trust (Waldman, 2018).

### 4.4.2 Implications for Regulators

As we saw, the presence of structural factors—such as platform economies, the attention economy, and surveillance capitalism—were defining features of double bind situations, especially those involving *Community*. Addressing these factors will likely require that individual value-centered privacy assistants, such as the VcPA, to be employed alongside regulation. For example, a smartphone VcPA could work alongside the Digital Market Act (DMA), which will ensure data interoperability between services (Digital Markets Act, 2022), including between different social media and messaging services. This could help with *Community*-related value tensions which, as the interviews showed, were largely due to market dominance and non-interoperability of social networks.

### 4.4.3 Implications for Smartphone Apps and Stores

App developers, especially those that develop social media apps and messaging services, will also need to be mindful of these values to resolve value tensions. In addition, it is likely in their interest to do so. Market research has demonstrated that products that best align with user values increase app uptake, user satisfaction, and brand loyalty (Stocchi et al., 2022). Here, we find that users want a return to a "true" social media platform, one that optimizes the value of *Community* over attention (P07). Optimizing social apps for connectivity, rather than engagement or attention, would better align with user values and likely help app developers cultivate a more stable user base.

These results also have implications for stores wishing to implement a value-centered approach to privacy via a privacy assistant such as a VcPA. Like the app developers, it may also behoove app stores to deploy a tool that will assist consumers in identifying apps most relevant to them—in this case, apps that best align with their values. However, a VcPA deployed in, say, the Apple App Store, and claiming to understand a user's "values," may cause concern among users. This would be especially true for those users who are concerned about what companies know about them and how they may use their data to influence their behavior (e.g., valuing *Control* (*power and choice*)). To mitigate these concerns, the logics for a VcPA system—such as their value profile and other data—could be stored locally on the user's phone, interacting with the app store only to gather information concerning what data the app-in-question will request access to. This, however, would need to be carefully balanced with memory and storage considerations to ensure that the system is not too large and retains usability. It is therefore more feasible for an app store to offer an integrated VcPA to users, rather than to store the system locally. To respect the data privacy choices of all users, we would therefore recommend that app stores implementing VcPAs ensure that: (1) the data obtained from the VcPA, such as a user's value profile, *exclusively* be used for the purposes of running the VcPA; (2) engagement with the VcPA on the app store should be fully voluntary (opt-in); and (3) all of this information should be clearly communicated when viewing one's profile.

### 4.4.4 Future Research

A critical area of future work will be to engage with more diverse stakeholders to construct a more inclusive picture of personal values in data privacy decision-making. In our study, participants were predominantly WEIRD (Western, Educated, Industrial, Rich, and Democratic) participants (Henrich et al., 2010). Some participants also had high levels of knowledge in fields relevant to this study (e.g., philosophy, psychology, technology). When studying values in technology, other scholars have pointed out the importance of understanding the values of underrepresented and marginalized user groups to avoid violating their values (e.g., Shams, 2022).

The same would hold true when designing for value-centered privacy via a VcPA. Engagement with diverse voices is critically required to ensure that those aiming to design for value-centered privacy choices do so in an inclusive manner.

## 4.5 Conclusion

In this work, we have identified six values that are motivationally related to data privacy in a primary context-dependent and individualized manner. We also identified three prevalent values (*Control*, *Use*, and *Community*) that were in frequent, and often irresolvable, tensions with one another. This suggests that these values are the most critical when creating environments that promote users' value-centered privacy decisions. Future privacy technologies, such as VcPAs, could therefore be designed with these values in mind. In addition, their frequent tensions, caused largely by structural hindrances and existing power dynamics, resembled double binds—situations where one is unable to act in accordance with two values they hold in high esteem. While the tension between *Control* and *Use* could be resolved using the "suggest alternatives" feature of a VcPA, the large influence of existing structures supports VcPA deployment complemented by legislation as a possible means of promoting value-centered privacy decisions. In addition, app developers and app stores could promote products better aligned with consumer values by considering *Control*, *Community*, and *Use* and their tensions, either in design or through deployment of a VcPA. Lastly, while these results suggest that the values of *Control*, *Use*, and *Community* are the most relevant to data privacy decisions, engagement with diverse voices will be a critical area of future research to ensure that those implementing a value-centered approach do so in a manner that is fully inclusive, beneficial, and ethical.

**Acknowledgments** The authors would like to thank Mathieu d'Aquin, Kathryn Cormican, Daniel Susser, Ed Curry, John Danaher, John Breslin, and Karen Young for their feedback on this work. This study described in this work was completed during author SEC's PhD, which was funded by the Science Foundation Ireland Centre for Research Training in Digitally-Enhanced Reality (d-real) under Grant No. 18/CRT/6224.

## References

Ben-Shahar, O., & Schneider, C. E. (2010). The failure of mandated disclosure. In *U of Chicago law and economics* (no. 516; Olin working paper, Vol. 159, issue 3). https://doi.org/10.2139/ssrn.1567284.

Braun, V., & Clarke, V. (2022). In A. Maher (Ed.), *Thematic analysis: A practical guide* (1st ed.). SAGE Publications.

Brignull, H. (n.d.). *Deceptive design*. Retrieved September 29, 2022, from https://www.deceptive.design.

Carter, S. E. (2021). Is downloading this app consistent with my values?: Conceptualizing a value-centered privacy assistant. In D. Dennehy, A. Griva, N. Pouloudi, Y. Dwivedi, I. Pappas, & M. Mäntymäki (Eds.), *Lecture notes in computer science (including subseries lecture notes in artificial intelligence and lecture notes in bioinformatics): Vol. 12896 LNCS* (pp. 285–291). Springer. https://doi.org/10.1007/978-3-030-85447-8_25

Carter, S. E. (2022). A value-centered exploration of data privacy and personalized privacy assistants. *Digital Society, 1*(27), 1–24. https://doi.org/10.1007/s44206-022-00028-w

Carter, S. E. (2023). A value-centered approach to data privacy decisions. PhD thesis, Discipline of philosophy, University of Galway, Ireland.

Carter, S. E., D'Aquin, M., Spagnuelo, D., Tiddi, I., Cormican, K., & Felzmann, H. (2023). The privacy-value-app relationship and the value-centered privacy assistant. ArXiv.

Cavoukian, A. (2009). Privacy by design—the 7 foundational principles—implementation and mapping of fair information practices. Information and Privacy Commissioner of Ontario, Canada, 5. https://doi.org/10.1007/s12394-010-0062-y.

Cox, A. L., Gould, S., Cecchinato, M. E., Iacovides, I., & Renfree, I. (2016). Design frictions for mindful interactions: The case for microboundaries. *Proceedings of the 2016 conference on human factors in computing systems (CHI).* (pp. 1389–1397). https://doi.org/10.1145/2851581.2892410.

Das, A., Degeling, M., Smullen, D., & Sadeh, N. (2018). Personalized privacy assistants for the internet of things: Providing users with notice and choice. *IEEE Pervasive Computing, 17*(3), 35–46. https://doi.org/10.1109/MPRV.2018.03367733

Davenport, T. H., & Beck, J. C. (2002). The attention economy: Understanding the new currency of Buisness (Revised Ad). Harvard Buisness Review Press.

Digital Markets Act. (2022). Pub. L. No. 2022/1925. https://eur-lex.europa.eu/legal-content/EN/TXT/?uri=CELEX%3A32022R1925&qid=1703171666542.

ePrivacy Directive. (2009). *Official Journal of the European Union* 11. https://eur-lex.europa.eu/LexUriServ/LexUriServ.do?uri=OJ:L:2009:337:0011:0036:en:PDF.

Garrido, G. M., Sedlmeir, J., Uludağ, Ö., Alaoui, I. S., Luckow, A., & Matthes, F. (2022). Revealing the landscape of privacy-enhancing technologies in the context of data markets for the IoT: A systematic literature review. *Journal of Network and Computer Applications, 207.* https://doi.org/10.1016/j.jnca.2022.103465

General Data Protection Regulation (GDPR). (2016). https://eur-lex.europa.eu/legal-content/EN/TXT/PDF/?uri=CELEX:32016R0679.

Glaser, B. G., & Strauss, A. L. (1999). *The discovery of grounded theory* (1st ed.). https://doi.org/10.4324/9780203793206

Goldhaber, M. H. (1997). The attention economy and the net. *First Monday, 2*(4). https://doi.org/10.5210/fm.v2i4.519

Gray, C. M., Kou, Y., Battles, B., Hoggatt, J., & Toombs, A. L. (2018). The dark (patterns) side of UX design. In A. Dey, E. Cutrell, & M. C. Shruefel (Eds.), *Proceedings of the conference on human factors in computing systems (CHI)* (pp. 1–14). ACM. https://doi.org/10.1145/3173574.3174108

Henrich, J., Heine, S. J., & Norenzayan, A. (2010). The weirdest people in the world? *Behavioral and Brain Sciences, 33*(2–3), 61–83. https://doi.org/10.1017/S0140525X0999152X

Heurix, J., Zimmermann, P., Neubauer, T., & Fenz, S. (2015). A taxonomy for privacy enhancing technologies. *Computers and Security, 53*, 1–17. https://doi.org/10.1016/j.cose.2015.05.002

Hoofnagle, C. J., & Urban, J. M. (2014). Alan Westin's privacy homo Economicus. *Wake Forest Law Review, 14*(1), 261–351. https://doi.org/10.31235/osf.io/ta2z3

Kahneman, D. (2011). *Thinking, fast and slow*. Penguin Books.

Killmister, S. (2017). *Taking the measure of autonomy: A four-dimensional theory of self-governance* (1st ed.). https://doi.org/10.4324/9781315204932

Liu, B., Andersen, M. S., Schaub, F., Almuhimedi, H., Zhang, S., Sadeh, N., Acquisti, A., & Agarwal, Y. (2016). Follow my recommendations: A personalized privacy assistant for Mobile app permissions. In M. E. Zurko, S. Consolvo, & M. Smith (Eds.), *Proceedings of the twelfth symposium on usable privacy and security (SOUPS)* (pp. 27–41). USENIX.

Malterud, K., Siersma, V. D., & Guassora, A. D. (2016). Sample size in qualitative interview studies: Guided by information power. *Qualitative Health Research, 26*(13), 1753–1760. https://doi.org/10.1177/1049732315617444

Mathur, A., Acar, G., Friedman, M. J., Lucherini, E., Mayer, J., Chetty, M., & Narayanan, A. (2019). Dark patterns at scale: Findings from a crawl of 11K shopping websites. In E. Gilbert & K. Karahalios (Eds.), *Proceedings of the ACM conference on computer supported cooperative work (CSCW)* (Vol. 3, pp. 81:1–81:32). ACM. https://doi.org/10.1145/3359183

Mittelstadt, B. (2017). From individual to group privacy in big data analytics. *Philosophy and Technology, 30*(4), 475–494. https://doi.org/10.1007/s13347-017-0253-7

Nissenbaum, H. (2004). Privacy as Contextual Integrity. *Wash. L. Rev., 79*, 119–158.

Obar, J. A., & Oeldorf-Hirsch, A. (2020). The biggest lie on the Internet: Ignoring the privacy policies and terms of service policies of social networking services. *Information, Communication & Society, 23*(1), 128–147.

Peters, D., Calvo, R. A., & Ryan, R. M. (2018). Designing for motivation, engagement and wellbeing in digital experience. *Frontiers in Psychology, 9*. https://doi.org/10.3389/fpsyg.2018.00797

Sandelowski, M. (1995). Sample size in qualitative research. *Research in Nursing & Health, 18*(2), 179–183. https://doi.org/10.1002/nur.4770180211

Sandhaus, H. (2023). Promoting bright patterns. *CHI '23: Designing technology and policy simultaneously workshop*. (pp. 1–9). https://brightpatterns.org/.

Schaub, F., Balebako, R., Durity, A. L., & Cranor, L. F. (2015). A design space for effective privacy notices. In L. F. Cranor, B. Robert, & S. Consolvo (Eds.), *Proceedings of the 11th symposium on usable privacy and security (SOUPS 2015)* (pp. 1–17). USEBIX Association. https://doi.org/10.1017/9781316831960.021

Shams, R. A. (2022). Operationalizing values in mobile applications: A mixed-methods empirical study on agriculture apps for Bangladeshi female farmers. PhD thesis, Faculty of Information Technology, Monash University, Australia.

Solove, D. J. (2021). The myth of the privacy paradox. *George Washington Law Review, 89*(1), 1–51. https://doi.org/10.2139/ssrn.3536265

Spiekermann, S., Grossklags, J., & Berendt, B. (2001). E-privacy in 2nd generation E-commerce: Privacy preferences versus actual behavior. In M. P. Wellman & Y. Shoham (Eds.), *Proceedings of the conference on electronic commerce (EC)* (pp. 38–47). ACM.

Stocchi, L., Pourazad, N., Michaelidou, N., Tanusondjaja, A., & Harrigan, P. (2022). *Journal of the Academy of Marketing Science, 50*(2), 195–225.

Story, P., Cranor, L. F., Smullen, D., Sadeh, N., Acquisti, A., & Schaub, F. (2020). From intent to action: Nudging users towards secure mobile payments. *Proceedings of the 16th symposium on usable privacy and security (SOUPS)*. (pp. 379–416).

Terpstra, A., Schouten, A. P., de Rooij, A., & Leenes, R. E. (2019). Improving privacy choice through design: How designing for reflection could support privacy self-management. *First Monday, 24*(6). https://doi.org/10.5210/fm.v24i7.9358

Utz, C., Degeling, M., Fahl, S., Schaub, F., & Holz, T. (2019). (un)informed consent: Studying GDPR consent notices in the field. In D. M. Freeman, A. Mitrokotsa, & A. Sinha (Eds.), *Proceedings of the ACM conference on computer and communications security* (pp. 973–990). ACM. https://doi.org/10.1145/3319535.3354212

Waldman, A. E. (2018). *Privacy as Trust: Information Privacy for an Information Age*. Cambridge University Press.

Warberg, L., Acquisti, A., & Sicker, D. (2019). Can privacy nudges be tailored to individuals' decision making and personality traits? In J. Domingo-Ferrer (Ed.), *Workshop on privacy and electronic society (WPES)* (pp. 175–197). ACM. https://doi.org/10.1145/3338498.3358656

Zuboff, S. (2019). *The age of surveillance capitalism: The fight for a human future at the new frontier of power* (1st ed.). Profile Books.

**Sarah E. Carter** BA in Biochemistry and Music, MSc in Biomedical Science, PhD in Philosophy. 2023–present Postdoctoral Researcher in AI Ethics, TU Delft. Research areas: Bioethics, AI ethics, information ethics, data privacy, and ethical design, with an emphasis on personal values, meaning, and autonomy.

**Heike Felzmann** Dipl. Psych and M.A. in Philosophy. PhD in Philosophy. Involved in European research projects ERASMUS+ project PROSPERO on education for the use of robots in the caring professions (2018–2021); ELS issues in Wearable Robotics and the COST Action on Wearable Robotics (2017–2021); COST Action RANCARE (CA14208) on Rationing in Nursing Care (2016–2019); H2020 project (MARIO) on assistive robotics for older persons with dementia (2015–2018); COST Action CHIPME (IS1303) on ELS issues in genetic testing (2015–2017). 2004–present: Lecturer at University of Galway, currently Senior Lecturer in Philosophy/Ethics and head of discipline (Philosophy). Research areas: Research ethics, health care ethics, information ethics, moral theory, and (feminist) bioethics.

# Chapter 5
# Are Technologies Worthless? Environmentalist Engineers in Quest of Sustainable Compromises

**Antoine Bouzin**

**Abstract** Since the mid-2010s, engineers in France have become increasingly involved in environmental activism. This commitment is giving rise to an appropriation of environmental causes, including climate change, the degradation of biodiversity and the risks engendered by pollution, and giving rise to critical moments, i.e., situations of doubt with regard to lived reality. In particular, the discursive expression of these environmentalist engineers is sharply critical of technological objects accused of contributing to ecological catastrophe, for example through their greenhouse gas emissions and the destruction of ecosystems linked to the extraction of the natural resources needed for their production. These environmentalist engineers therefore question the values generally attributed to technologies—efficiency, reliability, utility—by the professional engineering community. They thus engage in the elaboration of compromises, i.e., a work of combination between several spaces of justification, in order to identify criteria according to which technological objects could be granted new positive values. A compromise emerges between industrial and ecological justifications, valuing technologies that limit their impact on the environment. Based on low-tech theories, another compromise adds a democratic criterion to the environmental protection criterion, valuing technological objects that are accessible to the greatest number of individuals. Finally, environmentalist engineers refuse to compromise with market values deemed incompatible with environmental protection.

**Keywords** Engineering · Values · Worth · Technology · Justification · Critique · Compromise · Environmentalism

---

A. Bouzin (✉)
Centre Émile Durkheim, Bordeaux, France
e-mail: antoine.bouzin@u-bordeaux.fr



C. Didier et al. (eds.), *Engineering and Value Change*, Philosophy of Engineering and Technology 48, https://doi.org/10.1007/978-3-031-83549-0_5

## 5.1 Introduction

Usually discreet in French public life, the professional group of engineers has attracted considerable media attention since the early 2020s due to their environmental activism (Miller, 2021). Indeed, collective actions by engineers in support of environmental causes have been multiplying in recent years: pleas at graduation ceremonies, disruptions at corporate forums at engineering schools, or participation in social mobilizations of civil disobedience and site occupation. Moreover, a number of environmentalist organizations have recently been created on the initiative of engineers in France: Avenir Climatique in 2007, Les Shifters in 2014, Ingénieurs engagés in 2017, Together for Earth and Pour un réveil écologique in 2018, La Bascule in 2019, and Vous n'êtes pas seuls and Les Désert'heureuses between 2020 and 2022. While this momentum confirms the observation of a "multiplication of environmentalist collectives and commitments" (Comby & Dubuisson-Quellier, 2023, p. 5) during the 2010s, the environmentalist commitment of engineers raises several questions, notably about their relationship to technologies. The latter are indeed the subject of fierce criticism in several forms of expression—tribunes, manifestos—used by environmentalist engineers:

> We don't see ecological and social devastation as "issues" or "challenges" to which we, as engineers, should find "solutions". We don't see science and technology as neutral and apolitical. We believe that technological innovation and start-ups will save nothing but capitalism. We don't believe in sustainable development, green growth or "ecological transition", an expression that implies that society can become sustainable without getting rid of the dominant social order.[1]

> Ecological catastrophe is not a problem that requires technical solutions or new models that apply to everyone. It is the very industrialization of the world itself that constitutes the catastrophe. Our lifestyles are unsustainable, and the system must collapse.[2]

In these indictments, technologies are presented more as problems than as solutions to the ecological emergency. Technological objects—in all their dimensions, from production to use, maintenance and recycling—are accused of numerous ecological ill effects: their contribution to climate change through greenhouse gas emissions, their participation in the destruction of ecosystems through the extraction of materials from the soil and subsoil, not forgetting the environmental pollution and health risks they engender. These positions raise questions insofar as engineers are, in the common sense, inextricably linked to technology. Trained in science and technology, engineers are a professional group that is expected to participate in the

[1] Extract from a speech given by the engineering students' collective Les Agros qui bifurquent at the graduation ceremony of one of France's most renowned engineering schools, AgroParisTech, on April 30, 2022: https://www.youtube.com/watch?v=SUOVOC2Kd50.

[2] Excerpt from the manifesto Courage, fuyons! Lettres et autres idées pour un mouvement de désertion des ingénieur.e.s (Courage, let's run away! Letters and other ideas for an engineers' desertion movement) from the association Les Désert'heureuses: https://desertheureuses.noblogs.org/files/2022/05/2021-courage-fuyons-desertheureuses-lecture-.pdf.

design, use, operation, efficiency and dissemination of technological objects in society.

The relationship between engineers, technologies, and environmental sustainability has already inspired a great deal of research. From an educational science perspective, McLaughlan (2007) and Huntzinger et al. (2007) have analyzed the integration of sustainable development knowledge and skills into engineering training and the new learning methods associated with this. Other works, more rooted in anthropology and sociology, approach these engineering-technology-environment relationships from the angle of work. Hojem & Lagesen (2011), for example, have studied how consulting engineers take environmental concerns into account in their professional activities. Date and Chandrasekharan (2018), for their part, looked more specifically at engineers involved in the design of innovative, environmentally and socially sustainable objects. However, the case of engineers specifically involved in environmental activism remains a subject not yet fully explored in academic literature (Bouzin, 2024). It is precisely these environmentalist engineers who will be studied in this paper. This investigation is thus more in line with the research conducted by Smith and Wisnioski. Smith (2021) studied American engineers employed in the fossil fuel sector and examined the ethical and environmental issues raised. Wisnioski (2012) looked at the social mobilization of American engineers in the 1960s and 1970s, including their critique of technology and concern for environmental issues.

In this chapter, I'll be looking at how environmentalist engineers perceive technology. The aim is to understand the extent to which the value they attach to technological objects varies over the course of their activist involvement. I will first present my theoretical background, my working hypotheses and the methodology I used in my investigation. I will then look at the professional community of engineers in France, and describe the dominant representations and values shared with regard to technology. I will then outline the critical moments, i.e., the temporal sequences during which environmentalist engineers come to doubt these representations and values. Finally, I'll explain the diversity of compromises that these engineers may or may not have worked out in order to give new, positive values to technologies.

## 5.2 Theoretical Background, Hypotheses, and Methodology

My investigation of environmentalist engineers' perceptions of technology—in other words, the meanings and values attributed to it—is based on an analysis of activist discourses and actions. From a theoretical standpoint, I draw on the work of Boltanski and Thévenot, who are rooted in pragmatic sociology and political philosophy. The latter closely study situations of crisis and dispute involving individuals and objects, and "the critical operations that people carry out when they want to show their disagreement without resorting to violence, and the ways they construct, display, and conclude more or less lasting agreements" (Boltanski & Thévenot,

2006, p. 25). Both authors point out that, in a situation of disagreement, "people need to justify their actions" (ibid., p. 37). They thus identified six categories of justification in situations of disagreement. Each of these categories of justification articulates a specific modality of evaluation and the ideality of a social order, in other words, a ranking of the value of objects and beings in the world. According to these two authors, these spaces of justification constitute "Worlds", each endowed with a common higher principle, specific worth and mode of evaluation. Incompatible with each other, these "Worlds" on which justifications are based make it possible to measure the value of a person or an object, and also offer the possibility of resolving disagreement or dispute. Boltanski and Thévenot distinguish between The Inspired World, The Domestic World, The World of Fame, The Civic World, The Market World, and The Industrial World (ibid.). I will therefore use these theoretical tools to analyze the discourses of environmentalist engineers and the way in which they qualify technologies.

The work presented in this chapter is based on three hypotheses:

1. The majority of the professional engineering community subscribes to the Industrial World, its common higher principle and its values.
2. The environmentalist commitment creates a moment of crisis for engineers: disagreement emerges over the value to be placed on technologies that are accused of contributing to ecological problems.
3. Environmentalist engineers work to build different compromises in order to attribute a new value to technologies.

My investigation is based on a comprehensive sociological approach in line with Weberian principles. In particular, the aim is to grasp as precisely as possible the representations, doubts, certainties and critical real-life experiences, particularly in relation to technologies, as well as the different registers mobilized in terms of arguments and justification. I therefore conducted 75 biographical interviews between 2018 and 2023 with French environmentalist engineers, students and graduates, involved in various activist organizations. I was keen to vary several social characteristics in the choice of people I met: gender, age, disciplinary specialization in engineering school, and professional sector. In addition to these interviews, I also analyzed the discursive productions of the environmentalist organizations in which these engineers are involved: manifestos, tribunes, conferences, and meeting minutes.

## 5.3 The Advent of Critical Moments in Environmentalist Commitment

The strong criticism of technology and industry presented in the introduction seems particularly surprising given that the professional group of engineers has embodied the figure of progress in France since the nineteenth century (Jarrige, 2014). The identity of civil engineers and their social recognition have historically been based "on their power to invent, on their efficiency in solving technical problems" (Vérin, 1998, p. 19). Since the Industrial Revolution, the positive value accorded to technology by engineers has been based on the promise of economic and social progress, i.e., a means for human societies to emancipate themselves from natural determinisms and thus acquire a form of autonomy in terms of social, political, and economic organization (Catton & Dunlap, 1980). Historical studies show the extent to which scientific discoveries and technological innovations of the nineteenth century shaped the imagination of French Saint-Simonian engineers, who developed modern utopias of highly technicalized, industrialized, and interconnected modern societies (Picon, 2002).

Before analyzing the multiple ways in which engineers come to doubt and reevaluate technologies in the course of their environmentalist engagement, I would like to discuss commonly shared perceptions of technological objects. The first hypothesis I wish to test is that the values commonly shared within the professional engineering community correspond rather closely to those attached to the Industrial World identified by Boltanski and Thévenot. According to the two authors, the Industrial World is based "on the *efficiency* of beings, their *performance*, their *productivity*, and their *capacity* to ensure *normal operations* and to respond usefully to *needs*" (Boltanski & Thévenot, 2006, p. 204). The values promoted in this space of justification—and which can be called upon in the exchange of arguments—are therefore performance, productivity, optimization, utility, measurement, rationalization processes, efficiency, and reliability. These values are very similar to those promoted by the organizations and institutions representing engineers in France, which are also explicitly mentioned in the diversity of their discursive production (codes of conduct, ethics charters, white papers):

> The engineer strives to produce the best result, at the best cost, in the best conditions and within the given timeframe.[3]

> The engineer is a source of innovation and a driving force for progress.[4]

> Pushing the innovation process all the way through to application and market, foreseeing technological breakthroughs and their consequences, ordering them and situating them precisely in time cannot be done without an in-depth knowledge of the sciences and techniques involved. This is an approach in which the engineer or, even better, multidisciplinary teams of engineers and scientists play an essential role.[5]

---

[3] Extract from the *Code de déontologie de l'ingénieur* (Code of conduct for engineers) produced in 1996 by the Conseil national des ingénieurs et scientifiques de France (CNISF), quoted by Didier (2008, p. 79).

[4] Extract from the *Charte éthique de l'ingénieur* (Ethical charter for engineers) produced in 2001 by the CNISF, quoted by Didier (2008, p. 82).

[5] Extract from *Cahier blanc. Réussir le futur: jouons collectif* (White paper. Making the future a success: let's play together) produced in 2015 by the Société des ingénieurs et scientifiques de France (IESF, 2015, p. 3).

Technological objects are therefore highly valued within the professional engineering community, specifically for their efficiency, performance, and reliability. I'm not saying that French engineers form a homogeneous group structured exclusively by these Industrial World values. Rather, my aim is to draw up an ideal-type in the Weberian sense, i.e., to construct a generic portrait of engineers by highlighting relevant characteristics. It is also interesting to note that these Industrial World values are reaffirmed by engineers when they are surveyed about environmental issues. In their survey on environmental concerns, Didier & Talin (2015) show that 87% of engineers subscribe to the following proposition "Man's genius will enable the Earth to remain livable", while 51% of French people as a whole do. How then can we explain why engineers come to doubt technological objects and question their value?

I would now like to turn to my second hypothesis, according to which environmental activism plays an important role in changing perceptions of technology. In the course of their activist involvement, environmentalist engineers encounter critical moments, described by Boltanski & Thévenot as follows: "People, involved in ordinary relationships, who are doing things together – let us say, in politics, work, unionism – and who have to coordinate their actions, realize that something is going wrong; that they cannot get along anymore; that something has to change" (Boltanski & Thévenot, 1999, p. 359). Critical moments thus constitute "instances in which order breaks down, as evidenced by some moment of crisis, disequilibrium, critique, dispute, or contestation" (Boltanski & Thévenot, 2006, p. 25). Analysis of the interviews conducted with environmentalist engineers reveals a number of recurring patterns in the nature of the events during which these engineers realize that something is wrong—these events are also often triggers for the activist commitment itself. These events are often based on the discovery of phenomena such as climate change, the destruction of biodiversity or pollution—and the consequences these phenomena engender, such as global warming, the emergence of uninhabitable land areas, rising sea levels, the disappearance of animal and plant species or health risks for human populations. This acculturation of engineers to environmental issues generally occurs through the appropriation of scientific and technical information—documentaries, reports from scientific organizations such as the Intergovernmental Panel on Climate Change (IPCC), conferences by scientists or experts.[6]

This process of discovery, exploration and assimilation of environmental issues often provokes "moral shock" (Jasper, 1997), i.e., deep indignation and a desire for action. It is therefore through the experience of such situations that environmentalist engineers come to question the value of technologies insofar as their operation and use may involve greenhouse gas emissions (participation in climate change) and the extraction of raw materials (participation in the destruction of ecosystems). Perceived in the light of the above-mentioned environmental risks, technological

[6] Jean-Marc Jancovici's lectures are particularly popular with environmentalist engineers. Jean-Marc Jancovici is himself an engineer, a graduate of the École Polytechnique—the most famous and recognized engineering school in France—and a fervent advocate of the decarbonization of society, with a strong commitment to energy and climate issues.

objects are ultimately reevaluated and requalified as unsuitable, counterproductive, a source of nuisance, and deterioration—in other words, in total contradiction with the values tending to be defended by the professional engineering community. This reappraisal thus gives rise to a critical moment, a disagreement with the values previously attributed to technology. In the next section, I will look at how environmentalist engineers attempt to resolve these critical moments, and give new values to technological objects.

## 5.4 Overcoming Disagreement, Working Out Compromises

The challenge to technologies can be interpreted as an ordeal undergone by environmentalist engineers, in other words, a situation in which they "experience the vulnerability of the social order, by the very fact that they experience doubt about what reality is" (Lemieux, 2018, p. 39). The criticisms formulated by environmentalist engineers against technologies are thus based on a new space of justification that could be called the Environmental World, the Green World or the Ecological World (Lafaye & Thévenot, 1993). This Ecological World is built on a foundation of values oriented, according to Latour (1995), towards protecting and caring for terrestrial entities. I will now attempt to show the combined cognitive and practical work carried out by environmentalist engineers to overcome these trials—in other words, these critical moments of doubt—and attempt to restore meaning to lived reality. This work is based—and this is the core of my third hypothesis—on the elaboration of compromises: "A compromise suggests the possibility of a principle that can take judgments based on objects stemming from different worlds and make them compatible. It aims at a common good that transcends the two different forms of worth in presence by including both of them" (Boltanski & Thévenot, 2006, p. 278). For environmentalist engineers, this means ensuring the compatibility of technological objects (Industrial World) with ecological objects (Ecological World). More specifically, this involves protecting terrestrial entities from the degradation caused by technologies. I will begin by presenting the main compromise that I identify in the discourses and justifications of environmentalist engineers. I will then specify to what extent this compromise can be adapted, or not, to other spaces of justification, notably the Civic World and the Market World. The cognitive elaboration of compromises is also accompanied by the concrete production of objects. The function of these compromise-specific objects is to hold the compromise together, to solidify it, to give it a known and recognized material existence. I will therefore also indicate the objects specifically designed within the framework of the compromises shaped by environmentalist engineers.

### 5.4.1 Defining Sustainable Technologies

The first compromise shaped by environmentalist engineers involves two spaces of justification: the Ecological World and the Industrial World. The question is: under what conditions can technologies remain—or become (again)—effective, useful, efficient and optimized in a context of environmental urgency? One line of thought concerns the socio-technical framework (Flichy, 2003) of technological objects. Flichy defines the socio-technical framework of a technological object as the articulation between its operating framework, in other words, the set of "knowledge and know-how that are mobilized or can be mobilized" (ibid., p. 124), and its usage framework, which refers to the diversity of possible social uses of the object in question. The process of elaborating this industrial-ecological compromise therefore consists in establishing evaluation criteria according to which technologies, both in their operation and in their use, can be considered efficient, useful, and benign with regard to the environment. Publicly promoted by well-known French issue entrepreneurs, including engineer Jean-Marc Jancovici, the most common criterion is decarbonation, i.e., the reduction of greenhouse gas emissions linked to technological objects—in their design, production, use, maintenance, and recycling (Bouzin, 2023).

I will now present a concrete case to illustrate the form this industrial-ecological compromise can take. To do this, I am going to draw on the work carried out by the négaWatt Association. Co-founded by an engineer, and composed of many engineers, the association focuses on climate issues and produces energy transition scenarios. The purpose of these scenarios is to map out "the path to a desirable and sustainable future" and to describe "the solutions for achieving it" based "on technologies that are sufficiently mature for them to be deployed on a large scale" (Néga Watt, 2022, p. 5). Below is an extract (ibid.) from *La transition énergétique au cœur d'une transition sociétale. Synthèse du scénario négaWatt 2022* (The energy transition at the heart of a societal transition. Summary of the negaWatt 2022 scenario):

> Like its predecessors, the négaWatt 2022 scenario is based on the négaWatt approach, which consists of:
>
> - Prioritizing essential needs in individual and collective energy use through sobriety actions (eliminating waste, containing urban sprawl, opting for alternatives to the car, reducing packaging, etc.);
> - Reduce the amount of energy needed to satisfy the same needs through energy efficiency (insulating buildings, improving the efficiency of electrical appliances or vehicles, etc.);
> - Favoring renewable energies for their low environmental impact and inexhaustibility. These are flow energies, as opposed to stock energies, based on finite reserves of coal, oil, fossil gas and uranium.

The value of technological objects is thus based on their ability to satisfy these three criteria: sobriety in use, energy efficiency, and the use of renewable energy sources. In this respect, wind turbines, bicycles, and heat pumps are more valuable technologies than coal-fired power plants, sport utility vehicles, and oil-fired boilers. The environmentalist engineers involved in this compromise produce specific objects to consolidate this same compromise: for example, carbon footprints[7] or other tools to measure, control, and reduce the negative consequences of technologies on the environment in terms of greenhouse gas emissions, but also the consumption of natural resources, waste production, and pollution.

### 5.4.2 *Valuing Democratic Criteria*

The compromise between the Industrial and Ecological Worlds thus reconfigures the judgments held by environmentalist engineers with regard to technologies. However, an analysis of the discourses of environmentalist engineers allows us to characterize another form of compromise that calls on a new space of justification, the Civic World, whose values are based in particular on the pursuit of the general interest, solidarity, equality, and democratic expression (Boltanski & Thévenot, 2006). In this compromise, which I will call industrial-ecological-civic (because it brings together these three modes of justification), the value placed on technologies depends not only on their ability to limit their impact on the environment—the result of the industrial-ecological compromise—but also on their ability to express the general interest, to be accessible to the entire population and to meet the needs of the greatest number.

I'm going to illustrate a case of this industrial-ecological-civic compromise using an extract from an interview conducted with an environmentalist engineer, Alan. Alan is 36 years old and specializes in biochemical engineering. He has completed a doctoral thesis on life-cycle analysis,[8] works in a consultancy firm and is involved in an activist organization specializing in raising awareness and popularizing energy-climate issues. Alan is very interested in the low-tech movement theorized by Bihouix[9] in particular, according to which "instead of seeking topdown solutions to current environmental and societal challenges, instead of seeking ever more innovation, high technology, digitalization, competition, networking, growth – giving them names such as 'sustainable development', 'green growth', 'Economy 2.0' – we must instead direct ourselves, as quickly as possible, to a society based primarily on simpler technologies, undeniably cruder and more basic, maybe a little less powerful, but much more resource-efficient and locally controllable" (Bihouix, 2020, p. xiii-xiv). This is how Alan describes low-tech technologies:

> Low-tech means trying to answer the question: how can we democratize technology? Low-tech is low technology, resilient technology. The question is: what can we develop? What know-how do we already have that we can democratize, that the greatest number of people can appropriate? How can we appropriate technologies? So there's a very strong social and

[7] Carbon footprints are methods for measuring greenhouse gas emissions linked to human activities.

[8] A method for assessing the environmental impact of technological objects and production processes.

[9] Philippe Bihouix is a French engineer and graduate of the Ecole Centrale de Paris, one of France's most renowned engineering schools.

> sociological component, as well as an economic one. We are currently considering research projects on this subject, and indicators of this type: how many people on Earth could have access to this technology given the current state of know-how and available resources?

As we can see from Alan's discourse, the value of technological objects no longer depends solely on their limited effects on the environment, but also on their democratic characteristics, i.e., their capacity to be appropriated by the greatest number of individuals. A new social and political criterion is thus added to the evaluation process. In addition to technologies that contribute to environmental degradation, "there are products that are unaffordable in a low-tech world, by their very design but above all their societal uselessness, from yachts to limousines, from private planes to rockets for (future) space tourists" (Bihouix, 2020, p. 108). The consolidation of this industrial-ecological-civic compromise is based on the manufacture of low-tech objects, such as washing-machine bicycles, solar ovens and driers, and household waste composters. These objects are therefore more valuable than high-tech objects, which are deemed more resource-intensive, more expensive and, consequently, less accessible to the population as a whole.

### 5.4.3 Refusing to Compromise with the Market World

After identifying the industrial-ecological and industrial-ecological-civic compromises, I looked at the existence, within the discourse of environmentalist engineers, of a compromise with the Market World. Individual desires, the possession and accumulation of wealth, the pursuit of self-interest, competition, and competitiveness are the values promoted in this space of justification. Such a compromise of the industrial-ecological-market form is also identifiable in academic literature. First theorized in Germany in the late 1980s (Jänicke, 1990), the concept of "ecological modernization" aims "to link the drive for modernization in the developed market economies and the long-term requirement for an ex ante more environmentally friendly development through innovation in environmental technologies" (Jänicke, 2008, p. 557). More specifically, "the task is therefore to change the direction of technological progress and to put the compulsion for innovation at the service of the environment. The emphasis of this approach lies on the possibility of ecological-economic 'win-win' solutions that can be achieved, above all, through cost reduction and competition for innovation" (ibid., p. 558). In other words, in the industrial-ecological-market compromise formulated by the theories of ecological modernization, the value of technological objects is assessed on the one hand according to their minimal impact on the environment, and on the other hand according to their capacity to fit into the market economy, to integrate into a logic of so-called green economic growth and to become desirable and marketable goods.

However, I found no trace of this form of compromise in the analysis of interviews conducted with environmental engineers or in the discursive forms produced by the various activist organizations studied. On the contrary, the notion of green growth is often strongly criticized by the environmentalist engineers I met. Let me cite the case of Lyam. Lyam is 41 years old, has a degree in engineering, specializing in energy, and works as an independent consultant on ecological transition issues. He is also project manager for a think tank.

> I don't believe in it [green growth], it's just a way of doing the same thing without changing anything, well without changing the foundations, the root causes [of ecological problems]. It's a way of maintaining business as usual, it's a kind of marketing [...] Green growth goes against what we say about the fact that GDP [Gross Domestic Product] growth is inseparable from energy and therefore from greenhouse gas emissions.

The notion of green growth, which faithfully crystallizes this industrial-ecological-market compromise, is rejected using scientific and technical arguments: economic growth is said to be irremediably coupled with the growth of physical flows, one of the causes of ecological problems. Green growth is thus seen as a false solution that in no way solves ecological problems, and could even make them worse. This does not mean, however, that environmentalist engineers are indifferent to economic issues. The think tank where Lyam works, for example, is specialized in the decarbonization of the French economy, and provides sector-specific scenarios for reducing fossil fuel consumption, and hence greenhouse gas emissions.

## 5.5 Conclusion

In this chapter, I aim to show the extent to which engineers' involvement in environmental activism calls into question well-established interpretative schemas, particularly concerning the value placed on technological objects perceived as efficient, useful, and reliable. On the one hand, this activist commitment implies a new sociability, and on the other, the appropriation of environmental causes, which translates into the defense of ecological values based on the protection of the environment and ecosystems. Environmental activism thus gives rise to critical moments, doubts about the value of technologies in a context of ecological urgency manifested by phenomena such as climate change, biodiversity degradation and pollution. In these challenging situations, environmentalist engineers build compromises, cognitive arrangements between several spaces of justification—Boltanski and Thévenot's Worlds—in order to give new meaning to lived reality and attempt to resolve this disagreement over the value to be accorded to technological objects. A compromise between the Industrial World and the Ecological World emerges, enabling technologies to be valued once again, provided that environmental protection criteria such as decarbonization are met. This industrial-ecological compromise can also take another form by integrating Civic World. In this case, the value of technological objects includes, in addition to the criterion of environmental protection, a democratic criterion based on accessibility to the greatest number. Finally, the compromise with the Market World crystallized in academic theories of ecological modernization is strongly contested in the discourse of environmentalist engineers.

Many questions remain. We might, for example, ask whether this industrial-ecological compromise is defended only by environmentalist engineers, or whether it appears to be shared, on the contrary, by a larger fraction of the professional engineering community in France. We might also ask why environmentalist engineers integrate the Civic World—or, on the contrary, reject the Market World—in their work to elaborate a compromise. There are several stimulating paths worth exploring. The first concerns the politicization of engineers engendered by their involvement in environmental activism (Coutant & Bouzin, 2023). The appropriation of a political analytical grid for environmental problems could provide an explanation for the integration of Civic World. The second possibility concerns the relative downgrading of engineers in the economic field in terms of decision-making power, to the benefit of other professional categories such as managers and shareholders. The rejection of Market World could therefore be interpreted as a desire on the part of these environmentalist engineers to shift the balance of power in their favor (Coutant, 2020). Ultimately, this processual evolution in the perceptions and values accorded to technologies is accompanied by questions among environmentalist engineers regarding their professional activities, in particular their validity and relevance to the ecological emergency. Here we find one of Mitcham's conclusions that engineering does not in itself contain clear and determined objectives, which consequently leads to divergent negotiations and interpretations among engineers (Mitcham, 2019).

## References


Bihouix, P. (2020). *The age of low tech*. Bristol University Press.

Boltanski, L., & Thévenot, L. (1999). The sociology of critical capacity. *European Journal of Social Theory, 2*(3), 359–377.

Boltanski, L., & Thévenot, L. (2006). *On justification: Economies of worth*. Princeton University Press.

Bouzin, A. (2023). The language of ecology used by militant engineers: A cause suspended between technical reductionism and political consideration. *Socio, 17*, 139–160.

Bouzin, A. (2024). Negotiating engineering and activism: French environmentalist engineers conforming to, shifting, and overstepping professional boundaries. *Engineering Studies, 16*(2), 133–154.

Catton, W. R., Jr., & Dunlap, R. E. (1980). A new ecological paradigm for post-exuberant sociology. *American Behavioral Scientist, 24*(1), 15–47.

Comby, J.-B., & Dubuisson-Quellier, S. (2023). Vers une politisation des mobilisations écologiques contemporaines? In J.-B. Comby & S. Dubuisson-Quellier (Eds.), *Mobilisations écologiques* (pp. 5–17). Presses universitaires de France.

Coutant, H. (2020). Engineers' capitalism: Domestication of financialization in an aeronautics industrial group. *Sociologie, 11*(1), 73–89.

Coutant, H., & Bouzin, A. (2023). Ingénieurs écologistes. Reconfigurations par les enjeux environnementaux des luttes au sein du salariat de confiance du capitalisme. Paper presented at the study days "Ecology & Social Classes", Paris, France.

Date, G., & Chandrasekharan, S. (2018). Beyond efficiency: Engineering for sustainability requires solving for pattern. *Engineering Studies, 10*(1), 12–37.

Didier, C. (2008). *Penser l'éthique des ingénieurs*. Presses Universitaires de France.
Didier, C., & Talin, K. (2015). Attitudes et dynamiques environnementales des ingénieurs. Une entrée par la sociologie de l'environnement. Sociologies.
Flichy, P. (2003). *L'innovation technique. Récents développements en sciences sociales. Vers une nouvelle théorie de l'innovation*. La Découverte.
Hojem, T. S. M., & Lagesen, V. A. (2011). Doing environmental concerns in consulting engineering. *Engineering Studies, 3*(2), 123–143.
Huntzinger, D. N., Hutchins, M. J., Gierke, J. S., & Sutherland, J. W. (2007). Enabling sustainable thinking in undergraduate engineering education. *International Journal of Engineering Education, 23*(2), 218–230.
IESF (2015). Cahier blanc. Réussir le futur: jouons collectif, Paris, France.
Jänicke, M. (1990). State Failure. University Park: Pennsylvania State University Press.
Jänicke, M. (2008). Ecological modernisation: New perspectives. *Journal of Cleaner Production, 16*(5), 557–565.
Jarrige, F. (2014). *Technocritiques. Du refus des machines à la contestation des technosciences*. La Découverte.
Jasper, J. M. (1997). *The art of moral protest: Culture, biography, and creativity in social movements*. University of Chicago Press.
Lafaye, C., & Thévenot, L. (1993). Une justification écologique ? Conflits dans l'aménagement de la nature. *Revue française de sociologie, 34*(4), 495–524.
Latour, B. (1995). Moderniser ou écologiser. A la recherche de la Septième cité. *Écologie & politique, 13*, 5–27.
Lemieux, C. (2018). *La sociologie pragmatique*. La Découverte.
McLaughlan, R. G. (2007). Instructional strategies to educate for sustainability in technology assessment. *International Journal of Engineering Education, 23*(2), 201–208.
Miller, M. (2021). *La révolte. Enquête sur les jeunes élites face au défi écologique*. Seuil.
Mitcham, C. (2019). *Steps toward a philosophy of engineering. Historico-philosophical and critical essays*. Rowman & Littlefield International.
Néga, W. (2022). La transition énergétique au cœur d'une transition sociétale. Synthèse du scénario néga Watt 2022. Valence, France.
Picon, A. (2002). Les saint-simoniens. Raison, imaginaire et utopie. Paris: Belin.
Smith, J. M. (2021). *Extracting accountability. Engineers and corporate social responsibility*. MIT Press.
Vérin, H. (1998). Autour du mot "ingénieur". L'identité de "l'ingénieur": quelques repères historiques. *Recherche & Formation, 29*, 11–20.
Wisnioski, M. (2012). *Engineers for change. Competing visions of technology in 1960s America*. MIT Press.

**Antoine Bouzin** Hydraulics and fluid mechanics Eng., Master of Sociology, PhD Student in Sociology and Political Science (Centre Émile Durkheim, France). Co-founder of Collectif EPSI (Études Pluridisciplinaires sur l'Ingénierie—Multi-disciplinary engineering studies) and Collectif Biomasse & SHS (Biomass studies from human and social sciences). Research areas: Engineering studies, social movements, profession, public action and policies, environment, expertise.

# Part II
# Justice

# Chapter 6
# Justice and Smart Societies: Conceptual Foundations, Limitations, and Conditions of Algorithmizing Fairness

**Florian Richter**

**Abstract** Smart technologies should make cities, governments, and societies not only more efficient and sustainable but also more livable. First, the conceptual framework of smart cities and societies is reviewed. At first glance, it seems promising to implement AI-based systems to address the latent needs of citizens. However, which conception of justice this presupposes has not been sufficiently scrutinized. Second, different conceptions of justice are shortly discussed. It is argued that conceptions of justice are context-dependent, i.e., while in one context, one conception of justice could be considered ethically acceptable, in another context, it might be considered unacceptable. Third, the limits and ethical implications of operationalizing justice via algorithms are analyzed.

**Keywords** Justice · Political philosophy · Smart cities · Society 5.0

## 6.1 Introduction

The paradigm of classic technology is to control and regulate processes to ensure repeatability. In the so-called transclassic technology, we face an "intellectualization of the environment," for example, in ubiquitous computing or by virtualizing the environment, i.e., the access to the world is mediated via technology and intelligent systems (for example, via sensors, cameras, etc.) (Hubig, 2007, 43). Ethics of technology faces several other challenges, such as the complexity of intelligent systems that make it more difficult to operate and control them and the interaction with autonomous systems or virtual agents. Hence, technological strategies and procedures are part of the environment. The implementation of smart technologies into the environment takes place on a large scale in so-called "smart cities" or smart

F. Richter (✉)
Technische Hochschule Ingolstadt, Ingolstadt, Germany
e-mail: florian.richter@thi.de



C. Didier et al. (eds.), *Engineering and Value Change*, Philosophy of Engineering and Technology 48, https://doi.org/10.1007/978-3-031-83549-0_6

societies ("Society 5.0") (Hitachi-UTokyo Laboratory, 2018, 10/11). Smart technologies should make cities, governments, and societies more efficient, effective, and sustainable and, thus, more livable for humans (Bibri, 2018; Kumar et al., 2021; Zhong et al., 2021). So far, ethical aspects of smart cities, e.g., "surveillance," "data privacy," and "sustainability," have been examined (Ziosi et al., 2024). However, questions of justice have not been sufficiently scrutinized. Whereas Coeckelbergh discusses equality and distributive justice in his political philosophy of technology, he does not investigate justice concerning smart cities and societies (Coeckelbergh, 2022, 43–48).

If we assume that through the collection and algorithmic processing of large amounts of data, a more *personalized* conception of fairness would be possible, then we already commit ourselves to one conception of fairness. It is a form of moral desert that assumes that we deserve X because we need it to live a good life. These (latent) needs of the citizens could be predicted by, e.g., AI-based systems. (Needs are here taken in a general way by covering any aspect (physical, emotional, social, etc.) that one might need to achieve a certain kind of well-being.) I will argue that the conception of justice in smart cities or societies is not a realistic model for distributive justice; instead, another model is proposed that takes algorithmizing fairness as a starting point that allows the evaluation of different conceptions of fairness for different (smart) contexts. Since debates of political philosophy are also echoed in the formalization of fairness in machine learning (Binns, 2018), this topic needs further analysis in smart environments.

Not only that, but it might be particularly difficult for systems to address the (latent) needs of the citizens "perfectly" or even to detect them. First, there must be a general debate about the best way to cover basic needs: should it be a universal basic income, a social welfare state, or should it be based on competition (merit-based) and economic incentives (neoliberalism, forms of libertarianism)? Even a liberal state with a particular social welfare system faces the challenge of how "compensating benefits for everyone, and in particular for the least advantaged members of society" (Rawls, 1999, p. 13), can be distributed and how those benefits can be used by its citizens or how those primary goods can be converted "into achievements of well-being" (Sen, 1995, p. 27). *Consequently*, making certain conceptions of justice available and explicit is crucial to evaluate them for different contexts. It is thus proposed that conceptions of fairness must be located in (smart) contexts to assess the technological systems. It will make processes more transparent and, first and foremost, make it possible to discuss and examine which kind of conception of fairness might be adequate in a particular situation without imposing one.

The chapter makes three contributions. First, the literature about smart cities, smart government, and society 5.0 is reviewed. Questions about how a well-lived life is possible within smart environments are raised. Second, the main approaches in political philosophy are analyzed and systematized to contextualize the conceptions of justice within smart cities and societies; otherwise, the conceptions of justice remain abstract and are not operationalizable in smart environments. This framework has yet to be introduced into and applied to the field of societal justice within smart societies. Furthermore, the final aspect of the paper is that such a conceptual framework needs to be operationalized, i.e., located in specific contexts, so that the ethical claims can also be concretized. Based on the conceptual framework, limitations of algorithmizing fairness, and implications of smart societies can be outlined to yield a picture of their ethical acceptability.

## 6.2 Related Work on Smart Cities and Societies

Despite the vast amount of literature about topics like "smart city" or "smart government," the question of justice has not been sufficiently scrutinized within these topics. Using Internet of Things technology or AI-based systems is crucial for so-called smart cities. It can range from implementing energy efficiency for buildings and maintaining them ("smart building architecture") (Kumar et al., 2021), to the establishment of "an efficient monitoring and control network" for urban water infrastructure (Oberascher et al., 2022), to the general asset management of cities (Zhong et al., 2021). Smart cities are mainly discussed under the value of sustainability. For such a "smart sustainable city" an interdisciplinary and transdisciplinary approach is needed in research (Bibri, 2018). Smart cities characterized by an embedded cyber-physical critical infrastructure also face cyberattack threats (Grady et al., 2021). This embeddedness of cyberspace into the physical space results in technology weaving "into the fabric of everyday life until they are indistinguishable from it" (Weiser, 1991). This is seen as the solution to make cities smart and livable for humans, but the ethical and societal implications need further examination. The focus of this literature review is the conceptual frame of phenomena under the headings "smart city," "smart government," and "Society 5.0" and not the concrete technological implementations. The conceptual frame about the topics lacks, in general, a clarification of how to prioritize values, how to handle conflicts of value (aspects of systematization and contextualization), and how they can be made operationalizable.

### *6.2.1 Smart City*

Smart cities are characterized by (technological) innovation that increases the "competitiveness of local communities" "while increasing the quality of life for its citizens through better public services and a cleaner environment" (Appio et al., 2019, p. 1). To achieve this they rely on the one side "on state-of-the-art information technology (e.g., fiber optic networks, sensors and connected devices, open data analytics, internet of things, ICT-enabled participatory planning frameworks)" and on the other side they also need to rely "on human capital (e.g., research universities, knowledge-intensive companies and public institutions)" (Appio et al., 2019, p. 1). Consequently, smart cities need strategies that "will target the efficiency and technological advancement of the city's hard infrastructure systems (i.e. transport, water, waste, energy) or the soft infrastructure and the people of the city (i.e. social and human capital; knowledge, inclusion, participation, social innovation, social equity, etc.)" (Angelidou, 2014, p. 5). An example is the "city of London" that "has based its smart city initiative on four dimensions: a) technology innovation; b) open data and transparency; c) collaboration and engagement; d) efficiency and resource management" (Appio et al., 2019, p. 1).

Since "[m]ore than half of the World's population now lives in urban areas," and this "shift from a primarily rural to a primarily urban population is projected to continue for the next couple of decades" (Chourabi et al., 2012, p. 2289), it seems that there are significant challenges for cities and megacities concerning "waste management, scarcity of resources, air pollution, human health concerns, traffic congestions" (Chourabi et al., 2012, p. 2289). The concept of a smart city seems to offer a solution to such challenges because it is conceptualized "as an icon of a sustainable and livable city" (Chourabi et al., 2012, p. 2289). The management of smart cities surrounds the idea of ensuring livable conditions within the frame of rapid growth (Chourabi et al., 2012, p. 2289). "The connotation of a smart city represents city innovation in management and policy as well as technology" (Nam & Pardo, 2011, p. 185).

There are "[m]any different definitions in the literature that give importance to different issues: values, like sustainability, living quality, integration of technology and data collection (sensors, cameras, implanted medical devices), optimization of business processes, use of smart technologies used for infrastructure services" (Chourabi et al., 2012, p. 2290). Chourabi et al. present a "set of factors" as a "smart cities framework" "that are essential to understanding smart city initiatives and projects" (Chourabi et al., 2012, p. 2291). Those factors are: "(1) management and organization, (2) technology, (3) governance, (4) policy, (5) people and communities, (6) the economy, (7) built infrastructure, and (8) the natural environment" (Chourabi et al., 2012, p. 2291).

Smart cities are arranged or positioned around the well-being of their inhabitants. This is the "culmination of all the preceding layers," like the "smart environment" ("infrastructure") or "smart government" ("collaboration ecosystems"). These layers lead to some kind of "smart living," which is also a "result of local material conditions, quality of life, and sustainability" (Appio et al., 2019, p. 2 and 5). Nevertheless, what smart living is and how a well-lived life can be accomplished within a smart city must be asked; besides general claims about improving material conditions and enabling sustainable, economic, and social development growth of the citizens via technology.[1]

[1] "It must have positive environmental impacts, reduce vulnerability and improve safety. Quality of life also should rely on better jobs, housing, and infrastructure (material conditions) as much on the preservation and development of natural, economic, human and social capitals. All of these elements can be positively influenced by the use of information and communication technologies in smart cities" (Appio et al., 2019, p. 5).

Cardullo and Kitchin examine the "neoliberal ideals" underpinning smart cities. (Cardullo & Kitchin, 2019). The latter uses the envisioning of "future presents" of smart cities to create "a non-neoliberal, post-capitalist smart city" based on "technological sovereignty" (Kitchin, 2019). According to him, "instrumental rationality" underpins smart cities. Hence, ethical principles should be adopted. However, the discussion is mainly about the ethical value of privacy (Kitchin, 2016). Cities are not just technical entities but are embedded in social and political practices (Kitchin, 2018). He emphasizes the need for asking "fundamental, normative questions" as a starting point, particularly regarding the "many theories of social justice (e.g., egalitarianism, utilitarianism, libertarianism, communitarianism, contractarianism, etc." (Kitchin, 2018). Even though such questions about justice are seen as essential, they are not discussed more extensively. And it needs to be scrutinized how all of this can be measured and operationalized.

### 6.2.2 *Smart Government*

Implementing smart governments plays a crucial role in yielding a high quality of life within smart cities. Nevertheless, some have also argued that "smart government uses smart city as an area of practice" and "conclude that smart city is complimentary, part of larger smart government movement" (Anthopoulos & Reddick, 2016). Nevertheless, technological advances in smart cities "also have consequences for the state's work, altering governance practices and shifting the nature of governmentality and citizenship" (Kitchin, 2021). Kitchin has argued that "automated management facilitates and produces instrumental and technocratic forms of governance and government." Hence, it is a "rule-driven, top-down, autocratic means of managing how a system functions and how it processes and treats individuals within those systems" (Kitchin, 2018).

According to the review of Gil-Garcia et al., 14 components can be identified "to account for smartness in governments": "integration, innovation, evidence-based, citizen-centricity, sustainability, creativity, effectiveness, efficiency, equality, entrepreneurialism, citizen engagement, openness, resiliency, and technology savviness" (Gil-Garcia et al., 2016, p. 525). These components are according to them mainly "analyzed in city context" (Gil-Garcia et al., 2016, p. 525). It seems that the components, dimensions, and aspects that form the conceptual frame of smart governments and smart cities overlap or are interrelated. Consequently, similar frameworks are developed for both phenomena. Also, within these frameworks, there are interrelations. "While some may be seen as a means to an end, such as the data-driven or the technology dimension, others are a result of the value attained from a smart government, such as resiliency or efficiency" (Gil-Garcia et al., 2016, p. 525). However, values are interrelated, and when technologies are implemented, trade-offs are the most pressing issue (e.g., privacy vs. transparency, security vs. liberty). Operationalizing such general frameworks is much more complex than it seems from an abstract perspective. Even though smart governance might offer "efficient and effective management of city systems," "processes of co-creation and co-production between city administrations, companies and citizens" should be implemented (Kitchin, 2018). Furthermore, it is unclear how a framework could be elaborated that allows the distinction between instrumental and final values within societal or governmental contexts concerning questions of justice. Particularly to understand a "governmentality shifts from a disciplinary calculative regime in which people self-regulate behaviour based on the fear of surveillance and sanction, to control regimes in which people are corralled and compelled to act in certain ways" (Kitchin, 2021; Vanolo, 2014).

However, the contextualization or situatedness of the application of values and potential conflicts of values need to be spelled out more precisely. Otherwise, guidelines deteriorate to appeals or turgid speeches. Furthermore, to operationalize the general framework, for instance, the "value-criteria-indicator-observable model" (Hubig, 2016) could be helpful to specify the values or dimensions. Criteria allow an evaluation of the characteristics of an area of definition as good or bad: how can the values be fulfilled with regard to the factual circumstances? Criteria open a search space for indicators, which allow the evaluation of the fulfillment of the criterion via (quantitative or qualitative) observables (Hubig, 2016). While such general frameworks resemble the general nature of ethics guidelines more, it is crucial to establish criteria, indicators, and observables to concretize them. It is probably widely accepted that a value like sustainability is vital for smart cities. Still, there are different ways of looking at this value from an ecological, economic, or recreational perspective. Is a park that offers recreation and improves the air quality a better use of scarce plots of land, or should it be used for a smart office building that is more energy efficient than other buildings? Also, efficiency that provides "*more value to citizens with fewer resources*" in smart government (Gil-Garcia et al., 2016, p. 528) must be considered in different respects and contextualized. Optimizing technical processes can lead to possible threats by cyberattacks if fewer resources are mainly taken under monetary considerations. It can also mean that fewer people have to handle more tasks, which can lead to mistakes or psychological problems rooted in work overload. Under the value of equality, the "*use of information technologies, data, and appropriate strategies could help to reduce social exclusion and promote social justice*" (Gil-Garcia et al., 2016, p. 528). Data collection and its use are not value-free; in some cases, databases are one-sided, which can lead to discrimination.

### *6.2.3 Society 5.0*

The concept of Society 5.0 is fairly new and is part of a governmental plan in Japan. It contains similar aspects to the concepts of smart city and smart government, which are amplified in a societal context. "On January 22, 2016, the Government of Japan released the 5th Science and Technology Basic Plan" This plan proposes a vision of a so-called "Society 5.0" (Hitachi-UTokyo Laboratory, 2018, p. xi). The "Society 5.0" is defined in the following way:

> A human-centered society that balances economic advancement with the resolution of social problems by a system that highly integrates cyberspace and physical space. […] It follows the hunting society (Society 1.0), agricultural society (Society 2.0), industrial society (Society 3.0), and information society (Society 4.0). (Cabinet Office)

The aspiration embedded in a plan of Society 5.0 is to yield a "human-centered society," and this will be accomplished "through the high degree of merging between cyberspace and physical space," it "will be able to balance economic advancement with the resolution of social problems by providing goods and services that granularly address manifold latent needs regardless of locale, age, sex, or language to ensure that all citizens can lead high-quality, lives full of comfort and vitality" (Cabinet Office, 2017 as cited in (Hitachi-UTokyo Laboratory, 2018, p. xii)).

Such a vision comes along with two problematic issues that need further examination: (1) The merging of physical space and cyberspace is seen as the solution to several societal problems or at least as contributing to the solution. Still, the ethical and societal implications are not discussed. The profoundness of technological embeddedness leads to multifaceted challenges (see Sect. 6.4). (2) It seems that it might be possible to solve societal problems with technology and that the "manifold latent needs" can be addressed without considering differences. But does this lead to equality, and is it also possible to "guess" the people's latent needs without imposing certain ideologically imbued needs on them by the AI-based systems, institutions, etc., or by someone with more power and resources? This is also something that Hagendorff diagnoses about ethics guidelines that barely discuss "the possibility for *political abuse of AI systems* in the context of automated propaganda, bots, fake news, deepfakes, micro targeting, election fraud, and the like" (Hagendorff, 2020). Not only can the abuse of power be problematic, but the imposition of certain lifestyles or ideas of quality of life can be either ideologically loaded or arise from one-sided notions. The "Montréal Declaration for Responsible Development of Artificial Intelligence" (2018) and also the "AI Now 2019 Report" (2019) are, according to Hagendorff, "the only guidelines that explicitly prohibit[] imposing certain lifestyles or concepts of 'good living' on people by AI systems, as it is for example demonstrated in the Chinese scoring system" (Hagendorff, 2020). Another example is mentioned in the book *Society 5.0* about the capability approach of Amartya Sen: it is proposed that it might be better to teach a socially isolated person to ride a bike instead of giving her DVDs (Hitachi-UTokyo Laboratory, 2018, p. 139/140). Of course, such a proposition is highly paternalistic, even though it was made with good intentions. Such impositions would also threaten the citizens' autonomy to make their own decisions and the freedom to choose their idea of a good life as is possible in modern liberal societies. This conceptual background of liberalism must also be examined to scrutinize the implications of smart cities and societies to conceptualize justice. And, of course, technology should, in Society 5.0, only be used for the sake of humanity and the well-being of the people. But again, what is a well-lived life in a smart city or Society 5.0 without imposing ways of living?

The discussion about the so-called Society 5.0 is an excellent exemplar of identifying issues because it envisions incorporating technological advancements on a societal level and thus makes the consequences even more visible and pressing. The merging of physical and cyberspace (virtualization), hybridization, increase in complexity, and autonomization, i.e., the autonomous decision-making in cyberspace by AI-based systems, are all challenges of the ethics of technology mentioned in the beginning. Smart cities and societies seem thus laboratories for an ethics of technology. Such challenges must be confronted without falling into some kind of pessimism about technology that leaves no room for autonomy on the side of humans (as suggested by Jacques Ellul, for example (Ellul, 1965)). On the contrary, confronting the challenges implies systematizing, contextualizing, and operationalizing values like fairness within these contexts to preserve human autonomy. Otherwise, the promise that algorithmic formulations of fairness that are based on data processing within smart environments will lead to fairer societies is in vain.

## 6.3 Justice and Fairness

According to John Rawls, "distributive shares" should not correspond to a "moral desert." We can have "legitimate expectations as founded upon social institutions" that tell us what we are entitled to. *But* this "is not proportional to nor dependent upon" our "intrinsic worth" (Rawls, 1999, p. 273). Rawls believes that the "principles of justice" "regulate the basic structure and specify the duties and obligations of individuals" (Rawls, 1999, p. 273). David Schmidtz states, following Rawls, that "principles are meant to *guide*" our thinking and not to "*determine*" it.[2] Also, "they do not add up to a decision procedure" (Schmidtz, 2006, p. 227). However, such decision procedures or more concrete measures need to be implemented. Rawls thinks the "function of unequal distributive shares is to cover the costs of training and education, to attract individuals to places and associations where they are most needed from a social point of view, and so on" (Rawls, 1999, p. 277).

But who sets the standards for the most needed from a social point of view to attract them? Nurses are much needed, but neither are they paid well nor have good working conditions. Furthermore, Rawls thinks that under ideal circumstances, the entitlement to something and the moral desert of it would coincide (Rawls, 1999, p. 276/277). This is similar to Broome's formulation of "fairness-claims" that "*if* needed resources are being distributed, the person should have a share." For example, in the case of a research grant, an applicant has a claim to it if "her application is good enough," and others also receive one. Consequently, Broome discusses fairness under a formal and legal aspect that defines claims as

[2] Schmidtz thinks that a theory of justice is like a map that can consist of two principles (Rawls) or of four elements like he suggests (Schmidtz, 2006, p. 227).

not based on needs ("it seems implausible that a person has a right to whatever she needs")[3] (Broome, 1990/1991, p. 96/97). Nevertheless, much depends on how one determines what "good enough" means because decision procedures must become more concrete.

Apart from the decision procedure, the "assessment of justice" depends also on the type of freedom. Primary goods or resources are understood as "*means* to freedom," while the focus, according to Amartya Sen, should instead be "on the extent of the freedom a person actually has." In such a "capability-assessment of justice, individual claims are not to be assessed in terms of the resources or primary goods the persons respectively hold, but by the freedom they actually enjoy to choose the lives that they have reason to value" (Sen, 1995, p. 81). So, it is based on the options at hand that the individual has, based on her capabilities. A pregnant woman has different options at hand than a man, even though they might have the same income (Sen, 1995, p. 27).

To look at justice independently of freedom, background, and context seems only to produce a formal framework that is not adequate for evaluating the conditions of systems in which we participate and operate. Not only liberty but also other values like personal flourishing, health, security, etc., play an essential role as conditioning factors to determine justice. Even Rawls determines a starting point to evaluate a social system. ("We know from what position the social system is to be judged.") It is "to identify the least advantaged representative man" (Rawls, 1999, p. 79). However, it needs to be clarified regarding what the representative individual is the least advantaged, either concerning a value like health or security. Moreover, the values must be operationalized by criteria and indicators that open the search space for adequate (qualitative or quantitative) observables (Hubig, 2016). The purely quantitative measure of wealth by money income is problematic (see above the example from Sen).

Another common distinction with regard to justice is the one between equality of opportunity and equality of outcome. While the latter can be disregarded, the former is taken as crucial for discussing fairness. The Friedmans criticize equality of outcome by stating that governments that want to achieve "fair shares for all" will "reduce liberty," while governments that "promote personal equality and equality of opportunity enhance liberty." According to them, equality of outcome, achieved through rigorous taxation of the rich, leads to slower economic growth. However, such measures are going "against one of the most basic instincts of human beings"—to better their condition or well-being in a broad and not only material sense (Friedman & Friedman, 2007, pp. 53, 58). Nevertheless, Such a critique is far from dystopian visions about equality of outcome by equalizing everyone to a mere average, like in the short story Harrison Bergeron from Kurt Vonnegut, where the strong have to carry weights, for instance (Vonnegut, 2014). Besides the context of distributive shares, where the second principle of Rawls could also be regarded as

[3] Although need is one of the elements of justice according to Schmidtz, the distribution according to need might stop at one point, because this kind of distribution is not needed by the people anymore (Schmidtz, 2006, pp. 4, 166).

promoting a kind of equality of outcome, the concept does not make so much sense in other contexts. Binns takes security checks at airports as a context where we might favor equality of outcome so that some groups are not searched more thoroughly because few members of that group might be considered as posing a higher risk for air travel (Binns, 2018). But this is instead something like an equality in treatment rather than an equality of outcome.

In a nutshell, justice is not a question of addressing (latent) needs based, for example, on moral desert, but instead of entitlements to shares either from a more libertarian (Robert Nozick, Friedmans) or liberal (Rawls, Schmidtz) point of view.[4] It seems to me problematic to base shares on moral desert by either letting an individual's need or an individual's talent stand out because it leads either to paternalism (to tell people what they need) or to elitism, where some believe they are apt to make decisions for others, because of their fortunate positioning in society. Nevertheless, this does not imply that competition and a free market where some get more for their "hard" work should be suspended. Of course, some advocates of moral desert fear that equality of outcome will be implemented and effort might even be penalized so that it will lead to some kind of socialism. On the contrary, entitlements are the basis for a free market, and as normative or deontic concepts, they go beyond determining what people need, want, and are doing. Still, it is crucial to conceptualize how such a formal conception of justice can be operationalized by discussing its limitations and implications.

## 6.4 Limitations and Conditions

Operationalizing the distribution via algorithms could take different forms. (1) Personalization based on, for example, deep neural networks that collect large amounts of data and could thus be able to detect the (latent) needs of persons. (2)

[4] Communitarianism is usually taken as the antagonist of liberalism. Hence, it should communitarian ideas should be mentioned for a more complete picture (I am thankful to an anonymous reviewer who pointed this out): A rather simplistic conception of communitarianism is that values are part of the common tradition and are accepted by its members. In a certain way, the community is prior to the individual. This implies, however, that communitarianism cannot distinguish between "good" and "bad" communities (like criminal organizations) because in both the members might follow accepted values (or vices). Even in a communitarian approach where the common values of a specific practice should contribute to the well-being of patients, ends or values are presupposed and are, in the end, not being questioned (Santoni de Sio et al., 2021). (They assume that in care activities, human caretakers should be involved, and even simple care activities like taking urine samples should not be outsourced completely to technology. However, studies have shown that people who are affected, i.e., need care, are often willing to interact with robots (Schönmann et al., 2023). It seems rather paternalistic to construct such scenarios without considering the preferences of stakeholders.) Hence, the mediating role of technology seems to be mainly to nudge people to realize or follow the values and ends of the community. Here, notions of the good life are imposed. A problematic point that will be discussed further below regarding the social credit system.

Implementing fixed rules, as is already done in affirmative action policies where, for instance, a diverse student body is sought until a certain percentage of diversity is reached. Algorithms could help execute such rules without human (biased) interference. (3) Randomizing the access to goods or the distribution of resources like in a lottery. Much depends on the specific contexts. Affirmative action policies seem, for instance, fair for job applications but not for the selection to go on a dangerous mission (Broome, 1990/1991).

### *6.4.1 Limitations of Operationalizing Fairness*

While some have argued that randomization to distribute goods or opportunities in the face of scarce resources might make sense, such as access to prestigious universities (Sandel, 2021) or even in the case of distributing organs (Broome, 1990/1991, p. 99/100; Caplan, 2014), it is not acceptable to randomize the access to social welfare or health care. Beyond lines of thought that opt for an approach where the free market fixes almost everything (e.g., Milton Friedman), the problematic question arises of operationalizing the Rawlsian conception of building a system where the least advantaged can benefit. How should we measure who is the least advantaged? Is it by money income or rather real income? However, statistics are often profoundly political (Deaton, 2023, pp. 84–86). Furthermore, the critique of Sen (see above) shows that many circumstances and conditions play a role that need to be considered to understand the different options at hand of individuals. Which ones are relevant and need to be considered depends also on political deliberation.

To address these issues, the authors of *Society 5.0* have proposed using Sen's capability approach by, for instance, teaching a socially isolated person to ride a bike instead of giving her DVDs (Hitachi-UTokyo Laboratory, 2018, p. 139/140). However, such a position is a paternalism of ends and goals and is unacceptable in a liberal society.[5] A rather mild form would be a "paternalism of means" (Thaler & Sunstein, 2021, p. 7 and 312). Technology seems here to be an adequate means to nudge people's behavior in specific directions. Within the context of smart cities and societies, nudging might seem viable to optimize not only the use of technology but also human behavior. However, such a "paternalism of means" presupposes a specific understanding of the right thing to do to nudge people to these goals. Even though Sunstein and Thaler "only" appeal to some kind of common sense understanding of morality, it is, in the end, not only a paternalism of means but also of goals. It undermines democratic deliberation, liberty of citizens, and autonomy on the strategic and moral level. Operational autonomy is somewhat less critical because it can help implement security features like car assistant systems to prevent accidents, and they often function better than humans. However, implementing

[5] It should be mentioned that Sen's approach is not paternalistic, but how it is wrongfully used in the book *Society 5.0*. I am thankful to an anonymous reviewer for pointing out this issue to me.

systems that make decisions on a strategic or moral level must be accompanied by an assessment process of whether the stakeholders share the goals or not.

Another limitation of algorithmizing fairness is that values sometimes must be weighed. Of course, there are also cases where one value is accepted as having priority over others. Everyone must be, for instance, checked at the airport, and everyone's luggage is scanned. Safety is the top priority here and comes with restrictions on privacy and individual freedom. Artificial intelligence systems, for example, facial recognition, are also implemented under this priority to solve crimes. However, this comes with a qualitative leap because more data must be collected for the system's security. Consequently, it is not only for an external purpose, which is either considered as "good" (fight against crime) or "bad" (citizen surveillance). In many cases, the values (safety vs. freedom or privacy) must be weighed, and how they are implemented must be debated, as well as what is acceptable for the ones affected and what is considered fair.

Also, the surveillance of citizens cannot be algorithmized by randomization to avoid, for instance, discrimination, as in the stop-and-frisk program by the New York City Police Department, which led to discrimination against persons of African and Hispanic descent. Randomized controls could feel arbitrary and intransparent and lead to somewhat dystopian or Kafkaesque scenarios of bureaucracy exemplified, for instance, vividly in the novel *The Trial* written by Franz Kafka. Consequently, different conceptions of fairness and their operationalization via algorithmizing have limitations in the context of specific use cases. It must be scrutinized to see if they make sense or are acceptable.

### 6.4.2 Conditions for a Model of Technologically Mediated Justice[6]

If the analysis of information and the decision-making processes are delegated and outsourced into a cyber model or virtual space like it is suggested for a Society 5.0, then the reflection on the technological constitution of the lifeworld will need to be crucial, particularly in the case when the systems address the latent needs of the citizens almost "perfectly," because reflection will almost not be possible anymore. But what might the latent needs of the citizens be? There is also a general debate about which might be the best way to cover basic needs: should it be a universal basic income, a social welfare state, or should it be based on competition (merit-based) and economic incentives (neoliberalism, forms of libertarianism)? Even a liberal state that might have a specific social welfare system that always has certain "inequalities of wealth and authority" and will have thus "compensating benefits for

[6] I am thankful for the comment of an anonymous reviewer who pointed out that the final sections could serve as "presuppositions for the possible construction of another model of justice". I also added a conclusion to sketch such a model.

everyone, and in particular for the least advantaged members of society" (Rawls, 1999, p. 13) is faced with the challenge of how its citizens can use those benefits or how those primary goods can be converted "into achievements of well-being" (Sen, 1995, p. 27). Sen criticizes Rawls and emphasizes that personal diversity must be considered because the same income or primary goods may not lead to living well in the same way. A pregnant woman must overcome disadvantages that a man does not, even though they have the same income (Sen, 1995, p. 27). This means we cannot use the goods similarly because of our personal, socio-economic, and cultural background. This is a complicated issue that can neither be transferred to a system that suggests addressing specific latent needs by offering personalized counseling nor can it be delegated to systems to make decisions over citizens. Both ways would be threats to the citizens' autonomy.

**Condition I: Personal Identity**

One "must also ensure that the subjects can develop as autonomous subjects of evaluation. This includes maintaining the social structures that first and foremost allow the formation of personal identity" (Hubig, 2007, p. 142). Consequently, forms of "dogmatic upbringing and oppression" that "undermine the autonomy of decisions" must be disregarded. Hubig calls this dimension "legacy values" (Hubig, 2007, p. 142). This does not mean to preserve everything traditional, but to preserve the ones that "guarantee such structures (e.g., upbringing, education, experience of nature, and many more) that are the presuppositions that subjects learn to set themselves in a (if necessary, also negative) relation to them, thus learn to value or assess" (Hubig, 2007, p. 142). The "formation of value competence is a competence to set oneself in relation to ones wishes" and options (Hubig, 2007, p. 142). Consequently, the formation of critical judgment is crucial so that the citizens can participate in the discourse of policymaking and understand and identify their needs and how they can be converted into achievements of well-being without ideological impositions.

**Condition II: Not Imposing Ideas of the "Good"**

Another implication of delegating the decision-making process and not critically accompanying the generation of knowledge is that it can result in the classification of "good" and "bad" behavior by the systems. It would not then be transparent how the system arrived at this classification anymore because some AI-based systems (like deep neural networks) are opaque, and the vast amount of processed data cannot be reconstructed or made comprehensible. Are certain behaviors regarded as inefficient or not sustainable, or do they follow "good" or "bad" latent needs because they are selfish? These problematic issues need to be negotiated in a public discourse among citizens. If a system or a group of people establishes such a distinction between "good" and "bad" behavior, ideological pitfalls are always present. It might be seen as disciplining the people to a "morally good" or "tasteful" behavior that is only regarded as suitable by a powerful class. The social credit system in China functions like this. "Good" behavior that is listed on a so-called Redlist is rewarded with the praise of being a morally good person. In contrast, "negative" behavior that is listed on a so-called Blacklist is sanctioned either by being put on the list and publicly shamed or by other more profane punishments like being banned from air travel. Engelman et al. conclude their analysis by stating that there is no transparent behavior mapping to rewards or sanctions (Engelmann et al., 2019). Here, the opaqueness might be due to the system's imperfection or the unwillingness to provide transparency. Nonetheless, it is problematic to sanction or reward "morally good" behavior for three reasons. Two are more of a philosophical nature and one more of a political nature. (1) From a political (and legal) aspect, it is problematic because the good would be prior to the right. The common good would establish a certain kind of moral pressure to live in accordance with the established norms, even though some (especially oppressed minorities) will not or, in some cases, cannot even disagree. Here, it is problematic that it is unclear who establishes the norms and under which coercive circumstances they are set. Furthermore, not every morally wrong behavior is also legally sanctioned. Lying is generally regarded as morally wrong but does not always have legal repercussions. And not all legally sanctioned behavior is morally wrong (e.g., parking in the wrong spot). (2) The reward for "morally good" behavior devalues the behavior because it is then only done because of incentives or the good consequences that come along with it. This is, of course, a more Kantian view, and Utilitarians might have some objections. (3) Some can regard the classification as an imposition of a particular way of life that they do not share and that does not fit into their pursuit of happiness. These outliers are forced by sanctions to conform. A pluralistic society lives from the diversity of lifestyles and people and the degree to which it enables inclusion. That, of course, does not mean the toleration of any "lifestyles" like criminal ones.

## 6.5 Conclusion

In smart cities, governments, and societies, technology is embedded in the environment (sensors, cameras, etc.) to collect data, and it steers and controls processes (for instance, the traffic flow) to manage the lives of the citizens in an efficient and effective way. Strategic matters of prioritizing access and distributing resources are also outsourced to algorithms. Nevertheless, in the health sector, studies have shown that such systems can also discriminate against minorities (Obermeyer et al., 2019). Another example regarding fairness and discrimination is the system COMPAS to assess the risk of recidivism (Mehrabi et al., 2021). Kaminski, for instance, has argued that such systems are not adequate at all for the purposes of assessing guilt (Kaminski, 2020). It seems that the underlying problem of building fair algorithms is mainly that some kind of unbiased original position is presupposed that needs to be reached either by unbiasing data or algorithms (see the survey from (Mehrabi et al., 2021) for different attempts). However, as Green has argued such "formal algorithmic fairness" "leads to the impossibility of fairness and to models that exacerbate oppression." He opts for a "substantive algorithmic fairness" where "reforms can remediate the substantive inequalities identified." This is done by a "relational response" to "reduce dignitary and material disparities that reflect social hierarchies" and by a "structural response" to "alter the structure of decisions to reduce the extent to which decisions exacerbate social hierarchies" (Green, 2022). Such an approach does not promote affirmative action policies or equality of outcomes, as it is also proposed by (Wachter et al., 2021a, 2021b). The biases need to be identified and discovered in their social and political context (for instance, by a standard like "conditional demographic disparity") (Wachter et al., 2021a, 2021b).

Whether "algorithms can enhance or facilitate the reforms" (Green, 2022) is another question. And even though political philosophy is used for questions of algorithmic fairness, the "reverse – that novel insights from the algorithmic fairness literature are fed back into ethics and political philosophy – is far less established." Franke undertakes this by discussing "algorithmic group fairness" in relation to Nozick's philosophy (Franke, 2023). However, the interdependence between political philosophy and algorithmic fairness is much more difficult in practice. Franke tests the idea of Rawls' original position and difference principle for claimants in cases of loss to an insurance company. According to him, the ones who suffered a loss and get no compensation (false negatives) are the least advantaged group, but they would not be motivated to buy the insurance in the first place because it is an additional loss or cost for them (Franke, 2021). However, Rawls talks about the conditions for finding just principles, whereas Franke looks at the outcome. The idea of Rawls is that the least advantaged group gets some kind of benefit to have a chance to participate. If some of the disadvantaged groups do not receive compensation due to adverse circumstances, then this is more a failure of the system or the institutions and not of the conception of justice. Hence, it is about a distribution due to an event (case of damage) and not about setting conditions for a (fair) distribution; as a condition, one would stipulate that someone who has a justified claim would also be compensated. It is more a problem of implementation or realization. Instead, the conditioning factors for inequality need to be identified, and here, algorithms can play an essential role. It is a model that uses algorithmizing fairness as a starting point that allows the evaluation of different conceptions of fairness for different (smart) contexts by identifying the social conditions. Nevertheless, such a model also needs to consider that the formation of personal identity and critical judgment is a condition for participating in the shaping of a just society. Another condition is that notions of the good are not imposed and that they need to be negotiated within a public discourse to prevent technological implementations, systems, and algorithms from nudging and controlling citizens. Such a model can only be reached through a "process of mutual adjustment of principles and considered judgments" ("reflective equilibrium") (Rawls, 1999, p. 18) where use cases of technology are discussed and aligned with our principles. What would be the alternatives? Either models of algorithmic fairness that rely only on the functionalism of algorithms to make processes more efficient and effective (technological reductionism) or models of fairness that rely solely on dogmatic positions of principles, which lead to quarrels between different schools of thought about the right "-ism."

## References

Angelidou, M. (2014). Smart city policies: A spatial approach. *Cities, 41*, 3–11.

Anthopoulos, L., & Reddick, C. (2016). Smart City and smart government: Synonymous or complementary? *Proceedings of the 25th international conference companion on world wide web (WWW '16 companion).* (pp. 351–355).

Appio, F., Lima, M., & Paroutis, S. (2019). Understanding smart cities: Innovation ecosystems, technological advancements, and societal challenges. *Technological Forecasting and Social Change, 142*, 1–14.

Bibri, S. (2018). A foundational framework for smart sustainable city development: Theoretical, disciplinary, and discursive dimensions and their synergies. *Sustainable Cities and Society, 38*, 758–794.

Binns, R. (2018). Fairness in machine learning: Lessons from political philosophy. Conference on fairness, accountability and transparency. (pp. 149–159).

Broome, J. (1990/1991). Fairness. Proceedings of the Aristotelian Society, New Series, 91, pp. 87–101.

Cabinet Office. (2017). *Society 5.0.* Retrieved November 28th, 2021, from https://www8.cao.go.jp/cstp/english/society5_0/index.html.

Caplan, A. (2014). Bioethics of organ transplantation. *Cold Spring Harbor Perspectives in Medicine, 4*(3).

Cardullo, P., & Kitchin, R. (2019). Smart urbanism and smart citizenship: The neoliberal logic of 'citizen-focused' smart cities in Europe. *Environment and Planning C: Politics and Space, 37*(5), 813–830.

Chourabi, H., Nam, T., Walker, S., Gil-Garcia, J., Mellouli, S., Nahon, K., et al. (2012). Understanding smart cities: An integrative framework. In *Proceedings of the annual Hawaii international conference on system sciences* (pp. 2289–2297).

Coeckelbergh, M. (2022). *The political philosophy of AI: An introduction*. Polity.

Deaton, A. (2023). *Economics in America: An immigrant economist explores the land of inequality*. Princeton & Oxford.

Ellul, J. (1965). Technological society.

Engelmann, S., Chen, M., Fischer, F., Chingyu, K., & Grossklags, J. (2019). Clear sanctions, vague rewards: How China's social credit system currently defines 'good' and 'bad' behavior. *FAT* 2019—Proceedings of the 2019 conference on fairness, accountability, and transparency*. (pp. 69–78).

Franke, U. (2021). Rawls's original position and algorithmic fairness. *Philosophy and Technology, 34*, 1803–1817.

Franke, U. (2023). Algorithmic fairness, risk, and the dominant protective agency. *Philosophy and Technology, 36*(76).

Friedman, M., & Friedman, R. (2007). Free to choose. In M. Sandel (Ed.), *Justice: A reader* (pp. 49–60).

Gil-Garcia, J., Zhang, J., & Puron-Cid, G. (2016). Conceptualizing smartness in government: An integrative and multi-dimensional view. *Government Information Quarterly, 33*(3), 524–534.

Grady, C., Rajtmajer, S., & Dennis, L. (2021). When smart systems fail: The ethics of cyber–physical critical infrastructure risk. *IEEE Transactions on Technology and Society, 2*(1), 6–14.

Green, B. (2022). Escaping the impossibility of fairness: From formal to substantive algorithmic fairness. *Philosophy and Technology, 35*(90).

Hagendorff, T. (2020). The ethics of AI ethics: An evaluation of guidelines. *Minds & Machines, 30*, 99–120.

Hitachi-UTokyo Laboratory. (2018). *Society 5.0: A people-centric super-smart society*. SpringerOpen.
Hubig, C. (2007). Die Kunst des Möglichen II: Grundlinien einer dialektischen Philosophie der Technik. Bielefeld.
Hubig, C. (2016). *Indikatorenpolitik: Über konsistentes und kohärentes kommunikatives Handeln von Organisationen und in Unternehmen*. Retrieved from https://www.cssa-wiesbaden.de/fileadmin/Bilder/B%C3%BCcher_Brosch%C3%BCren/Papers-cssa/cssa-paper_Indikatorenpolitik_2_2016.pdf.
Kaminski, A. (2020). Gründe geben. Maschinelles Lernen als problem der Moralfähigkeit von Entscheidungen. In K. Wiegerling, M. Nerurkar, & C. Wadephul (Eds.), *Datafizierung und big data* (pp. 151–174). Springer.
Kitchin, R. (2016). The ethics of smart cities and urban science. *Philosophical Transactions of the Royal Society A: Mathematical, Physical and Engineering Sciences, 374*(2083).
Kitchin, R. (2018). Reframing, reimagining and remaking smart cities. In C. Colleta, L. Evans, L. Heaphy, & R. Kitchin (Eds.), *Creating smart cities* (pp. 219–230). Routledge.
Kitchin, R. (2019). Toward a genuinely humanizing smart urbanism. In *In the right to the smart city* (pp. 193–204). Emerald Publishing Limited.
Kitchin, R. (2021). Decentering the smart city. In S. Flynn (Ed.), *Equality in the city: Imaginaries of the smart future*. Bristol.
Kumar, A., Sharma, S., Goyal, N., Singh, A., Cheng, X., & Singh, P. (2021). Secure and energy-efficient smart building architecture with emerging technology IoT. *Computer Communications, 176*, 207–217.
Mehrabi, N., Morstatter, F., Saxena, N., Lerman, K., & Galstya, A. (2021). A survey on bias and fairness in machine learning. *ACM Computing Surveys, 54*(6), 1–35.
Nam, T., & Pardo, T. (2011). Smart city as urban innovation: Focusing on management, policy, and context. In *Proceedings of the 5th international conference on theory and practice of electronic governance (ICEGOV '11)* (pp. 184–195).
Oberascher, M., Rauch, W., & Sitzenfrei, R. (2022). Towards a smart water city: A comprehensive review of applications, data requirements, and communication technologies for integrated management. *Sustainable Cities and Society, 76*.
Obermeyer, Z., Powers, B., Vogeli, C., & Mullainathan, S. (2019). Dissecting racial bias in an algorithm used to manage the health of populations. *Science, 366*(6464), 447–453.
Rawls, J. (1999). *A theory of justice* (revised ed.). Cambridge, MA.
Sandel, M. (2021). *The tyranny of merit: What's become of the common good*. Penguin Books.
Santoni de Sio, F., Almeida, T., & van den Hoven, J. (2021). The future of work: Freedom, justice and capital in the age of artificial intelligence. *Critical Review of International Social and Political Philosophy*, 1–25.
Schmidtz, D. (2006). Elements of justice.
Schönmann, M., Bodenschatz, A., Uhl, M., & Walkowitz, G. (2023). The care-dependent are less averse to care robots: An empirical comparison of attitudes. *International Journal of Social Robotics, 15*, 1007–1024.
Sen, A. (1995). Inequality reexamined. *New York/Cambridge (Mass.)*.
Thaler, R., & Sunstein, C. (2021). *Nudge: The final edition*. Penguin Books.
Vanolo, A. (2014). Smartmentality: The Smart City as disciplinary strategy. *Urban Studies, 51*(5), 883–898.
Vonnegut, K. (2014). Harrison Bergeron. In K. Vonnegut (Ed.), *Welcome to the Monkey House*.
Wachter, S., Mittelstadt, B., & Russell, C. (2021a). Bias preservation in machine learning: The legality of fairness metrics under EU non-discrimination law. *West Virginia Law Review, 123*(3).
Wachter, S., Mittelstadt, B., & Russell, C. (2021b). Why fairness cannot be automated: Bridging the gap between EU non-discrimination law and AI. *Computer Law & Security Review, 41*.

Weiser, M. (1991, September). The computer for the 21st century. Specialized elements of hardware and software, connected by wires, radio waves, and infrared, will be so ubiquitous that no one will notice their presence. Scientific American Special Issue on Communications, Computers, and Networks.
Zhong, C., Ng, S., & Skitmore, M. (2021). An interdependent infrastructure asset management framework for high-density cities. *Proceedings of the Institution of Civil Engineers: Municipal Engineer, 174*(3), 180–190.
Ziosi, M., Hewitt, B., Juneja, P., Taddeo M., & Floridi. (2024). Smart cities: Reviewing the debate about their ethical implications. *AI & Soc, 39*, 1185–1200.

**Florian Richter** MA Philosophy and History, PhD Philosophy (TU Darmstadt 2015). 2017–2022 Teaching Professor of Ethics and Philosophy, Tecnológico de Monterrey, Mexico. Since 2022, Postdoctoral Researcher at the Technical University Ingolstadt of Applied Sciences, Germany. Research areas: Technology ethics, philosophy of technology.

# Chapter 7
# Applied Trustworthy Data and Artificial Intelligence Governance in Precision Agriculture

**Artur Bogucki**

**Abstract** This chapter critically examines the ethical, legal, and socio-economic (ELSE) challenges and opportunities associated with data and AI in precision agriculture. By conducting an extensive literature review spanning publications from 2016 to 2023, the study systematically analyzes key arguments and trends, focusing on data governance issues. The review includes a meticulous analysis of over 505 distinct sources, revealing a significant shift in thematic focus towards challenges such as data ownership, security, privacy, and labor market disruptions. The findings underscore the predominance of ethical challenges, highlighting critical issues like discrimination, bias, privacy breaches, and the need for robust legal frameworks.

Building on the literature review, the chapter addresses the complexities of agri-data governance, providing strategic recommendations for improving data management in the sector. The analysis emphasizes the importance of establishing a centralized European Ag-Data System to consolidate diverse datasets, enforce interoperability standards, and ensure transparent data governance. This approach aims to balance market dynamics, enhance data accessibility, and foster trust among farmers and stakeholders while ensuring adherence to the EU values in practice.

Further, the chapter evaluates the European Commission's Ethics Guidelines for Trustworthy AI using the ALTAI framework through three case studies in Spain, Finland, and Serbia/Lithuania. The assessment highlights improvements in transparency, privacy, and data governance, yet identifies ongoing challenges in technical robustness and accountability. The chapter advocates for continuous ethical evaluations and the use of tools like model cards and datasheets to enhance transparency and accountability. The findings underscore the necessity for robust data governance frameworks and ethical evaluations to ensure the responsible deployment of AI in agriculture, balancing innovation with the protection of stakeholders' rights and interests.

A. Bogucki (✉)
Centre for European Policy Studies (CEPS) | Warsaw School of Economics, Warsaw, Poland
e-mail: artur.bogucki@ceps.eu



C. Didier et al. (eds.), *Engineering and Value Change*, Philosophy of Engineering and Technology 48, https://doi.org/10.1007/978-3-031-83549-0_7

**Keywords** Data governance · Artificial intelligence governance · AI ethics · Digital law · Technology law

## 7.1 Introduction

Digital precision agriculture, or "smart farming," employs digital methods to monitor and optimize farm processes, transforming agriculture through technologies like Internet of Things (IoT), remote sensing, and machine learning (Kritikos, 2017; Al-Amin et al., 2023; Cammarano et al., 2023). These tools gather and use data to enhance productivity, streamline operations, and perform autonomous tasks (Ayoub Shaikh et al., 2022). Driven by global population growth and climate change, smart farming faces pressures to meet increasing demand with fewer resources and a shrinking labor force (Baudron & Giller, 2014; Schuh, 2019). Advances in mechanization and automation, such as robotic milking and automated climate controls, have revolutionized farm management (Bhakta et al., 2019; Botta et al., 2022; Cisternas et al., 2020). However, this digital transformation generates vast amounts of data, posing challenges in data governance, ownership, security, privacy, and technology literacy (Acosta et al., 2020; Niedbała et al., 2023; Lorencowicz & Uziak, 2023). This chapter reviews the ethical, legal, and socio-economic (ELSE) factors of smart farming, specifically analyzing data governance issues and applying the Assessment List for Trustworthy Artificial Intelligence (ALTAI) framework to assess the achievement of trustworthy Artificial Intelligence (AI) in practice.

To provide a comprehensive understanding of the ELSE challenges and opportunities associated with data and AI in precision agriculture, this chapter begins with an extensive literature review. The literature review spans publications from 2016 to 2023 and systematically examines key arguments and trends in these areas. This review employed a rigorous methodology, involving a systematic search of key journals and databases using specific keywords, followed by meticulous analysis and categorization of arguments from over 505 distinct sources. The review was conducted twice to capture evolving perspectives, with significant shifts observed in thematic focus, particularly an increase in discussions on challenges such as data governance and concerns about data ownership, security, privacy, and labor market disruptions. Furthermore, the results reveal a predominance of challenges over opportunities, especially in the ethical domain, highlighting critical issues like discrimination, bias, privacy breaches, and the need for robust legal frameworks to address liability and accountability (Gabriel & Gandorfer, 2023). These findings underscore the necessity for targeted, sector-specific research and solutions to effectively manage the integration of AI and data in agriculture.

Following the literature review, the chapter first addresses agri-data governance, delving into the complexities and providing strategic recommendations for improving data management in the sector. Precision agriculture, with its reliance on advanced technologies like IoT and sensors, faces significant security threats and risks associated with data storage and use (Akhtar et al., 2016). Farmers increasingly express concerns over data ownership and its use, impacting their decision-making and leading to suboptimal socio-economic outcomes (Clark, 2016; Kamble et al., 2020). The European Union's initiatives, such as the Code of Conduct for agricultural data sharing, aim to clarify roles and processes, thereby enhancing productivity and profitability (Renda et al., 2019). However, challenges persist in defining clear values and principles for data access, ownership, and governance, which are crucial for ensuring fair competition and protecting farmers' interests (Kaur et al., 2022; Atik, 2022). To mitigate these issues, the passage recommends establishing a centralized European Ag-Data System to consolidate diverse datasets, enforce interoperability standards, and ensure transparent data governance. This approach, supported by sector-specific regulations and a neutral public authority, aims to balance market dynamics, enhance data accessibility, and foster trust among farmers and stakeholders in the agricultural ecosystem.

Lastly, building on the relationship between data governance and AI implementation within the agricultural sector, the Sect. 7.4 examines the European Commission's Ethics Guidelines for Trustworthy AI and their application using the ALTAI framework. The guidelines underscore the need for AI to be lawful, ethical, and robust, with principles such as human oversight, technical robustness, privacy, transparency, fairness, and accountability (AI-HLEG, 2019). This part assesses the implementation of these principles in three experimental case studies in Spain, Finland, and Serbia/Lithuania.

The ALTAI framework, a comprehensive tool developed by the High-Level Expert Group on Trustworthy AI (HLEGAI), was used to evaluate the trustworthiness of AI systems in these pilots at two stages of the project. Initial assessments during the first year of the project identified the need for further implementation of recommended measures, while subsequent evaluation in year three highlighted significant improvements in transparency, privacy, and data governance. However, challenges remain in ensuring technical robustness, safety, and accountability, particularly given the complexity of integrating multiple AI systems within unmanned ground and aerial vehicles. The assessment also discusses the limitations of using ALTAI for ongoing research projects and the necessity for staged assessments to capture evolving ethical considerations.

The detailed assessments revealed notable advancements in human oversight, with mechanisms like privacy-by-design tools and emergency stop buttons enhancing safety and transparency. Despite improvements, areas such as diversity and non-discrimination require further attention to prevent bias and ensure fair decision-making in agricultural AI applications. Additionally, the use of model cards and datasheets is recommended to enhance transparency and accountability by providing structured documentation of AI models and datasets. This holistic approach aligns with the European Commission's Ethics Guidelines for Trustworthy AI and is essential for assessing the broader implications of AI technologies.

Finally, the findings underscore the importance of robust data governance frameworks and continuous ethical evaluations to ensure the responsible deployment of AI in agriculture, balancing innovation with the protection of stakeholders' rights and interests.

## 7.2 Literature Review

The literature review has the objective to provide an overview of key ethical, legal, socio-economic and environmental (ELSE) challenges and opportunities linked to data and AI in precision agriculture and related sectors from literature between 2016 and 2023. Other reviews on data and AI in agriculture focus on emerging trends from AI (Rejeb et al., 2022), impacts of agricultural robotics (Sparrow & Howard, 2021), as well as ethical and social impacts of AI and data (Ryan, 2022). This review attempts to give a systematic overview of the key arguments involved.

### *7.2.1 Methodology*

To synthesize the ELSE literature review process for the FlexiGroBots project, a systematic method was employed to examine key journals and databases using specific keywords via Google Scholar and direct searches. This strategy ensured a comprehensive collection of insights from both scholarly and practical perspectives. Individual sources were meticulously analyzed, extracting salient arguments related to the challenges and opportunities presented by robotics and AI in agriculture. These were meticulously recorded in an Excel spreadsheet, ensuring each unique argument from the diverse range of sources, including academic papers, industry reports, and high-quality blogs, was categorized and documented. Further classification of arguments into overarching categories such as AI, robotics, or both, along with metadata like the relevant ELSE factor, refined the review, resulting in a standardized compilation of the main arguments from 505 distinct sources. This process was conducted twice, first time in 2021 and another time in 2023 to discern the impact of AI's heightened media presence, marked by milestones such as the release of DALL.E 2 and ChatGPT, on the nature of discourse within this field.

### *7.2.2 Results*

The systematic follow-up review extended the literature review from January 2022 to September 2023. The second batch added 322 arguments from 78 new publications, culminating in 1074 arguments sourced from 505 publications, compared to the first batch's 747 arguments from 426 publications. Notably, Table 7.1 depicts a shift in the thematic focus of the literature: the second batch saw an increase in environmental and societal arguments to 16.46% and 29.19%, respectively, while the first batch had a stronger emphasis on economic and ethical issues. The second batch also indicated a trend away from discussions on discrimination, bias, privacy, and explainability towards transparency and data governance in ethical discourse, and a pivot from inequality and physical safety to (cyber)security risks and labor

**Table 7.1** Percentual recorded ELSE factors for data from 2016 to 2021 and 2022 to 2023

| ELSE factor | First batch | Second batch |
|---|---|---|
| Economic | 34.44% | 25.47% |
| Environmental | 11.30% | 16.46% |
| Ethical | 30.19% | 22.36% |
| Legal | 8.24% | 6.52% |
| Societal | 15.82% | 29.19% |
| **Grand total** | **100%** | **100%** |

market disruptions on the societal front. This suggests a recent literature tilt towards societal challenges. Despite these shifts, there were no substantial changes in the overall literature landscape on AI and robotics in agriculture during 2022 and 2023, with academic discourse largely reflecting on historical observations and the inapplicability of advancements in generative AI to the tangible application of agricultural robotics.

### 7.2.3 Quantification and Main Arguments

The dataset analysis indicates a significant academic contribution, with 68% of arguments on AI and robotics in agriculture originating from academic sources. Figure 7.1 illustrates the distribution between challenges and opportunities across ELSE factors, highlighting a concentration of opportunities in economic and environmental categories (40% of the total), while ethical, legal, and societal categories predominantly feature challenges. The ethical domain shows a notable imbalance, with a singular opportunity versus 299 challenges. Additionally, AI topics tend to attract more challenges (513) than opportunities (269), in contrast to robotics where opportunities (126) outnumber challenges (104). Mixed systems, which encompass both AI and robotics, show a balanced dialogue with slightly more opportunities (37) than challenges (25). Overall, the discourse is shaped by contributions from various sectors, with the majority of arguments stemming from academia, complemented by insights from the private sector, public sector, and civil society.

### 7.2.4 Challenges of Smart Farming

Identified legal challenges outline that AI and robotics systems lack an adequate legal framework. The gaps in legal frameworks range from issues of data ownership, and intellectual property (Meltzer, 2021), to copyright (Dwivedi et al., 2021), privacy (Kaplan & Haenlein, 2020) and more. Advanced edge devices can monitor video input in real-time (Dhanya et al., 2022) and surveillance practices are becoming more mainstream (UNESCO, 2023). The most frequently raised specific challenges concern liability and accountability. Under current law, robots cannot be held

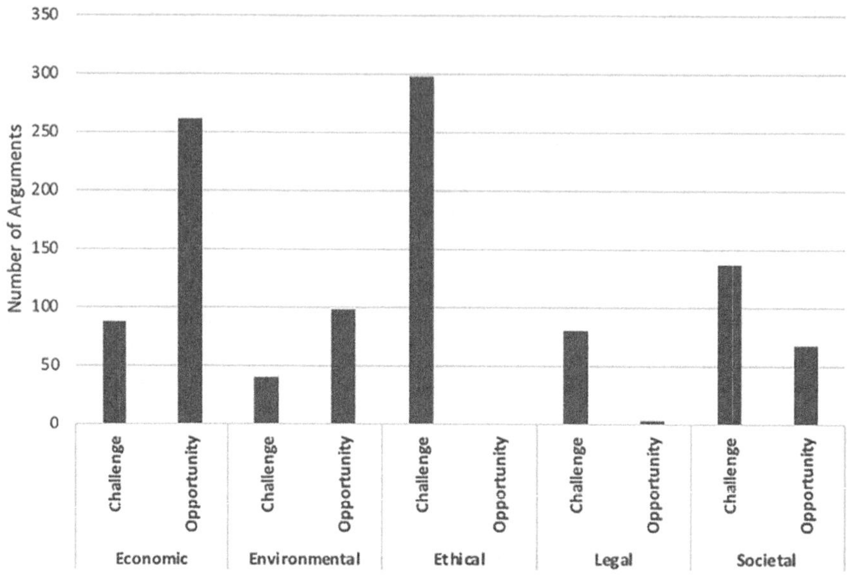


**Fig. 7.1** Number of arguments for ELSE factors, distinguishing between challenges and opportunities

liable for their actions (*O*'Sullivan et al., 2019) and there are many different stakeholders in the AI development process (Giuffrida, 2019). Moreover, a lack of explainability in deep learning makes it difficult to understand why systems fail (Samek et al., 2019). Note that the legal challenges listed in the figure above are not exhaustive and they partly overlap with ethical challenges on privacy, explainability and transparency (Fig. 7.2).

Identified ethical challenges rarely acknowledge solely the agricultural sector nonetheless most literature either addresses discrimination and bias in sectors such as healthcare and finance. Furthermore, Ennaji et al. (2023) make the case for bias in AI and agriculture. There exist many misleading publications from researchers that either do not have the proper soil and plant nutrient knowledge or the right machine learning (ML) expertise and select the wrong parameters in their models. Li (2023) addresses the concerns for biased data towards economic growth over environmental protection as an ethical issue. Autonomous systems will likely only be affordable for affluent countries and decrease the need for low-skilled labor, leading to reinforced gender and global inequalities (Guenat et al., 2022) (Fig. 7.3).

Questions around accountability and privacy are also relevant in the ethical domain (Ma, Zhang & Zhang, 2018; Dogru & Keskin, 2020; Currie & Hawk, 2021). Many data used for AI or ML-based solutions provide real-time information about people (Mark & Anya, 2019) and facial recognition can hamper privacy (Borenstein & Howard, 2021). Although transparency challenges are apparent in health care (Quinn et al., 2021) and public defense (Sharma, 2021), they are rarely discussed in the agriculture technology literature.

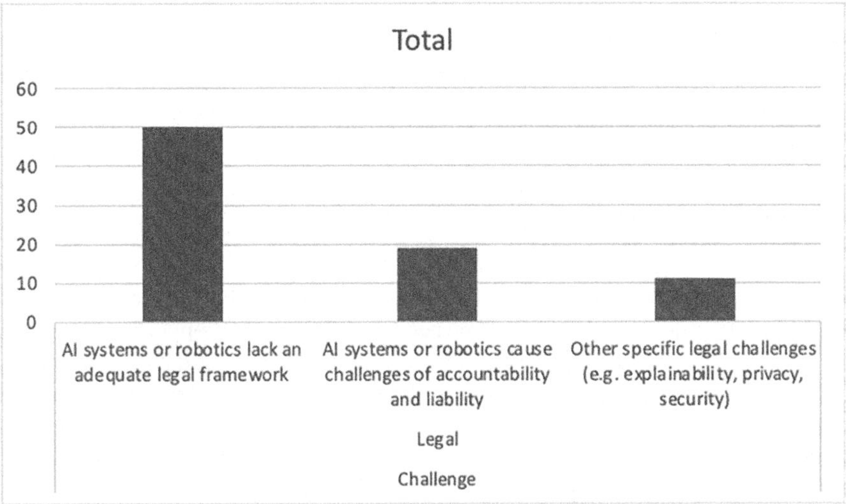


**Fig. 7.2** Main legal challenges recorded for data from 2016 to 2021 and 2022 to 2023

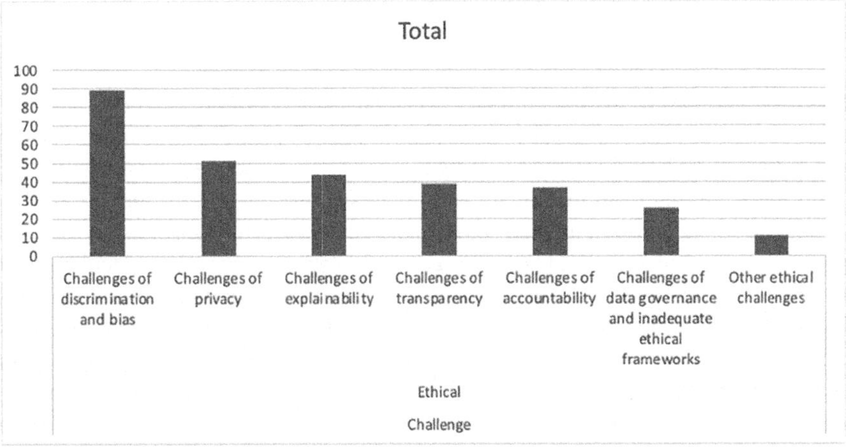


**Fig. 7.3** From the wide range of ethical challenges, discrimination and bias stand out

The literature review highlights that the integration of AI and robotics in agriculture presents a complex array of ethical and legal challenges. Ethically, the main issues include potential discrimination, embedded biases, privacy breaches, and a lack of transparency in AI operations, which mirror broader societal concerns about inequality, surveillance, and labor market disruptions. Legally, the primary obstacle is the absence of an adequate legal framework to govern these emergent technologies, with specific issues around ownership, privacy, liability, and accountability needing urgent attention. These challenges are particularly pertinent in the agricultural sector, where the use of data-driven technologies is growing rapidly. Addressing these issues requires sector-specific research and solutions, particularly in the realm of data governance, which is crucial for ensuring the responsible and equitable use of agricultural data.

## 7.3 Agri-Data Governance

Given the critical importance of data governance highlighted in the literature review, this subchapter delves deeper into the specific challenges and opportunities associated with data governance in precision agriculture. Effective data governance is essential for managing the vast amounts of data generated by advanced agricultural technologies, ensuring data security, ownership, and privacy. By examining these issues, I aim to provide a comprehensive understanding of how robust data governance frameworks can support the ethical and legal deployment of AI in agriculture with consideration for EU values in the process.

### *7.3.1 Data Governance Challenges*

Precision agriculture, with its reliance on advanced technologies like IoT and sensors, introduces various security threats and risks associated with data storage and use. Farmers are becoming increasingly aware of these risks, especially as the data collected—ranging from machinery sensors to satellite data—transfers out of their control once shared (Akhtar et al., 2016). The data's security then rests on the company in possession and could be used against the farmers in market speculations or by tech providers to sell new products. Concerns about data ownership and its use impact farmers' decisions and can lead to suboptimal socio-economic outcomes. Thus there's a recognized need for clear principles regarding data exploration (Clark, 2016). Furthermore, the European Union has been discussing data privacy and security for some time now with the European Data Strategy (Ryan et al., 2024). A critical issue is determining the access rights to data from farming machinery, which aids in multiple operational facets, and allowing data access to various service entities could boost competition in the sector (Kamble et al., 2020).

In 2018, the EU launched a Code of Conduct to streamline "agricultural data sharing, stemming from a collaboration among numerous farming organisations and significant industrial entities across the EU." The code's main thrust is to demystify the roles and processes surrounding agricultural data sharing. By offering clear guidelines on contracts and the reuse of data, it aims to enhance the efficiency with which farm data is used, thereby boosting both productivity and profitability. Hence, as the digital era advances, cloud computing is playing an increasingly pivotal role in agriculture (Renda et al., 2019).

### 7.3.2 *Agriculture Data and Stakeholders Ecosystem*

The data employed in precision agriculture comes in various shades (Castle et al., 2015). Under the EU Code of Conduct, these are categorized into several types. Personal data pertains to information that can pinpoint an individual (Kaur et al., 2022, p. 4). Anonymized data is data purged of any identifying characteristics (Kaur et al., 2022, pp. 6–7). Publicly available data is open for general use and distribution, while raw data is collected in its most unrefined form (Punt & Snijkers, 2020). Metadata gives insight about other sets of data (Urdu et al., 2023). Primary data transforms raw data into a format that is easily understandable by humans (Gnip & Kafka, 2018). Finally, aggregated data is a synthesis from numerous sources (Žuraulis & Pečeliūnas, 2023). Furthermore, agricultural data is specifically tied to farming operations and can be further divided into specific segments. This includes data about the farm's operations and management, covering aspects like plant and animal production, and data that revolves around machinery operations, which is meticulously protected against any external interference. Albeit data associated with machine maintenance and repair falls under service data. Then there's agri-supply and agri-service data: the former delves into agricultural inputs, and the latter pertains to services aiming to benefit farmers (Muangprathub et al., 2019).

The ecosystem is not complete without mentioning the users of farm data. This encompasses contractual partners such as farmers and manufacturers of machines and equipment, and a host of contractors and service providers. This last group is broad, covering entities ranging from software companies to research institutions (Kalmar et al., 2022, p. 281).

### 7.3.3 *Contracting and Control of Data*

The new-found value that can be derived from data has given rise to debates surrounding its ownership and control. While farmers are at the forefront of generating this data through AgriTech, it is predominantly the developers of these technologies who assert ownership (Atik, 2022). This predominant flow from farmers to AgriTech companies underscores the pressing need for governance structures centered around farmers. Such structures are not just essential for safeguarding farmers' rights (Kaur et al., 2022) but also play an instrumental role in defining the revenue models of AgriTech enterprises (Tatge, 2017).

Contracts, which are meant to protect and define data rights, often fall short in providing robust protections for farmers (Lorencowicz & Uziak, 2023). The role of contracts in the agricultural tech world is pivotal (EPRS, 2016). Often presented as click-wrap agreements, they dictate the terms of farm data usage. These data licenses, which are intricate by nature, tend to heavily favor corporate objectives (Kaur et al., 2022). It is common for agribusinesses to treat data as their intellectual property, creating additional layers of complexity in data sharing (Chapman &

Sherman, 2018). One of the glaring issues in this domain is the lack of transparency. Farmers, the primary data creators, are frequently left in the dark regarding the full extent of data usage, sharing, and trade. This opacity is even more pronounced when AgriTech providers are based in foreign countries, leading to potential trust issues and uncertainties regarding data governance (Jakku et al., 2019).

Once the data is out of the exclusive control of farmers, its usage becomes largely dictated by these contracts (Wang et al., 2022). Further complicating the matter is the ambiguous legal status of farm data. Neither classified distinctly as intellectual property nor as trade secrets, there is a gray area when it comes to its rights and protections (Kaur et al., 2022, p. 6; Jouanjean et al., 2020). The World Trade Organization's TRIPS agreement, for instance, does not address the rights of those who contribute the data. While some data collections might be shielded under copyright laws, raw data is often left unprotected (Chapman & Sherman, 2018). And even when protections exist, contractual terms might overshadow them.

Furthermore, the environment that dictates agricultural data governance is multifaceted, encompassing societal norms, policies, infrastructure, and various oversight mechanisms (Anidu & Dara, 2021). Factors like digital literacy and the prevailing regulatory environment play a pivotal role in shaping the governance of agricultural data (Lin et al., 2023; Hasan et al., 2022). With the global rise of AgriTech, there is an evident need for cohesive data policy interventions. Given the diversity in data protection laws across the globe, a unified approach to farm data protection remains elusive. Many farmers, despite being the primary stakeholders, remain uninformed of their rights (Clark, 2016). This scenario is exacerbated when AgriTech companies operate across borders, introducing jurisdictional complexities.

### 7.3.4 Data Ownership

The prominent mentioning of empowering farmers and other data producers with rights, including data ownership rights, might not help mitigate data-related issues of precision agriculture. Ownership of data has been touted as a possible legal mechanism for farmers, allowing them to use their agricultural data with various technology providers (Kritikos, 2017). This concept suggests that granting farmers explicit data ownership rights could alleviate concerns about data lock-ins and foster trust in data-sharing decisions (Kerber, 2016). However, data ownership doesn't fundamentally change the underlying contractual relationships surrounding the use, processing, and sharing of data (Drexl, 2017). Moreover, it doesn't address the inherent disparity in bargaining power between parties (Sykuta, 2016, p. 69). Even if farmers had inherent ownership rights, this would primarily benefit them in scenarios where contracts become nullified—a rather limited advantage, especially when considering potential risks (Van der Burg et al., 2020). Similarly, these dynamics might also apply to other types of agricultural data beyond farm-specific data.

Owning data may not be a panacea because the "right to transfer" data is a double-edged sword. While it offers flexibility, it also poses risks of loss of control. Historical examples show that property rights, over time, tend to gravitate towards those who value them most. In the context of data, this could mean stronger entities with more bargaining power gain even more control, exacerbating power disparities (Drexl, 2016). Transferring data ownership to dominant AgriTech entities could heighten their control, making it harder for farmers to switch providers and deepening their dependency.

The potential for data fragmentation and limited access presents another challenge (Kerber, 2016, p. 761). Ownership rights, including the "right to encumber," can stifle broader access to and innovation with data sets. First movers in the agricultural sector may hoard data, inhibiting competitors and other noncompeting entities from accessing it (Drexl, 2016). This can hamper innovation (Drexl, 2020). For instance, if an irrigation company withholds crucial soil data, other connected sectors might find it challenging to offer competitive solutions. By the same token, exclusive ownership can obstruct access to critical data, curtailing data-driven innovation (Kerber, 2016). If only a few powerful entities control data, we might see a monopolistic or oligopolistic sector emerge, further stagnating growth and innovation. While initially, data ownership might boost farmers' confidence in digital technologies, in the long run, the concentration of data rights among a few major players could erode this trust. The potential monopolization of data by a handful of agricultural giants could disillusion farmers, hindering the European Union's goal of enhancing digital adoption in the sector (Atik, 2022, p. 706). Given these complexities, it would be prudent to tread carefully with the concept of data ownership in agricultural regulations, especially when considering trust issues.

### 7.3.5 Recommendations for Agri-Data Governance

To address data fragmentation and access issues in agriculture, a comprehensive strategy involves creating a centralized data platform. This platform would consolidate diverse datasets from various company silos, enhancing visibility and accessibility. The proposed European Ag-Data System would serve as an infrastructure to support diverse stakeholder data needs, promoting innovation through improved data accessibility. The Common European Agricultural Data Space (CEADS) is envisioned to set sectoral data standards and ensure interoperability, with a central hub enforcing data access rights and encouraging format alignment (Ryan et al., 2024). CEADS aims to develop a transparent data hub with clear regulations, helping to balance market dynamics and giving farmers the flexibility to choose between smart farming options. This system would shift the competitive focus to service quality and data analytics.

However, there are challenges, such as enforcing direct interoperability across various machines and services, which the recent Data Act's provisions on interoperability do not fully address. Sector-specific interventions are suggested to ensure interoperability between different machines and digital services by setting technical standards for manufacturers and data space operators. This would also involve enforcing data portability within the centralized European Ag-Data System.

To keep pace with rapid digital service evolution and potential regulatory limitations, it's advised not to pursue overly ambitious tech-neutral standards for agricultural data regulation. A pragmatic approach includes setting clear criteria for third-party data access and reuse, guided by the Principles for a Data Economy from the American Law Institute and the European Law Institute (ELI-ALI principles). A sector-specific authority could manage these aspects and the Common European Agricultural Data Space (CEADS), adjusting access terms as needed and enforcing data rights. This authority could be backed by the recent Data Act, which allows Member States to appoint competent authorities for oversight.

Furthermore, to cultivate trust among farmers, a neutral public authority could oversee the CEADS and enforce sectoral regulations, possibly involving farmer representatives in governance or auditing roles. The Data Act's provisions for competent authorities, including raising user awareness, handling violations, and imposing penalties, could support trust in regulatory measures if aligned with the agricultural sector's specific requirements. Centralizing a European Ag-Data authority might be essential to ensure consistent governance and penalties across Europe.

The complexities of data governance in precision agriculture underscore the necessity for comprehensive frameworks that address security, ownership, and privacy concerns. Establishing clearer principles and regulations for data management can mitigate the risks identified and support the ethical use of agricultural data. However, data governance is only the first aspect of the broader ethical landscape of AI in agriculture. To fully address these challenges, one must also consider the overarching ethical guidelines and assessment tools designed to ensure trustworthy AI in practice.

## 7.4 The Ethics Guidelines for Trustworthy AI

AI technology has brought ethical considerations to the forefront of policy debates in the EU, addressing challenges such as labor and creative production impacts (Bordot, 2022; Frey & Osborne, 2017), increased surveillance (Kalluri et al., 2023), perpetuated biases (Buolamwini & Gebru, 2018), and the epistemic "black box" dilemma complicating responsibility assignment (Coeckelbergh, 2020). The European Commission's strategy, "Artificial Intelligence for Europe" (COM(2018) 237 final), underscores a commitment to fostering trust and accountability within the AI landscape. This dedication is further evidenced by the establishment of the High-Level Expert Group on AI (AI-HLEG) in June 2018, comprising 52 members from European universities, industry, and civil society organizations (AI-HLEG, 2019). The AI-HLEG's mandate includes developing Ethics Guidelines for Trustworthy AI, Policy and Investment Recommendations, the Assessment List for

Trustworthy AI (ALTAI), and Sectoral Considerations on these recommendations (Smuha, 2019), thereby reinforcing the EU's commitment to ethical AI governance.

### *7.4.1 Trustworthy AI and Its Applicability to the Agricultural Use Cases*

In 2019, the AI-HLEG published the Ethics Guidelines for Trustworthy AI, defining an AI system as "trustworthy" if it is lawful, ethical, and robust (AI-HLEG, 2019, p. 2). These guidelines stress the importance of enhancing human capabilities through AI, advocating for governance structures such as human-in-the-loop (HITL), human-on-the-loop (HOTL), and human-in-command (HIC) to ensure proper human oversight (AI-HLEG, 2019). In agriculture, maintaining human agency is vital to counter automation bias and de-skilling risks, with experts arguing for the irreplaceability of subjective expertise (Legun et al., 2023; Sparrow & Howard, 2021). Ensuring technical robustness and safety in AI involves building resilience against changes and adversarial interactions, with agricultural robotics particularly concerned about physical harm and cybersecurity threats (ENISA, 2023; Simpson, 2024). Even though privacy and data governance in agriculture focus on nonpersonal data, they still face significant challenges related to bias and data quality, which are crucial for building trust in AI systems (AI-HLEG, 2019). Transparency requires AI systems to be traceable and explainable, and current literature indicates a gap in research on the transparency of AI in agriculture (Ryan, 2022). Throughout AI development, diversity, nondiscrimination, and fairness are critical, as biases in agriculture can lead to harmful ecological impacts (Galaz et al., 2021). Environmental and societal well-being, though not initially framed as an ethical principle, are essential for the sustainability and societal impact of AI, reflecting the complexity of these technologies in agriculture (AI-HLEG, 2019). Lastly, accountability covers responsibility for AI systems and their outcomes, with ethical considerations extending to contractual agreements within the agricultural sector (AI-HLEG, 2019).

### *7.4.2 Trustworthiness Assessment*

The "Assessment List for Trustworthy Artificial Intelligence" (ALTAI), developed by the European Commission in collaboration with the AI-HLEG, serves as the foremost instrument for evaluating AI trustworthiness within the European Union (AI-HLEG, 2019; Stahl & Leach, 2023). ALTAI functions as a comprehensive questionnaire that embodies the seven essential requirements for trustworthy AI as identified by the AI-HLEG. It is intended as an operational checklist for organizations seeking to validate the trustworthiness of their AI products.

In the context of the FlexiGroBots project, this tool was employed during two interview sessions with project partners: initially during months 8–9 and subsequently in the project's final stages during months 33–34. These sessions aimed to ensure compliance with ethical standards in the development of the project's pilot programs. This section presents the findings from the second phase of this assessment.

### 7.4.3 Using ALTAI for Agri AI Assessments

An initial assessment of the project's use cases was conducted during months 8–9. However, it was determined that it was too early for many of the ALTAI-recommended mitigation measures to be implemented. Consequently, a more comprehensive reassessment took place towards the project's conclusion in September and October 2023. This reassessment involved interviews with pilot team members, an ethics expert, and a note-taker, utilizing the ALTAI web tool and supplementary notes for enhanced depth and clarity. Due to the complexity of the questions, a final meeting was necessary with the project's platform team specifically to address mitigation measures.

It is essential to note that these assessments provide a snapshot of an evolving research project, where the developers or co-creators are also the primary users, thus requiring no additional training. The perspectives and requirements of eventual end-users, who will interact with the fully functional system once it is market-ready, remain speculative at this stage.

Another significant limitation of applying the ALTAI to the FlexiGroBots pilots lies in the complexity of the use cases. Each pilot comprises multiple unmanned ground vehicles (UGVs) and unmanned aerial vehicles (UAVs), partially or fully integrated into the Mission Control Centre and platform. Furthermore, each UGV and UAV might incorporate multiple AI systems within one device. Consequently, assessing all AI systems in one ALTAI interview necessitates generalization across technologies, diminishing the overall meaningfulness of the assessment. Given that the ALTAI can include up to 154 questions, it is not feasible to assess every single AI model. Such fragmentation of the product would also complicate the assessment of the holistic impacts of the AI systems, which are included in the ALTAI.

Nevertheless, certain aspects of the ALTAI should be considered even before a project begins, such as understanding the technology's impact on environmental and societal well-being. If these impacts pose a disproportionate risk, the development and implementation process should be reconsidered entirely. The limitations of applying the ALTAI to ongoing research projects have also been highlighted by Stahl and Leach (2023). The mismatch in timelines—requiring a complete and functional AI system versus considering the project conception—makes it challenging to utilize the ALTAI effectively. A combination of multiple methods or staged assessments should be employed for ethical technology assessments.

#### 7.4.3.1 Results from Grapevine Farming Use Case

The Spanish pilot, centered on grapevine farming, involves a collaborative robot that works in close conjunction with human workers, thereby emphasizing the significance of labor conditions and safety in AI applications. The ALTAI assessment indicated overall enhancement across all trustworthiness dimensions, with transparency achieving the highest score due to the clear user awareness of AI interaction. Notable progress was observed in privacy and data governance, particularly through the implementation of a privacy-by-design tool in unmanned ground vehicles (UGVs) to anonymize faces in video recordings. Nonetheless, broader data governance challenges remain to be addressed beyond the scope of ALTAI (Fig. 7.4).

Accountability remained relatively stable, attributed to the public availability of datasets and codes from the project's outset. The greatest challenge was ensuring technical robustness and safety, especially in terms of cybersecurity and potential system misuse, areas that had not been thoroughly examined at the project's current stage (Fig. 7.5).

During the assessment discussions, the development team emphasized the importance of distinguishing between various technological components, such as unmanned ground vehicles (UGVs) and unmanned aerial vehicles (UAVs), and clarifying the extent of AI's role within the pilot's technology suite. There was also debate over the strictness of requirement interpretations, specifically whether the human capacity to recognize autonomous UGV operations was sufficient for compliance or if additional measures were necessary.

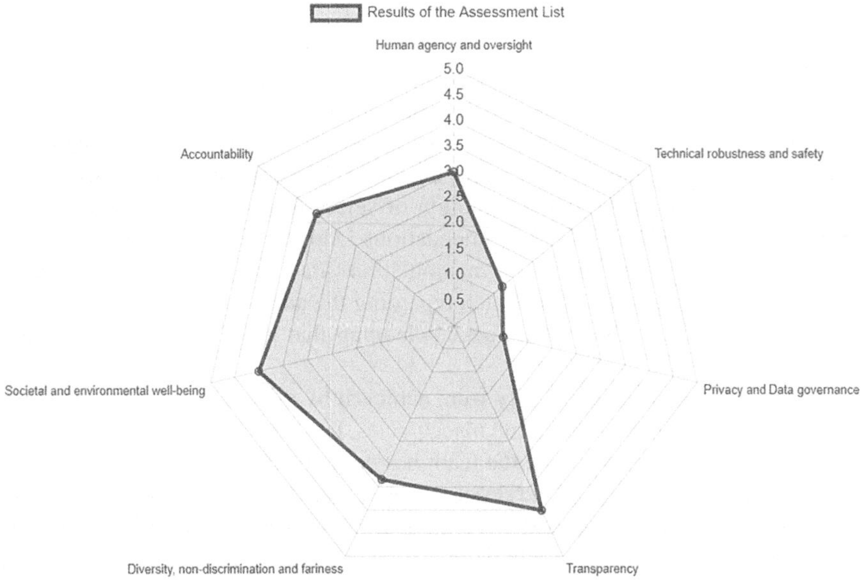


**Fig. 7.4** Pilot 1 ALTAI Assessment (M8) (See more https://flexigrobots-h2020.eu/library/deliverables)

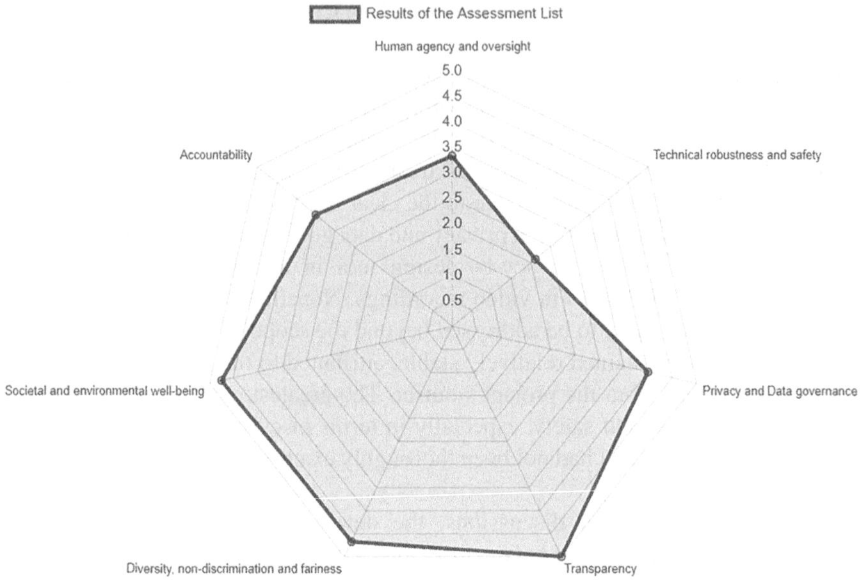


**Fig. 7.5** Pilot 1 ALTAI Re-assessment (M34) (See more https://flexigrobots-h2020.eu/library/deliverables)

#### 7.4.3.2 Results from Crop Farming Use Case

The Finish pilot, which focuses on crop farming, employs large machinery, including an autonomous tractor, where physical safety is of paramount concern (Fig. 7.6).

Similar to the Spanish pilot, this use case experienced significant advancements in transparency, privacy, and data governance, alongside notable improvements in accountability and human agency and oversight. These improvements were particularly evident with the implementation of safety features like emergency stop buttons, in line with ALTAI recommendations. Although risks related to human autonomy, such as attachment or manipulation, are considered less relevant to this pilot, there is a cautious approach regarding the adequacy of measures to prevent over-reliance on the system, acknowledging that it remains in development (Fig. 7.7).

Enhancements in accountability were achieved by establishing mechanisms for reporting system vulnerabilities and inaccuracies. Despite a lower score in diversity, nondiscrimination, and fairness, the team has committed to adhering to accessible design principles before market release. Although the technology targets a specific user group—farmers—it aims to accommodate the accessibility needs within this group as outlined in the Ethics Guidelines for Trustworthy AI.

Transparency considerations are currently shaped by the informed status of the pilot users regarding system limitations, with additional measures planned for commercial use.

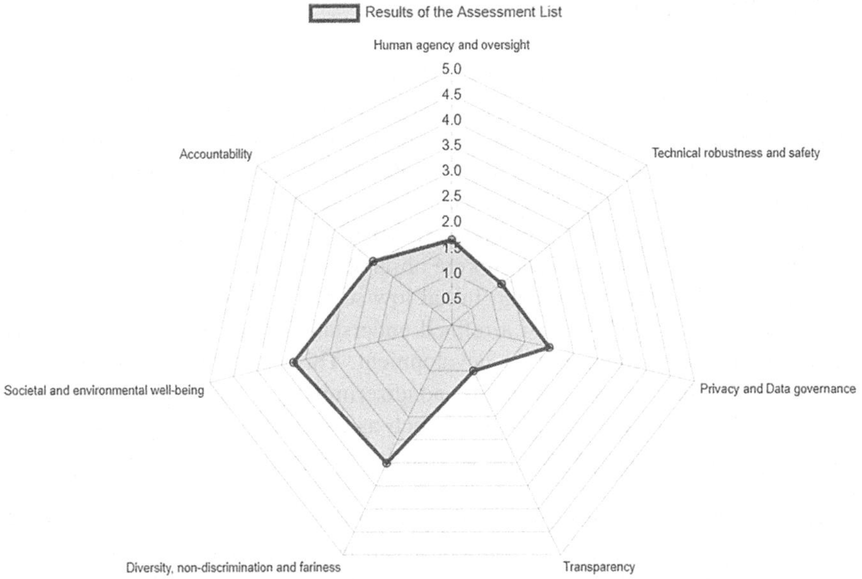


**Fig. 7.6** Pilot 2 ALTAI Assessment (M8) (See more https://flexigrobots-h2020.eu/library/deliverables)

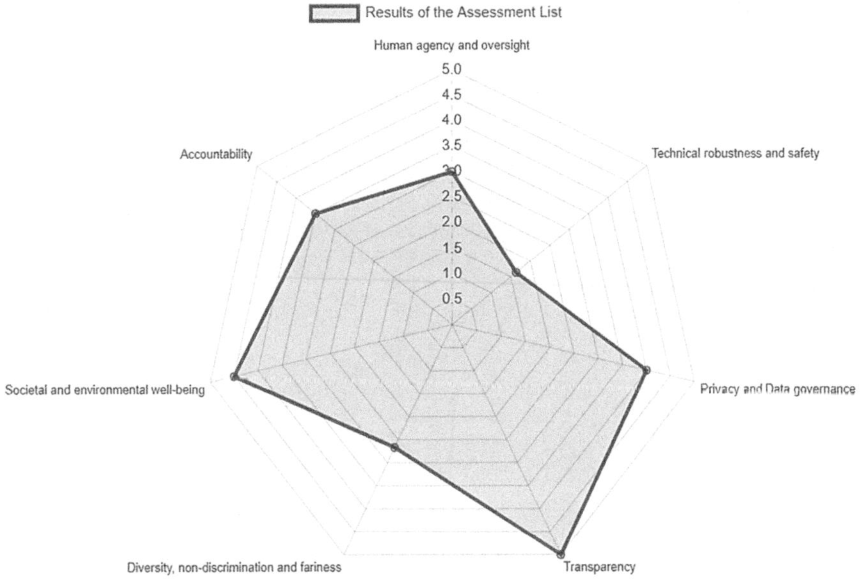


**Fig. 7.7** Pilot 2 ALTAI Re-assessment (M33) (See more https://flexigrobots-h2020.eu/library/deliverables)

#### 7.4.3.3 Results from Blueberry Use Case

The Serbian pilot, focused on blueberry farming in Serbia and Lithuania, is notable for its engagement with stakeholders, particularly a socio-economic partner, to address the specific needs of blueberry farmers. Significant progress in trustworthiness was observed across all areas, especially in the complex operations of the unmanned ground vehicles (UGVs) used for soil sampling near plant roots, which contributed substantially to these advancements (Fig. 7.8).

The most notable enhancements were observed in transparency, as well as societal and environmental well-being. Initial low scores in these areas saw considerable improvement, likely due to a refined understanding of the pilot's impact and active farmer involvement during development. This engagement also positively influenced the dimensions of diversity, nondiscrimination, and fairness. The inclusion of explainable AI (XAI) practices, such as uncertainty modeling, has significantly heightened transparency by clarifying the system's confidence levels for users (Fig. 7.9).

Human autonomy is safeguarded in the current design, ensuring that the technology does not pose a risk of over-reliance and that humans retain final decision-making capabilities. This assurance, along with the matured capabilities of the soil sampler and other robotic applications, led to a threefold improvement in this area.

Privacy and data governance were bolstered by integrating privacy-by-design principles in the unmanned aerial vehicles' (UAVs) flight parameters, setting altitudes that prevent human identification.

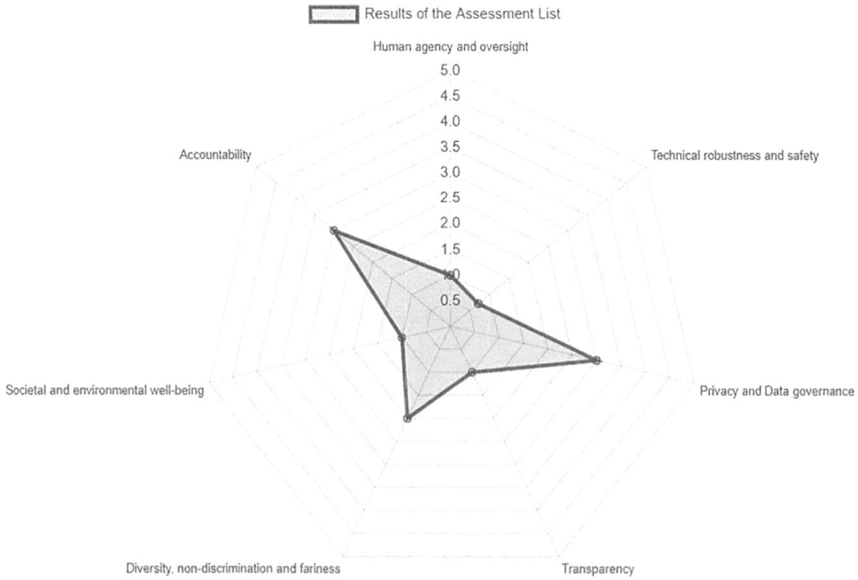


**Fig. 7.8** Pilot 3 ALTAI Assessment (M8) (See more https://flexigrobots-h2020.eu/library/deliverables)

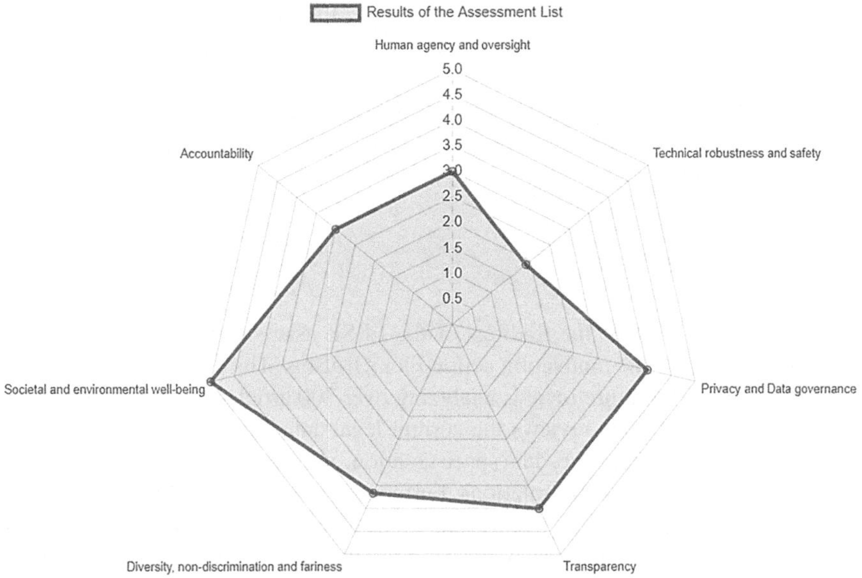


**Fig. 7.9** Pilot 3 ALTAI Re-Assessment (M34) (See more https://flexigrobots-h2020.eu/library/deliverables)

## 7.4.4 *Enhancing Transparency with Model Cards and Datasheets*

Model cards and datasheets are tools that enhance AI transparency by providing structured documentation of AI models and datasets' characteristics (Brynjolfsson et al., 2018; Bender et al., 2021). They ensure thorough documentation by developers and clarify the potential and limitations of AI technologies for external users. These instruments are crucial for preventing misuse in unintended contexts and have been widely adopted by organizations like Microsoft and Google. Since the start of the project, "system cards" standards have emerged, as seen with OpenAI's release of GPT-4 (OpenAI, 2023), which extend the concept to document the entire system encompassing the model. This holistic approach is advocated by the Ethics Guidelines for Trustworthy AI (AI-HLEG, 2019) and is essential for assessing the broader implications of AI technologies.

For the FlexiGroBots project, the need for model cards and datasheets was identified at the project's inception, following the first ALTAI questionnaire round and a literature review. As a result, custom templates were created for model cards and datasheets, combining insights from the literature and the specific needs of the FlexiGroBots project. The complete templates are available on the project website. Developers filled in separate documents for each model or dataset and uploaded them to the public domain.

In the interviews for the ALTAI reassessments, FlexiGroBots partners concurred that model cards, datasheets, and system cards contribute to fulfilling the requirements of transparency and accountability. However, they also recognized that these measures alone are not sufficient. Throughout the project's duration, partners continuously added the necessary documentation to every dataset and model published.

## 7.5 Conclusion

This chapter synthesizes the findings of the ELSE assessment of the AI and data challenges faced by the innovation action FlexiGroBots. The document begins with a comprehensive literature review, analyzing over 700 articles to identify key arguments and trends. It then reviews the current legal data landscape and offers guidance on agri-data governance, data space creation, and data ownership within farmer and AgriTech provider (ATP) relations. Furthermore, the chapter builds upon previous analyses and verifies their applicability in practice by applying the European Ethics Guidelines for Trustworthy AI to agriculture. It assesses the project's pilots and technologies using the ALTAI and suggests the use of model cards, datasheets, and system cards for transparent reporting.

## References


Acosta, M., van Wessel, M., van Bommel, S., Ampaire, E. L., Twyman, J., Jassogne, L., & Feindt, P. H. (2020). What does it mean to make a 'Joint' decision? Unpacking intra-household decision making in agriculture: Implications for policy and practice. *The Journal of Development Studies, 56*(6), 1210–1229. https://doi.org/10.1080/00220388.2019.1650169

Akhtar, P., Tse, Y. K., Khan, Z., & Rao-Nicholson, R. (2016). Data-driven and adaptive leadership contributing to sustainability: Global agri-food supply chains connected with emerging markets. *International Journal of Production Economics, 181*, 392–401. https://doi.org/10.1016/j.ijpe.2015.11.013

Al-Amin, A. K. M. A., Lowenberg-DeBoer, J., Franklin, K., & Behrendt, K. (2023). Economics of field size and shape for autonomous crop machines. *Precision Agriculture, 24*(5), 1738–1765. https://doi.org/10.1007/s11119-023-10016-w

Anidu, A., & Dara, R. (2021). A review of data governance challenges in smart farming and potential solutions. In *2021 IEEE international symposium on technology and society (ISTAS)* (pp. 1–8). https://doi.org/10.1109/ISTAS52410.2021.9629169

Atik, C. (2022). Towards comprehensive European agricultural data governance: Moving beyond the "Data Ownership" debate. *IIC - International Review of Intellectual Property and Competition Law, 53*(5), 701–742. https://doi.org/10.1007/s40319-022-01191-w

Ayoub Shaikh, T., Rasool, T., & Rasheed Lone, F. (2022). Towards leveraging the role of machine learning and artificial intelligence in precision agriculture and smart farming. *Computers and Electronics in Agriculture, 198*, 107119. https://doi.org/10.1016/j.compag.2022.107119

Baudron, F., & Giller, K. E. (2014). Agriculture and nature: Trouble and strife? *Biological Conservation, 170*, 232–245.

Bender, E. M., Gebru, T., McMillan-Major, A., & Shmitchell, S. (2021, March). On the dangers of stochastic parrots: Can language models be too big?. In *Proceedings of the 2021 ACM conference on fairness, accountability, and transparency* (pp. 610–623).

Bhakta, I., Phadikar, S., & Majumder, K. (2019). State-of-the-art technologies in precision agriculture: A systematic review. *Journal of the Science of Food and Agriculture, 99*(11), 4878–4888. https://doi.org/10.1002/jsfa.9693

Bordot, F. (2022). Artificial intelligence, robots and unemployment: Evidence from OECD countries. *Journal of Innovation Economics & Management, 37*(1), 117–138.

Borenstein, J., & Howard, A. (2021). Emerging challenges in AI and the need for AI ethics education. *AI and Ethics, 1*(1), 61–65. https://doi.org/10.1007/s43681-020-00002-7

Botta, A., Cavallone, P., Baglieri, L., Colucci, G., Tagliavini, L., & Quaglia, G. (2022). A review of robots, perception, and tasks in precision agriculture. *Applied Mechanics, 3*(3) Article 3. https://doi.org/10.3390/applmech3030049

Brynjolfsson, E., Mitchell, T., & Rock, D. (2018, May). What can machines learn and what does it mean for occupations and the economy?. In *AEA papers and proceedings* (Vol. 108, pp. 43–47). American Economic Association.

Buolamwini, J., & Gebru, T. (2018, January). Gender shades: Intersectional accuracy disparities in commercial gender classification. In *Conference on fairness, accountability and transparency* (pp. 77–91). PMLR.

Cammarano, D., van Evert, F. K., & Kempenaar, C. (2023). *Precision agriculture: modelling*. Springer Nature.

Castle, M., Lubben, B., & Luck, J. (2015). Precision agriculture usage and big agriculture data. *Cornhusker Economics*. https://digitalcommons.unl.edu/agecon_cornhusker/725

Chapman, S., & Sherman, B. (2018). Finding a place for agriculture in intellectual property law. *IIC - International Review of Intellectual Property and Competition Law, 49*(7), 759–762. https://doi.org/10.1007/s40319-018-0753-8

Cisternas, I., Velásquez, I., Caro, A., & Rodríguez, A. (2020). Systematic literature review of implementations of precision agriculture. *Computers and Electronics in Agriculture, 176*, 105626. https://doi.org/10.1016/j.compag.2020.105626

Clark, P. (2016). Opinion: Growers shouldn't need a law degree to understand their data terms agreements | Agri-Pulse Communications, Inc. https://www.agri-pulse.com/articles/7259-opinion-growers-shouldn-t-need-a-law-degree-to-understand-their-data-terms-agreements.

Coeckelbergh, M. (2020). Artificial intelligence, responsibility attribution, and a relational justification of explainability. *Science and engineering ethics, 26*(4), 2051–2068.

Currie, G., & Hawk, K. E. (2021). Ethical and legal challenges of artificial intelligence in nuclear medicine. *Seminars in Nuclear Medicine, 51*(2), 120–125. https://doi.org/10.1053/j.semnuclmed.2020.08.001

Dhanya, V. G., Subeesh, A., Kushwaha, N. L., Vishwakarma, D. K., Nagesh Kumar, T., Ritika, G., & Singh, A. N. (2022). Deep learning based computer vision approaches for smart agricultural applications. *Artificial Intelligence in Agriculture, 6*, 211–229. https://doi.org/10.1016/j.aiia.2022.09.007

Dogru, A. K., & Keskin, B. B. (2020). AI in operations management: Applications, challenges and opportunities. *Journal of Data, Information and Management, 2*(2), 67–74. https://doi.org/10.1007/s42488-020-00023-1

Drexl, J. (2016). *Designing competitive markets for industrial data—between propertisation and access* (SSRN Scholarly Paper 2862975). https://doi.org/10.2139/ssrn.2862975.

Drexl, J. (2017). On the future EU legal framework for the digital economy: A competition-based response to the 'Ownership and Access' debate. In Trading data in the digital economy: Legal concepts and tools.

Drexl, J. (2020). *Connected devices—an unfair competition law approach to data access rights of users* (SSRN Scholarly Paper 3746163). https://papers.ssrn.com/abstract=3746163.

Dwivedi, Y. K., Hughes, L., Ismagilova, E., Aarts, G., Coombs, C., Crick, T., Duan, Y., Dwivedi, R., Edwards, J., Eirug, A., Galanos, V., Ilavarasan, P. V., Janssen, M., Jones, P., Kar, A. K., Kizgin, H., Kronemann, B., Lal, B., Lucini, B., et al. (2021). Artificial intelligence (AI): Multidisciplinary perspectives on emerging challenges, opportunities, and agenda for research, practice and policy. *International Journal of Information Management, 57*, 101994. https://doi.org/10.1016/j.ijinfomgt.2019.08.002

Ennaji, O., Vergütz, L., & El Allali, A. (2023). Machine learning in nutrient management: A review. *Artificial Intelligence in Agriculture, 9*, 1–11. https://doi.org/10.1016/j.aiia.2023.06.001

EPRS. (2016). Precision agriculture and the future of farming in Europe: Scientific foresight study. Publications Office European Parliament. Directorate General for Parliamentary Research Services. https://data.europa.eu/doi/10.2861/020809.

European Union Agency for Cybersecurity (ENISA). (2023). *ENISA Threat Landscape 2023*. Retrieved from https://www.enisa.europa.eu/sites/default/files/publications/ENISA%20Threat%20Landscape%202023.pdf

Frey, C. B., & Osborne, M. A. (2017). The future of employment: How susceptible are jobs to computerisation?. *Technological forecasting and social change, 114*, 254–280.

Gabriel, A., & Gandorfer, M. (2023). Adoption of digital technologies in agriculture—an inventory in a European small-scale farming region. *Precision Agriculture, 24*(1), 68–91. https://doi.org/10.1007/s11119-022-09931-1

Galaz, V., Centeno, M. A., Callahan, P. W., Causevic, A., Patterson, T., Brass, I.,... & Levy, K. (2021). Artificial intelligence, systemic risks, and sustainability. *Technology in society, 67*, 101741.

Giuffrida, I. (2019). Liability for AI decision-making: Some legal and ethical considerations. *Fordham Law Review, 88*, 439–456.

Gnip, P., & Kafka, S. (2018). Using technology of data collection and data processing in precision farming. *Agricultural Economics (Zemědělská Ekonomika), 49*, 419–426. https://doi.org/10.17221/5426-AGRICECON

Guenat, S., Purnell, P., Davies, Z. G., Nawrath, M., Stringer, L. C., Babu, G. R., Balasubramanian, M., Ballantyne, E. E. F., Bylappa, B. K., Chen, B., De Jager, P., Del Prete, A., Di Nuovo, A., Ehi-Eromosele, C. O., Eskandari Torbaghan, M., Evans, K. L., Fraundorfer, M., Haouas, W., Izunobi, J. U., et al. (2022). Meeting sustainable development goals via robotics and autonomous systems. *Nature Communications, 13*(1), 3559. https://doi.org/10.1038/s41467-022-31150-5

Hasan, K., Masriadi, Muchlis, Aftah, R. M., & Syakban, M. N. (2022). Digital skills in the optimization of agricultural technology among Milenial 2022 (description study on agricultural students at Malikussaleh University). *Proceedings of international conference on social science, political science, and humanities (*ICoSPOLHUM), 3, 00007–00007. https://doi.org/10.29103/icospolhum.v3i.149.

High-Level Expert Group on Artificial Intelligence (AI-HLEG). (2019). *Ethics guidelines for trustworthy AI*. Brussels: European Commission. Retrieved from https://ec.europa.eu/newsroom/dae/document.cfm?doc_id=60419

Jakku, E., Taylor, B., Fleming, A., Mason, C., Fielke, S., Sounness, C., & Thorburn, P. (2019). "If they don't tell us what they do with it, why would we trust them?" Trust, transparency and benefit-sharing in smart farming. *NJAS - Wageningen Journal of Life Sciences, 90–91*, 100285. https://doi.org/10.1016/j.njas.2018.11.002

Jouanjean, M.-A., Casalini, F., Wiseman, L., & Gray, E. (2020). Issues around data governance in the digital transformation of agriculture: The farmers' perspective. *OECD*. https://doi.org/10.1787/53ecf2ab-en

Kalluri, P., Agnew, W., Cheng, M., Owens, K., Soldaini, L., & Birhane, A. (2023). The Surveillance AI Pipeline. https://arxiv.org/abs/2309.15084

Kalmar, R., Rauch, B., Dörr, J., & Liggesmeyer, P. (2022). Agricultural data space. In B. Otto, M. ten Hompel, & S. Wrobel (Eds.), *Designing data spaces: The ecosystem approach to competitive advantage* (pp. 279–290). Springer. https://doi.org/10.1007/978-3-030-93975-5_17

Kamble, S. S., Gunasekaran, A., & Gawankar, S. A. (2020). Achieving sustainable performance in a data-driven agriculture supply chain: A review for research and applications. *International Journal of Production Economics, 219*, 179–194. https://doi.org/10.1016/j.ijpe.2019.05.022

Kaplan, A., & Haenlein, M. (2020). Rulers of the world, unite! The challenges and opportunities of artificial intelligence. *Business Horizons, 63*(1), 37–50. https://doi.org/10.1016/j.bushor.2019.09.003

Kaur, J., Hazrati Fard, S. M., Amiri-Zarandi, M., & Dara, R. (2022). Protecting farmers' data privacy and confidentiality: Recommendations and considerations. *Frontiers in Sustainable Food Systems, 6*. https://www.frontiersin.org/articles/10.3389/fsufs.2022.903230.

Kerber, W. (2016). *A new (intellectual) property right for non-personal data? An economic analysis* (SSRN scholarly paper 2858171). https://papers.ssrn.com/abstract=2858171.

Kritikos, M. (2017). Precision agriculture in Europe: Legal, social and ethical considerations. *European Parliamentary Research Service*. https://data.europa.eu. https://doi.org/10.2861/278

Legun, K., Burch, K. A., & Klerkx, L. (2023). Can a robot be an expert? The social meaning of skill and its expression through the prospect of autonomous AgTech. *Agriculture and Human Values, 40*(2), 501–517.

Li, R. (2023, May 8). The environmental impact of AI. https://insights.grcglobalgroup.com/the-environmental-impact-of-ai/.

Lin, D., Fu, B., Xie, K., Zheng, W., Chang, L., & Lin, J. (2023). Research on the improvement of digital literacy for moderately scaled tea farmers under the background of digital intelligence empowerment. *Agriculture, 13*(10) Article 10. https://doi.org/10.3390/agriculture13101859

Lorencowicz, E., & Uziak, J. (2023). Selected problems on data used in precision agriculture. In S. Pascuzzi & F. Santoro (Eds.), *Farm machinery and processes management in sustainable agriculture* (pp. 217–226). Springer. https://doi.org/10.1007/978-3-031-13090-8_23

Ma, L., Zhang, Z., & Zhang, N. (2018). Ethical dilemma of artificial intelligence and its research Progress. *IOP Conference Series: Materials Science and Engineering, 392*(6), 062188. https://doi.org/10.1088/1757-899X/392/6/062188

Mark, R., & Anya, G. (2019). Ethics of using smart city AI and big data: The case of four large European cities. *The ORBIT Journal, 2*(2), 1–36.

Meltzer, J. (2021). The impact of artificial intelligence on international trade: Evidence from Google neural machine translation. *Journal of Technological Advancements, 1*(1), 1–20. https://doi.org/10.4018/JTA.20210101.oa6

Muangprathub, J., Boonnam, N., Kajornkasirat, S., Lekbangpong, N., Wanichsombat, A., & Nillaor, P. (2019). IoT and agriculture data analysis for smart farm. *Computers and Electronics in Agriculture, 156*, 467–474. https://doi.org/10.1016/j.compag.2018.12.011

Niedbała, G., Piekutowska, M., & Hara, P. (2023). New trends and challenges in precision and digital agriculture. *Agronomy, 13*(8) Article 8. https://doi.org/10.3390/agronomy13082136

O'Sullivan, S., Nevejans, N., Allen, C., Blyth, A., Leonard, S., Pagallo, U., Holzinger, K., Holzinger, A., Sajid, M. I., & Ashrafian, H. (2019). Legal, regulatory, and ethical frameworks for development of standards in artificial intelligence (AI) and autonomous robotic surgery. *The International Journal of Medical Robotics and Computer Assisted Surgery, 15*(1), e1968. https://doi.org/10.1002/rcs.1968

OpenAI. (2023). *GPT-4 System Card*. Retrieved from https://cdn.openai.com/papers/gpt-4-system-card.pdf

Punt, T., & Snijkers, G. (2020). Exploring precision farming data: A valuable new data source? A first orientation.

Quinn, T. P., Senadeera, M., Jacobs, S., Coghlan, S., & Le, V. (2021). Trust and medical AI: The challenges we face and the expertise needed to overcome them. *Journal of the American Medical Informatics Association, 28*(4), 890–894. https://doi.org/10.1093/jamia/ocaa268

Rejeb, A., Rejeb, K., Zailani, S., Keogh, J. G., & Appolloni, A. (2022). Examining the interplay between artificial intelligence and the agri-food industry. *Artificial Intelligence in Agriculture, 6*, 111–128. https://doi.org/10.1016/j.aiia.2022.08.002

Renda, A., Reynolds, N., Laurer, M., & Cohen, G. (2019). Digitising agrifood: Pathways and challenges. In *CEPS papers* (25701; CEPS Papers). Centre for European Policy Studies. https://ideas.repec.org/p/eps/cepswp/25701.html.

Ryan, M. (2022). The social and ethical impacts of artificial intelligence in agriculture: Mapping the agricultural AI literature. *AI & Society*. https://doi.org/10.1007/s00146-021-01377-9

Ryan, M., Gürtler, P., & Bogucki, A. (2024). Will the real data sovereign please stand up? An EU policy response to sovereignty in data spaces. *International Journal of Law and Information Technology, 32*(1), eaae006.

Samek, W., Montavon, G., Vedaldi, A., Hansen, L. K., & Müller, K.-R. (Eds.). (2019). *Explainable AI: Interpreting, explaining and visualizing deep learning* (Vol. 11700). Springer. https://doi.org/10.1007/978-3-030-28954-6

Schuh, G. E. (2019). Out-migration, rural productivity. Migration and the labor market in developing countries, 161.

Sharma, M. (2021). The ethical challenges of AI in defence. *Analytics India Magazine*. https://analyticsindiamag.com/the-ethical-challenges-of-ai-in-defence/

Simpson, C., (2024). JRC Highlights Report 2023. In Bonjean, I., Fornara, M., Jonkers, K., Klaassen, T., Lehto, S., Soldatova, L., Thielen Del Pozo, J., and Westra Van Holthe, M. (Eds.), *Publications Office of the European Union*. Luxembourg. https://doi.org/10.2760/947345, JRC136884

Smuha, N. A. (2019). The EU approach to ethics guidelines for trustworthy artificial intelligence. *Computer Law Review International, 20*(4), 97–106.

Sparrow, R., & Howard, M. (2021). Robots in agriculture: Prospects, impacts, ethics, and policy. *Precision Agriculture, 22*(3), 818–833. https://doi.org/10.1007/s11119-020-09757-9

Stahl, B. C., & Leach, T. (2023). Assessing the ethical and social concerns of artificial intelligence in neuroinformatics research: An empirical test of the European Union Assessment List for Trustworthy AI (ALTAI). *AI and Ethics, 3*(3), 745–767.

Sykuta, M. (2016). Big data in agriculture: Property rights, privacy and competition in Ag data services. *International Food and Agribusiness Management Review, 19*(A) https://econpapers.repec.org/article/agsifaamr/240696.htm.

Tatge, T. (2017, June 8). Data is the new cash crop: Understanding the market for farm data. Medium. https://dirtforfarmers.com/data-is-the-new-cash-crop-understanding-the-market-for-farm-data-59056dda285d.

UNESCO. (2023, April 21). Artificial Intelligence: Examples of ethical dilemmas. https://www.unesco.org/en/artificial-intelligence/recommendation-ethics/cases.

Urdu, D., Goense, D., Booij, J., Graumans, C., & Kempenaar, C. (2023). Analysis of metadata standards for the exchange of image datasets and algorithms in the agricultural domain: A metadata-oriented approach to identify minimum interoperability mechanisms for image data and deep learning algorithms that is used for vision-based applications in agriculture. Sprint Robotics Project PL4.0 WP7. https://doi.org/10.18174/632203.

van der Burg, S., Oosterkamp, E., Bogaardt, M. J., Regan, A., Popa, E. O., Tabeau, E., Brunori, G., Favilli, E., & Wattel, C. (2020). Futures of farm data sharing practices; perspectives of European farmers, researchers and agri-tech businesses (Report IOF2020). https://doi.org/10.13140/RG.2.2.21562.41924.

Wang, Y., Huber, R., & Finger, R. (2022). The role of contractors in the uptake of precision farming—a spatial economic analysis. *Q Open, 2*(1), qoac003. https://doi.org/10.1093/qopen/qoac003

Žuraulis, V., & Pečeliūnas, R. (2023). The architecture of an agricultural data aggregation and conversion model for smart farming. *Applied Sciences, 13*(20) Article 20. https://doi.org/10.3390/app132011216

**Artur Bogucki** MS Law, MS Finance, LL.M. Law & Economics (EMLE), Ph.D. Law. Associate Researcher, Centre for European policy Studies (CEPS), Assistant Professor of Law and Economics, Warsaw School of Economics (SGH). Lecturer, European Master in Law & Economics program. Member, AI Lab, SGH. Leader, CIVICA Europe Revisited group, SGH. Visiting Lecturer, LUISS University. Visiting Researcher, ETH Zurich. Fellow, University of Glasgow. Board Member, Polish Association of Law and Economics. Author of several papers, chapters, and monographs including Behavioural Law and Economics and Inner Morality of Law (CH. Beck, 2023). Research areas: Legal philosophy, legal theory, theories of regulation, empirical legal studies, and technology law.

# Chapter 8
# Scientists and the Sovereigns: The Distributive Justice Implications of Digital Sequence Information Governance Under the Convention on Biological Diversity

**Bob Kreiken and Adam McCarthy**

**Abstract** This chapter presents an insight into current negotiations of the UN Convention on Biological Diversity (CBD) regarding the digitalization of genetic resources, from an ethics of technology perspective. The chapter gives an overview of the transition towards data-centric biological research and open science, and its implications for Access and Benefit-Sharing (ABS) policies, which consist of mechanisms that ensure fair access to genetic resources and the fair and equitable benefit-sharing from their utilization in research and development. Here, we examine value conflicts at the knowledge-governance interface, conceptualizing the CBD negotiations as a site that can be analyzed to provide a better understanding of the relationship between science, technology, and innovation policies, values, and distributive justice. In particular, we suggest that values regarding utilitarian notions of open science clash with sovereignty claims over so-called Digital Sequence Information (DSI) on genetic resources. We then apply a distributive justice lens to current debates about the modalities of a benefit-sharing system for DSI, illustrating how different normative approaches to distributive justice and criteria for the fair allocation of benefits underlie states' and stakeholders' political positions. The analysis concludes by discussing the need to consider the unequal scientific capabilities in the production of knowledge and calls for deeper reflection on value divergence in international Science, Technology, and Innovation policy-making processes.

**Keywords** Distributive justice · Open science · Biodiversity · Technology and innovation governance · Data values

---

B. Kreiken (✉)
Delft University of Technology, Delft, The Netherlands
e-mail: b.e.kreiken@tudelft.nl

A. McCarthy
Manchester Institute of Innovation Research, University of Manchester, Manchester, UK
e-mail: A.Mccarthy@manchester.ac.uk



C. Didier et al. (eds.), *Engineering and Value Change*, Philosophy of Engineering and Technology 48, https://doi.org/10.1007/978-3-031-83549-0_8

## 8.1 Introduction

At first glance, Nature appears untouched by contemporary digital society. Yet, we continuously generate vast amounts of data from it, from the canopy above to the microbes below, mainly for conservation but also for a range of other uses. Here, we focus on genetic sequence data that are produced by scientists to advance our understanding of and ability to utilize genetic diversity. Technological advances in data-driven biological Science, Technology, and Innovation (STI), such as the decreasing cost of sequencing technologies, have made the molecular scale, concerning but not limited to genetic sequences, proteins, and metabolites, increasingly accessible to the biosciences, which constitute disciplines concerned with the biological aspects of humans, plants, and animals. These data serve as input to fields like biodiversity genomics, conservation genetics, biotechnological research, and synthetic biology (Carbonell et al., 2019; Dixon et al., 2020; Leonelli, 2016). This "digital revolution" in the biosciences is supported by genetic sequence databases which provide scientists and companies with open access to data, a fundamentally important resource used to inform and perform observations, compare genetic traits across species, conduct experiments, and produce innovations.

Because new technologies in unequal societies tend to disproportionately benefit those with the resources to exploit them (Woodhouse & Sarewitz, 2007), it is important to reflect on how this ongoing and accelerating STI development affects society. Distributive justice is a useful lens to explore how benefits, opportunities, risks, and burdens arising from technological transformations are distributed in society, and we propose that digitalization in the biosciences is a useful case to examine in this context. This perspective states that the "datafication" of nature carries distributive effects, such as differential capacities to control and make use of data and benefit from its outcomes (Geary & Bubela, 2019; Pritchard et al., 2022). Because distributive justice is also concerned with the redistribution of scientific agency, these considerations are not only pertinent to the economic outputs of STI production but also to STI processes as sites of knowledge production. This relates to how benefits from STIs are shared in society and which STIs are pursued. In this vein, insights from the Science and Technology Studies field reveal that both knowledge generation and its outcomes are influenced by the norms and values of the society that produces them (Jasanoff, 2004). This points to the importance of examining STIs beyond their technical aspects and focusing on the socio-economic, cultural, and political contexts in which they exist. As a result, approaches like Ethical Legal and Social Implications (ELSI) and Responsible Research and Innovation have been applied to explore the broader implications of STI, including recently for biodiversity sequencing technologies (Laird & Wynberg, 2016; Sherkow et al., 2022). Thereby, we can raise critical questions such as "Who decides what STI is pursued" and "Who benefits from their use in conservation" (Arts et al., 2015; Heeks et al., 2014).

At the same time, distributive aspects of technologies are shaped by Science, Technology, and Innovation Policy (STIP), which plays a key role in influencing the "directionality" of STI. This influence arises from constraints on scientific activities, such as prohibitions on certain types of research, or funding decisions made in alignment with policy objectives (recently articulated through mission-oriented innovation policy, for example) (Gupta & Falkner, 2006; Kishi, 2020; Mazzucato, 2018). Furthermore, publicly funded programs for capacity-building and incentives or obligations for technology transfer often aim to broaden access to the benefits of STI and to include marginalized groups in its design and production. However, the impact of STIP on the distribution of the outcomes of STI and agency in data-driven research is less explored, particularly at an international level. Therefore, we argue that STIP processes are a key site to explore how different values associated with STI manifest and play out and that a distributive justice lens is useful to illuminate and discuss these values in more detail.

In this chapter, we focus on attempts to govern the complex issues raised by the bioscience digital revolution, such as the right to consent of Indigenous peoples about access to and use of their traditional knowledge and genetic resources on their lands, compensation of biodiverse developing states as "data providers", and a growing global technological divide. We take the negotiations of the Convention on Biological Diversity (CBD) on Digital Sequence Information as a case study of an international STIP process that seeks to advance distributive justice by addressing global disparities in the capacities to generate, access, and utilize genetic sequence data in the biosciences. Digital Sequence Information (DSI) is an undefined policy term that broadly refers to genetic sequence data but can be interpreted to encompass associated information and metadata, and other types of biochemical data. The CBD negotiations over DSI can be characterized as a site of science for policy and policy for science, as negotiators require knowledge inputs from the scientific community and scientific practice is also affected by the regulation of DSI. The analysis is informed by literature on ABS, DSI, and distributive justice in STIP, reading of policy documents, and in-person observation of the DSI negotiations, including the negotiations of the open-ended working group on DSI in Geneva in March 2022, the 15th Conference of the Parties (COP15) in Montreal in December 2022, and of the open-ended working group on DSI in Geneva in November 2023. We begin the chapter with a discussion on competing values over genetic sequence data, tracing these back to the turn towards the data driven biosciences. Then, we move on to explore how the different policy proposals for the allocation of monetary and non-monetary benefits from the use of DSI are rooted in different approaches to distributive justice. The chapter concludes that explicitly recognizing value conflicts and facilitating discussions where these are foregrounded, may lead to more effective decision-making processes.

## 8.2 Background: The Convention on Biological Diversity

Before we explore the value conflicts and distributive justice dimensions of the negotiations over DSI, it is important to understand the origin and aims of Access and Benefit-Sharing policies under the CBD. The CBD is a framework convention for biodiversity governance adopted in 1992 at the Rio Earth Summit. Its three objectives are the conservation of biodiversity, its sustainable use, and fair access to genetic resources and the fair and equitable sharing of the benefits arising from their utilization, in short, Access and Benefit-Sharing (ABS) (UNCED, 1992). By incorporating ABS as an objective, the CBD established a new legal regime for the governance of genetic resources by bringing them under the sovereignty of states. This meant that companies and scientists as "users" of genetic resources were obliged to work with governments and Indigenous Peoples and Local Communities (IPLCs) as "providers" to obtain their Prior and Informed Consent (PIC) and set Mutually Agreed Terms (MAT) regarding the conditions of access to their genetic resources and benefit-sharing resulting from their use. Although the function of this system has been criticized based on its failure to generate a large amount of benefits (Michiels et al., 2021), ABS provides a potential lever for distributive justice in STI based on the sharing of benefits derived from biological STI (Wynberg, 2023). The MAT have also been noted for their potential to produce more equitable co-production of knowledge in international scientific and commercial relationships because they can contain provisions about collaboration, capacity-building technology transfer, and shared authorship of scientific publications or even ownership of intellectual property. In sum, ABS was originally conceived as a win-win policy that would promote biodiversity conservation while reducing inequality in STI through benefit-sharing based on legally certain criteria (Laird et al., 2020; Kate & Laird, 1999).

The rules and norms of ABS were operationalized in the legally binding Nagoya Protocol (SCBD, 2010), which entered into force in 2014. But, as the regulations applied only to the transfer of physical genetic resources, the informational component of genetic resources, accessible through increasingly affordable sequencing technologies and open access genetic sequence databases, created a loophole for users of such "digitized" genetic resources. Because scientific publications that use; sequences include an average of 44 from a variety of states, regulating DSI under a bilateral system like the Nagoya Protocol was considered impractical by many stakeholders (Scholz et al., 2021). Therefore, efforts to regulate DSI under the CBD culminated in a decision at COP15 in Montreal in 2022 to establish a multilateral benefit-sharing mechanism for DSI, although some ambiguity remained over a potential "hybrid system" involving some bilateral arrangements for DSI and genetic resources (SCBD, 2022b; Van Vooren, 2022). With this decision, commercial and noncommercial users of DSI from states that are Party to the CBD will likely have varying duties for financial compensation, technology transfer, scientific cooperation, and capacity-building.

The way policies are designed to meet competing demands could have significant implications for the distributional aspects of biological STI. Although Parties to the CBD have heeded the call of companies and scientists to maintain open access to DSI and "*not hinder*" biological STI, integrating processes for equitable co-production of knowledge in existing scientific infrastructures, while at the same time strengthening rights to benefit-sharing from DSI undoubtedly involves trade-offs (SCBD, 2022a, 2022b). This dilemma, which will be discussed in more detail, is summarized in the often-used phrase to govern data "*as open as possible, as closed as necessary*" (European Union, 2019; Klünker & Richter, 2022).

## 8.3 Scientists and the Sovereigns

### *8.3.1 The Scientific Globalization of the Gene*

The open access model of the biosciences is a culmination of decades of value-based decisions. In the 1960s, biologists first began using computers to decipher and digitize DNA sequences, a development that fundamentally reshaped biological STI (Hagen, 2011). By the 1980s, research institutes worldwide established major databases, which later converged into national and regional databases. The USA, Europe, and Japan began standardizing data formats to facilitate data sharing, leading eventually to the formation of the International Nucleotide Sequence Database Collaboration (INSDC) in 1987. Some years later, when the international community was working towards sequencing and assembling the human genome, the Bermuda Principles were adopted which called for the pre-publication of sequence data following its generation. Hilgartner (2017), in his analysis of the Human Genome Project, suggests that this shift in the control of knowledge, driven by the mass proliferation of data, made the previous regime of publishing sequence data in journals untenable due to fears of data being "scooped" before publication and led to the emergence of open access norms. This highlights how, alongside arguments about the societal benefits of open access, bioscientists' norms, practices, and ownership claims shifted towards openness in concert with the digital transformation of STI.

This international regime based on open access to digitalized biological information involved the construction of genetic sequence data as "transnational digital objects" with an *abstractable*, *sharable*, and *mobile* character (Stevens, 2018). The facilitation of international "data journeys", such as is the case in the use of DSI, further involves the decontextualization and standardization of a sequence from the context of the genetic resource it was generated from, like a living being, field station, or laboratory, to enable its recontextualization in the specific research setting where the sequence is accessed and utilized (Leonelli, 2016, p. 30).

An illustrative example of this is the Earth Biogenome Project, which took inspiration from the Human Genome Project and aims to generate genome sequences for all Earth's eukaryotes with the collaboration of regional and national biogenome projects (Lewin et al., 2018). Doezema (2023) uses this project as a case study to explore how the global character of genetic sequence data is paired with the construction of biodiversity as a "global object" in biodiversity sequencing efforts of global concern that transcends cultural, ethnic, national, geopolitical, economic, and technological differences between states. The construction by the scientific community of biological sequence data as *global* resources to conserve biodiversity as a *global* object is important because it normalizes STIPs that prioritize the scientific and evidential value of DSI. This "scientific data value" refers to the attention and care scientific institutions give to ensuring sequence data can support knowledge claims (Leonelli, 2016).

Although the Earth Biogenome Project provides one example, the institutionalization of scientific data values for validating knowledge claims is well established in the open science movement which seeks to make research including data, samples, articles, and technologies more accessible. The foremost principles in open science are the FAIR data standards, meaning data is Findable, Accessible, Interoperable, and Reusable to increase scientific output and facilitate the replication of studies. The FAIR data standards have been successful in articulating an ideal for international scientific collaboration and for automating database integration. Additionally, they have gained increasing political traction, with promotion by both the OECD and UNESCO (OECD, 2020; UNESCO, 2023).

However, proposals for unlimited openness have been critiqued for failing to address existing inequalities related to the differential abilities to comply with and benefit from open data policies (Leonelli, 2018). The implementation of the FAIR data standards in research institutions involves significant human and technological resources that low- and middle-income countries may not have (Bezuidenhout, 2020). Additionally, there are arguments that these proposals neglect the rights and interests of groups in society, such as those related to privacy and ownership (Tsosie, 2019). In the context of the DSI negotiations, the CARE principles for Indigenous Data Governance, which stand for Collective Benefit, Authority to Control, Responsibility, and Ethics, form a notable addition to the FAIR principles. The CARE data principles aim for "*value-based relationships that position data approaches within Indigenous cultures and knowledge systems to the benefit of Indigenous Peoples* (Castellano, 2004; Anderson et al., 2003). *This shift ultimately promotes equitable participation*" (Carroll et al., 2021). Data standard setting as a site of STI policy therefore involves questions of distributive justice, along with a diversity of associated values, illustrating the need for further examination and discussion of the values underpinning data standards and the open science movement.

Despite these issues, several policy briefs, position statements, and publications suggest that there is general agreement among biological scientists worldwide that the open sharing of data should continue for the sake of scientific progress and conservation (Prathapan et al., 2018). In particular, in recent policy briefs, members of the DSI Scientific Network and the International Chamber of Commerce call for a simple multilateral benefit-sharing system for DSI that ensures legal certainty and harmonization of ABS policies of various UN fora while aligning with open access norms in research and innovation (Halewood et al., 2023; Michiels et al., 2021; Scholz et al., 2022). The partial uptake of this argumentation from the Geneva negotiations into the later COP15 decision, outlined in language that any decision on DSI should "*(e) Not hinder research and innovation; (f) Be consistent with open access to data*" suggests that the scientific data value has been emphatically and successfully argued (SCBD, 2022a, 2022b). However, we identify a potential clash between these criteria and norms and values of sovereignty, outlined explicitly in the criterion "*(i) Take into account the rights of indigenous peoples and local communities, including with respect to the traditional knowledge associated with genetic resources that they hold;*", but also implicitly via the potential extension of the ABS regulations to DSI use. This conflict is the subject of the next section.

### 8.3.2 Data Sovereignty as a Competing Set of Values

It could be argued that COP *decision 15/9* implicitly recognizes that the sovereignty of states over genetic resources will extend to DSI on those resources (although some countries like Brazil and South Africa already had national ABS legislation in place for DSI (Bagley et al., 2020)). This decision will likely have significant implications for the data-driven biological STI regime, not only in terms of data accessibility and standards but also in terms of professional responsibility to share benefits with states and IPLCs. However, the terms of the multilateral benefit-sharing system are yet to be determined, and we propose that clashes between scientific data values and data sovereignty values form significant, unresolved fault lines in the debate that should be made more explicit.

While the regulation of a key resource for biological STI under an environmental treaty may seem misplaced, states and IPLCs hold data sovereignty over DSI as a key principle, reflected in their written submissions and position statements. Data sovereignty is an emergent, multidimensional concept that is mobilized by individuals and societies to denote rights, control, and power over data (Von Scherenberg et al., 2024). The concept promotes values like transparency, inclusion, representation, deliberation, beneficence, and justice in how data are collected and used (Hummel et al., 2021). In addition, digital sovereignty, mostly applied in relation to Big Data and privacy, describes the efforts of governments to develop national technological identities, responding to the increasing power of multinational technology companies and concentration of digital control through state capitalism (Couture & Toupin, 2019). The DSI negotiations should be considered in this wider context of data geopolitics and science diplomacy.

Technological constructs like DSI can shape, challenge, and reinforce sovereignty over genetic resources (Timmers, 2022), necessitating an exploration of the role and meaning of sovereignty within the CBD as it relates to DSI. We do this based on Berry's (2022) distinction between de jure and de facto sovereignty over genetic resources. The de jure conception holds that the highest international legal order ought to be followed. That means that sovereignty is derived from the obligations that follow from the CBD's founding text and its consequent legally binding COP decisions. However, others argue that this idealistic conception ignores the political dimensions underpinning interactions between actors, including states. From this perspective, sovereignty is constituted by the power to decide and enforce laws, meaning the actor or state that exercizes control over genetic resources has sovereignty. This de facto interpretation of sovereignty is particularly important to determining sovereignty over viruses, migratory species and species, and traditional knowledge found across international and community borders (Berry, 2022).

The effects of the "dematerialization" of genetic resources through DSI on these conceptions of sovereignty are so far unclear. However, we can ascertain at least that honoring data sovereignty is likely to clash with notions of open access based on scientific data values, and combined with uncertainties about assigning sovereignty and responsibility, these form unresolved issues in the DSI debate. Firstly, even though the COP15 decision may implicitly recognize sovereignty over DSI due to the extension of the ABS regime, there is not yet a binding instrument that defines the specific rights and obligations of Parties concerning DSI under the CBD. Secondly, the determination of sovereignty is challenging because the decisions over DSI are dispersed over its lifecycle. Should we recognize as sovereign the state of the data's origin, where the database is located, or where the data are reproduced and utilized? From this perspective, multiple actors constantly exert their influence over DSI and its derivatives across various locations[1] (Klünker & Richter, 2022). Conversely, an actor lacking control over infrastructures to store and utilize DSI cannot assert de facto sovereignty, despite de jure rights.

This complexity is often used to justify the defense of open access, but states and IPLCs are unlikely to "drop" their data sovereignty claims for the sake of STI. In fact, data sovereignty in relation to DSI and the biosciences is increasingly reflected in regional and national bioeconomy strategies. For example, the valorization of biodiversity is key to the nation-building process underlying the Colombian bioeconomy strategy (Aparicio, 2022). Likewise, Indigenous data sovereignty is pursued to retain control over data and prevent misuse, misrepresentation, and unacknowledged use of Indigenous-held genetic resources and traditional knowledge (McCartney et al., 2022). This is based on the argument that imposing open access obligations on IPLCs without additional protections risks undermining their specific rights and values associated with the data (McCartney et al., 2022).

In this debate, the obligation to obtain free, prior, and informed consent (PIC) before acquiring DSI and the methods that increase the visibility of IPLCs in data infrastructures may provide a pathway for equitable data-driven biological STI. However, Prior Informed Consent (PIC) seems to be marginalized in the DSI debate compared to its strong articulation in the Nagoya Protocol, due to arguments about the challenges of tracking and tracing DSI, as well as the complexities related to sovereignty across both species and data lifecycles. This can be contrasted with "bottom-up" initiatives, such as Traditional Labels, and discussions about the importance of Indigenous provenance metadata. Such metadata may enable communities to assert their rights over digitized genetic resources and increase their visibility whether a multilateral system explicitly grants sovereignty over them or not (Golan et al., 2022).

---

[1] In the context of discussions on property and contractual rights on data, a de facto property right of the data producer has been discussed in legal literature, for example in Eckardt and Kerber (2024).

Furthermore, different assertions of data sovereignty are not aligned per se and can lead to further clashes. A telling example is that national genomics initiatives aimed at leveraging human genetic diversity for STIs as public goods have previously led to the inequitable dispossession of Indigenous peoples' DNA (Benjamin, 2009). The importance of data in national and IPLC agendas differs significantly: while nations may view DSI as a resource input for economic development based on bioeconomy strategies, many IPLCs describe genetic resources from relational perspectives, considering them a sacred part of Mother Earth and allowing their use based on values of reciprocity (Espinosa, 2014; Kreiken & Arts, 2024). Therefore, the issue of sovereignty in relation to DSI presents a complex and contested area, with different values underlying each position having implications for benefit-sharing modalities.

## 8.4 Distributive Justice in Biological STI

Revisiting the data governance dilemma of "*open as possible, closed as necessary*", we find a complex and potentially competing set of demands on the multilateral mechanism for DSI. This section outlines how this value diversity relates to approaches to distributive justice in STI. This chapter does not aim to simplify or resolve the complexity of sovereignty concerning DSI. Instead, it highlights the need to recognize and discuss these issues within the debate. Drawing from previous analyses of competing demands in STIP, when such demands are incommensurable, it is usually the most powerful actors whose demands are met (Parker & Crona, 2012). This gives a distributive justice lens further relevance for analyzing power dynamics. This section is based on the analysis of policy options put forward in the document "Synthesis of views pursuant to decision 15/9" (SCBD, 2023) alongside developments at CBD negotiations over DSI in Geneva, November 2023.

A first avenue for distributive justice is the variety of allocation criteria for the benefits generated from the multilateral mechanism. These will be analyzed using De Jonge's (2011) distinction between the principles of *desert*, *equity*, and *need*. *Desert* holds that actors that contribute to biodiversity conservation and the generation of DSI deserve compensation in accordance with their relative contribution. The principle of *need* demands that benefits accrue to those who need them most, for example, states with higher biodiversity richness, in line with the aims of the CBD. *Equity*, then, accounts for morally relevant differences between actors, such as development status, meaning that low- and middle-income countries with shortcomings in conservation and scientific budgets would be prioritized.

These three principles are reflected with regard to one major issue in the negotiations, namely the link between access and utilization to monetary and nonmonetary benefits being achieved through tracking and tracing of Parties and IPLC's DSI. This option, however, would involve mass collection of data for database managers and extra administrative burdens to users of DSI.[2] Such time, cost, and regulatory constraints, could arguably outweigh generated benefits and should be considered for a fair distribution of burdens and benefits (Collins et al., 2020). Another option is that benefits are allocated based on available information on the geographic origin of the sequences or underlying genetic resources in the complete database.

The above options reflect *desert*-based redistribution of benefits proportional to a state's share of DSI in the database. However, several Parties argue that this disbursement unfairly affects regions with fewer capacities to generate DSI, specifically identified by Scholz et al. (2021) as Latin America, Africa, and Eastern Europe. Other proposals, which do not link DSI back to its origin, emphasize the principles of *need* and *equity*, for example by allocating benefits to low- and middle-income countries and IPLCs with financial, technological, scientific, and self-identified needs. Thus, following need- and equity-based reasoning, states with low generation of DSI should instead be made beneficiaries. Such multilateral solutions, while technologically and regulatorily speaking are more pragmatic, counterintuitively erode data sovereignty over DSI because actors will not be able to control how DSI from their jurisdiction is being used in STI, yet receive benefits from its use. Moving towards a single multilateral system for both DSI and genetic resources (an option preferred by many users of DSI), also means that mechanisms under the Nagoya Protocol that currently enhance equity in research and development, such as PIC and MAT, potentially become obsolete without replacement. While access to DSI and its utilization may become separated for practical reasons, such as efficiency, current proposals do not detail how mechanisms for equity, such as PIC and MAT are reimagined within the multilateral system. Another proposal suggests that benefits should be enhanced for Parties opting into the multilateral mechanism. This argument suggests that if many states opt into the multilateral system, the last ones that 'cling' to national legislation will receive less scientific interest (SCBD, 2023). Hence, by putting DSI on the agenda of the CBD and asserting data sovereignty over it, states may have conversely introduced a stressor to their sovereignty over their genetic resources, highlighting how technological changes in the can biosciences reshape sovereignty through control over key resources for STI (Timmers, 2022).

Another reason why this loss of control over DSI's use in STI is relevant is that the current pathways for DSI narrow possibilities for the redistribution of scientific agency. We discuss this point by distinguishing between commutative and distributive justice in ABS rationales ( Korthals & De Jonge, 2009). Whereas commutative justice governs fair compensation in exchanges *downstream* of the STI pipeline, distributive justice entails discussions over STI priorities *upstream* so the needs of low- and middle-income countries and IPLCs are recognized and integrated into STIP priorities. Currently, we find that ABS remains in the commutative justice paradigm because most negotiations focus more on the modalities of a

[2] In adjacent ABS discussions over DSI on marine genetic resources held under the High Seas Treaty, Langlet and Dunshirn (2023) have discussed the strengths and weaknesses of several policy options for traceability.

benefit-sharing system (*Who gets What*) for DSI rather than the needs that the system and users of DSI must address (*the Why*).

With the maintenance of open access in the multilateral system, the so-called "decoupling" of access from benefit-sharing, benefit-sharing is only effectuated when benefits have materialized, for example in publications and patents. However, a key unresolved issue of the multilateral system is that it raises the impression that users of DSI fulfill their moral benefit-sharing duties if the databases they use are compliant with the new CBD rules. It is questionable whether this fairness in exchange of DSI actually meets the aforementioned needs or whether they would support the other aims of the CBD (Sirakaya, 2021). In this vein, Laird et al. (2020) argue that stimulation of equitable research processes in biological STI could achieve more than another dysfunctional ABS mechanism, their critique suggesting a failure of the commutative paradigm of justice. This is echoed by Gaffney et al. (2020) who state: "*Greater equity in use of genetic sequence data will not come from ABS schemes but through greater cooperation and capacity building*." These statements point to the need for scientists and companies to proactively take responsibility and engage stakeholders in the direction, design, and production of benefits and not face such questions when the research has already been conducted.

Therefore, the ideal of fair and equitable benefit-sharing from DSI necessitates a move away from a commutative to a distributive approach to justice. Distributive justice is not a unitary concept and harbors varying interpretations, though we limit our discussion to a few of them. In the submission statements on DSI, we broadly identify utilitarian and communitarian approaches to equity in STIP (Cozzens, 2007). Utilitarianism recognizes STIPs as fair when they produce the highest societal well-being with the support of the market. Inequalities are not considered morally wrong if the maximum utility situation demands this. Communitarianism, on the other hand, supports policies that build community bonds and values social responsibilities in STI.

Utilitarian arguments against the current ABS system are increasing, challenging its original conception as a win-win due to concerns about efficiency and effectiveness. Although there are no definitive figures on the benefits generated under the Nagoya Protocol, the regulatory and implementation costs involved with ABS possibly outweigh any financial benefits generated. There is recognition that the multilateral benefit-sharing system must avoid this situation (SCBD, 2022b). The utilitarian claims are supported by the argument that the direct impact of ABS on conservation is uncertain, while it indirectly hinders the progress of biodiversity research (Neumann et al., 2018). This has led to calls for reflection on the purposes of the ABS system, its desired outcomes, and whether it can effectively produce them (Wynberg, 2023). Utilitarian arguments, underpinned by scientific data values, point to the societal value of STI derived from open and unrestricted access to DSI. Rather than systemic inequalities, the users of DSI point towards capacity gaps as the key cause of the global STI divide. According to this reasoning, DSI policies should not restrict open science but rather support scientific capacity-building, technology transfer, and international collaboration that will help with "*leveling the DSI playing field*" (Scholz et al., 2021).

The provision of technological and scientific capacity according to a "catch-up" innovation approach, however, risks overlooking a key challenge (Cirera & Maloney, 2017). While utilitarianism assumes other redistributive mechanisms operating outside STIPs ensure the least well-off also benefit, there would be no such redistributive mechanisms from the use of DSI if allocation is conditional and based on notions of *desert* rather than *need* or *equity*, because countries already unable to generate DSI would be excluded from benefits (Cozzens, 2007). Furthermore, non-monetary benefits from open science, such as knowledge products and new technologies, have only minimally "trickled down" to the Global South, as was highlighted by the unequal access to COVID-19 vaccines (Reidpath & Allotey, 2019). This perspective suggests that overcoming technological divides involves not only providing access to technologies and information and strengthening STI capacities but also building the capabilities of the least advantaged for DSI and its associated technologies meaningfully, reflecting the concept of Inclusive Innovation (Heeks et al., 2014; Himma, 2007). Fieldwork in laboratory settings suggests that low-resourced scientists, even those with access to data face challenges that relate to personal, financial, infrastructural, and institutional constraints to generating, curating, sharing, and using data (Bezuidenhout et al., 2017). A similar issue is found in North-South capacity-building projects that focus on developing the technical competencies of scientists without strengthening the political and socio-economic structures that sustain scientific research (Mormina, 2019). Utilitarian ideas about the simple transfer of technologies and technical expertise to low- and middle-income countries overlook the learning process, the local situatedness of STI, and the need for strengthening institutional and policy capacities (Shen & Williams, 2005). Moreover, STI-oriented policy proposals for the allocation of benefits for projects on a "competitive" basis (SCBD, 2023) raise questions about who will receive funding and whose epistemic norms are promoted. Overall, as utilitarian approaches to DSI are suggested, policymakers and stakeholders should consider how policies will meaningfully strengthen individual and societal capabilities to benefit from DSI.

If the scientific community aims to advance equity in DSI use independently of sovereign states, it must move beyond focusing solely on the regulatory hurdles posed by ABS. A communitarian approach suggests that DSI users might embrace the spirit and principles of ABS and apply these in their own unique research and technological context. That is particularly important considering new technological developments that could outpace still-to-be-established DSI regulations, for example, the use of artificial intelligence in sequence predictions and self-driving synthetic biology labs (Holland et al., 2024). The "datafication" of Nature, in combination with such techniques and the predominance of scientific data values, might also make biodiversity issues to users of DSI less visible and decrease their concern for in-situ genetic diversity.

In communitarianism, the focus shifts towards how STI is done rather than on its outcomes, and thus closer to the *upstream* conceptualization of distributive justice that Korthals and De Jonge (2009) conceptualized, and aligning with notions of Responsible Research and Innovation (Stilgoe et al., 2013). Despite the lack of participatory processes for DSI users under a multilateral benefit-sharing system, users can learn from new organizations that proactively pursue benefit-sharing. One example is Base Camp Research, a company that is building a biological dataset for AI and setting up partnerships with local researchers and communities to valorize their biodiversity, though this model does not preserve open access. Another organization, Wise Ancestors, promotes co-production in biotechnological STI by mediating Western and Indigenous knowledge systems, strengthening conservation leadership by and data sovereignty for IPLCs. Adopting benefit-sharing responsibilities as part of professional practice extends beyond DSI users to other stakeholders in the STI ecosystem, such as journals, funding agencies, and patent offices, which play major roles in driving ethical conduct and fostering Responsible Research and Innovation in the biosciences. Take, for example, the journal *Molecular Ecology* which revised its data accessibility statement to invite authors to list any type of nonmonetary benefit that were generated and shared in the research (Marden et al., 2021). These emerging avenues address criticisms of the ABS system and offer potential models for communitarian relationships in biological STI. However, it remains to be seen how these might be integrated into a practical multilateral mechanism. Opening the conversation about fairness and equity in biological STI, while examining the values that underpin positions is therefore of paramount importance.

## 8.5 Discussion

Biological STI finds itself in a technological acceleration, most recently with the emergence of artificial intelligence-driven bioscience and synthetic biology (Holland et al., 2024). This chapter highlights that policy responses to technological change are driven by different values, and such changes can either challenge or reinforce the data sovereignty of states and IPLCs. In this context, the principle of benefit-sharing for DSI seems to involve the task of incorporating additional values, such as sovereignty and notions of distributive justice, alongside scientific data values. While some proposals may promote a utilitarian vision based on the scientific value of DSI, others show signs of evolving ideas about benefit-sharing from DSI in a communitarian sense without impeding STI. These biases may be caused by inadequate representation of companies and research institutions from low- and middle-income countries at the DSI negotiations, or by the lack of strategic discussions over how ABS policies could function best to advance STI and biodiversity conservation.

Based on this discussion, we advocate that engaged actors recognize and openly discuss the value-laden nature of STIP. In this sense, is important to acknowledge diverse values and whether they are served when weighing policy options (Stirling, 2010). Additionally, sovereignty over DSI seems to be critical for economic and political aspirations beyond biodiversity conservation and STI, and discussions should be more explicit about trade-offs that may be necessary to support states and

IPLCs to achieve self-determined goals. The place of ABS discussions under the CBD limits the discussion to a cost-benefit analysis for biodiversity conservation, while the procedural rights and recognition and scientific agency that are strengthened by ABS do not lend themselves easily to financial calculation.

Literature on distributive justice and STIP remains limited, especially regarding international STIP instruments like ABS. The international dimension of the DSI negotiations thus invites further research on how international STIPs are related to distributive justice. This article formed an introduction to some value conflicts in the DSI negotiations. Many aspects are left unexplored, such as the relationship between data values and nature values,[3] values of further observer groups, and the conflicts or synergies between values, such as efficiency and equity. Many of the DSI policy outcomes, such as the Geneva principles from March 2022 suggest value conflicts remain implicit in the negotiations, yet are represented in the list of criteria for a multilateral benefit sharing system for DSI. Possible theoretical frameworks to ground further value analyses are responsible and inclusive innovation, twin commons, neogeography in digital systems, and epistemic value theory in relation to the digital divide (Arts et al., 2015; Bruynseels, 2020; Fallis, 2007).

## 8.6 Conclusion

Examining the DSI negotiations as a site of value conflict in STIP reveals how data values in STI and in society relate to one another. Value-based discussions which also recognize different concepts of distributive justice could assist the DSI negotiations at least in understanding values inherent in the relationships with, access to, and utilization of DSI, and in proposals for greater equity in the biosciences.

**Acknowledgments** The authors gratefully acknowledge helpful comments from participants and discussants at the fPET conference 2023. The authors also benefited from feedback and discussions provided by colleagues from the Interdisciplinary DSI (iDSI) research group, in particular Irma Kluenker, and by the anonymous reviewers of this paper. The authors remain solely responsible for the content of the paper.

**Funding** AM received additional support from the Economic and Social Research Council (ESRC) [Grant number: ES/P000665/1; project 2488392].

## References

Anderson, I., Griew, R., & McAullay, D. (2003). Ethics guidelines, health research and indigenous australians. *New Zealand Bioethics Journal, 4*(1), 20–29.

[3] The IPBES Values Assessment (2022) contains a comprehensive overview of the variety of values people and groups hold towards nature.

Aparicio, A. (2022). The road ahead: Narratives and imaginaries of the value of biodiversity in shaping bioeconomy policy in Colombia. *Tapuya: Latin American Science, Technology and Society, 5*(1), 2059137. https://doi.org/10.1080/25729861.2022.2059137

Arts, K., van der Wal, R., & Adams, W. M. (2015). Digital technology and the conservation of nature. *Ambio, 44*(S4), 661–673. https://doi.org/10.1007/s13280-015-0705-1

Bagley, M., Karger, E., Ruiz Muller, M., Perron-Welch, F., Thambisetty, S., De Souza, L., Frere, T., Frison, C., Humphries, F., Khalaf, N., Lawson, C., Cabrera Madaglia, J., Meyer, H., Nair, P., & Rourke, M. (2020). Fact-finding study on how domestic measures address benefit-sharing arising from commercial and non-commercial use of digital sequence information on genetic resources and address the use of digital sequence information on genetic resources for research and development. Report no. CBD/DSI/AHTEG/2020/1/5. Available at: https://www.cbd.int/doc/c/428d/017b/1b0c60b47af50c81a1a34d52/dsi-ahteg-2020-01-05-en.pdf. Accessed June 21, 2024.

Benjamin, R. (2009). A lab of their own: Genomic sovereignty as postcolonial science policy. *Policy and Society, 28*(4), 341–355. https://doi.org/10.1016/j.polsoc.2009.09.007

Berry, T. (2022). The CBD's term "Sovereign Right(s)" does not necessarily mean sovereignty. In C. Lawson, M. Rourke, & F. Humphries (Eds.), *Access and benefit sharing of genetic resources, information and traditional knowledge* (pp. 12–24). Routledge. https://doi.org/10.4324/9781003301998

Bezuidenhout, L. (2020). Being fair about the design of FAIR data standards. *Digital Government: Research and Practice, 1*(3), 1–7. https://doi.org/10.1145/3399632

Bezuidenhout, L. M., Leonelli, S., Kelly, A. H., & Rappert, B. (2017). Beyond the digital divide: Towards a situated approach to open data. *Science and Public Policy, 44*(4), 464–475. https://doi.org/10.1093/scipol/scw036

Bruynseels, K. (2020). When nature goes digital: Routes for responsible innovation. *Journal of Responsible Innovations, 7*(3) Article 3. https://doi.org/10.1080/23299460.2020.1771144

Carbonell, P., Radivojevic, T., & García Martín, H. (2019). Opportunities at the intersection of synthetic biology, machine learning, and automation. *ACS Synthetic Biology, 8*(7), 1474–1477. https://doi.org/10.1021/acssynbio.8b00540

Carroll, S. R., Herczog, E., Hudson, M., Russell, K., & Stall, S. (2021). Operationalizing the CARE and FAIR principles for indigenous data futures. *Scientific Data, 8*(1) Article 1. https://doi.org/10.1038/s41597-021-00892-0

Castellano, M. B. (2004). Ethics of aboriginal research. *International Journal of Indigenous Health, 1*, 98–114.

Cirera, X., & Maloney, W. F. (2017). *The innovation paradox: Developing-country capabilities and the unrealized promise of technological catch-up*. World Bank. https://doi.org/10.1596/978-1-4648-1160-9

Collins, J. E., Sirakaya, A., Vanagt, T., & Huys, I. (2020). Developing a methodology to balance benefit-sharing: Application in the context of biodiversity beyond National Jurisdiction. *Genetic Resources, 1*(1), 24–39. https://doi.org/10.46265/genresj.2020.1.24-39

Couture, S., & Toupin, S. (2019). What does the notion of "sovereignty" mean when referring to the digital? *New Media & Society, 21*(10), 2305–2322. https://doi.org/10.1177/1461444819865984

Cozzens, S. E. (2007). Distributive justice in science and technology policy. *Science and Public Policy, 34*(2), 85–94. https://doi.org/10.3152/030234207X193619

Dixon, A. T., Curach, C. N., & Pretorius, I. S. (2020). Bio-informational futures: The convergence of artificial intelligence and synthetic biology. *EMBO Reports, 21*(3), e50036. https://doi.org/10.15252/embr.202050036

Doezema, T. (2023). The promise of ELSI: Coproducing the future of life on earth. *Science as Culture, 32*(4), 461–485. https://doi.org/10.1080/09505431.2023.2182189

Eckardt, M., & Kerber, W. (2024). Property rights theory, bundles of rights on iot data, and the eu data act. *European Journal of Law and Economics, 57*(1), 113–143. https://doi.org/10.1007/s10657-023-09791-8

European Union. (2019). Directive (EU) 2019/1024 of the European Parliament and of the Council of 20 June 2019 on open data and the re-use of public sector information. *Official Journal of the European Union, 172*, 56–83.

Fallis, D. (2007). Epistemic value theory and the digital divide. In E. Rooksby & J. Weckert (Eds.), *Information technology and social justice* (pp. 29–46). IGI Global). https://doi.org/10.4018/978-1-59140-968-7.ch002

Gaffney, J., Tibebu, R., Bart, R., Beyene, G., Girma, D., Kane, N. A., Mace, E. S., Mockler, T., Nickson, T. E., Taylor, N., & Zastrow-Hayes, G. (2020). Open access to genetic sequence data maximizes value to scientists, farmers, and society. *Global Food Security, 26*(100411). https://doi.org/10.1016/j.gfs.2020.100411

Geary, J., & Bubela, T. (2019). Governance of a global genetic resource commons for non-commercial research: A case-study of the DNA barcode commons. *International Journal of the Commons, 13*(1). https://doi.org/10.18352/ijc.859

Golan, J., Riddle, K., Hudson, M., Anderson, J., Kusabs, N., & Coltman, T. (2022). Benefit sharing: Why inclusive provenance metadata matter. *Frontiers in Genetics, 13*(1014044). https://doi.org/10.3389/fgene.2022.1014044

Gupta, A., & Falkner, R. (2006). The influence of the Cartagena protocol on biosafety: Comparing Mexico, China and South Africa. *Global Environmental Politics, 6*(4). https://doi.org/10.1162/glep.2006.6.4.23

Hagen, J. B. (2011). The origin and early reception of sequence databases. In M. Hamacher, M. Eisenacher, & C. Stephan (Eds.), *Data mining in proteomics* (Vol. 696, pp. 61–77). Humana Press. https://doi.org/10.1007/978-1-60761-987-1_4

Halewood, M., Bagley, M. A., Wyss, M., & Scholz, A. H. (2023). New benefit-sharing principles for digital sequence information. *Science, 382*(6670), 520–522. https://doi.org/10.1126/science.adj1331

Heeks, R., Foster, C., & Nugroho, Y. (2014). New models of inclusive innovation for development. *Innovation and Development, 4*(2), 175–185. https://doi.org/10.1080/2157930X.2014.928982

Hickel, J., Dorninger, C., Wieland, H., & Suwandi, I. (2022). Imperialist appropriation in the world economy: Drain from the global south through unequal exchange, 1990–2015. *Global Environmental Change, 73*(102467). https://doi.org/10.1016/j.gloenvcha.2022.102467

Hilgartner, S. (2017). *Reordering life: Knowledge and control in the genomics revolution*. The MIT Press.

Himma, K. E. (2007). The information gap, the digital divide, and the obligations of affluent. *The International Review of Information Ethics, 7*, 63–76. https://doi.org/10.29173/irie7

Holland, C., McCarthy, A., Ferri, P., & Shapira, P. (2024). Innovation intermediaries at the convergence of digital technologies, sustainability, and governance: A case study of AI-enabled engineering biology. *Technovation, 129*(102875). https://doi.org/10.1016/j.technovation.2023.102875

Hummel, P., Braun, M., Tretter, M., & Dabrock, P. (2021). Data sovereignty: A review. *Big Data & Society, 8*(1). https://doi.org/10.1177/2053951720982012

Jasanoff, S. (2004). *States of knowledge: The co-production of science and social order*. Routledge.

Kishi, N. (2020). How does policy focus influence scientific research? *Science and Public Policy, 47*(1), 114–124. https://doi.org/10.1093/scipol/scz051

Klünker, I., & Richter, H. (2022). Digital sequence information between benefit-sharing and open data. *Journal of Law and the Biosciences, 9*(2). https://doi.org/10.1093/jlb/lsac035

Kreiken, B. E. & Arts, B. J. M. (2024). 'Disruptive data: How access and benefit-sharing discourses structured ideas and decisions during the convention on biological diversity negotiations over digital sequence information from 2016 to 2022' Global Environmental Change, 87 p. 102892. DOI: https://doi.org/10.1016/j.gloenvcha.2024.102892 Available at: https://www.sciencedirect.com/science/article/pii/S0959378024000967.

Kupferschmidt, K. (2018). Plan to amend biopiracy rules would 'smother research,' biologists warn. Science https://doi.org/10.1126/science.aau6619.

Laird, S. A., & Wynberg, R. P. (2016). Locating responsible research and innovation within access and benefit sharing spaces of the convention on biological diversity: The challenge of emerging technologies. *NanoEthics, 10*(2), 189–200. https://doi.org/10.1007/s11569-016-0268-z

Laird, S., Wynberg, R., Rourke, M., Humphries, F., Muller, M. R., & Lawson, C. (2020). Rethink the expansion of access and benefit sharing. *Science, 367*(6483) Article 6483. https://doi.org/10.1126/science.aba9609

Langlet, A., & Dunshirn, P. (2023). Traceability options for marine genetic resource from areas beyond national jurisdiction. *High Seas Alliance*. Available at: https://highseasalliance.org/resources/paper-traceability-options-for-marine-genetic-resource-from-areas-beyond-national-jurisdiction/. Accessed June 20, 2024

Leonelli, S. (2016). *Data-centric biology: A philosophical study*. The University of Chicago Press.

Leonelli, S. (2018). Without urgent action big and open data may widen existing inequalities and social divides. *LSE Impact Blog*. Available at: https://blogs.lse.ac.uk/impactofsocialsciences/2018/02/14/without-urgent-action-big-and-open-data-may-widen-existing-inequalities-and-social-divides/. Accessed June 20, 2024

Lewin, H. A., Robinson, G. E., Kress, W. J., Baker, W. J., Coddington, J., Crandall, K. A., Durbin, R., Edwards, S. V., Forest, F., Gilbert, M. T. P., Goldstein, M. M., Grigoriev, I. V., Hackett, K. J., Haussler, D., Jarvis, E. D., Johnson, W. E., Patrinos, A., Richards, S., Castilla-Rubio, J. C., et al. (2018). Earth BioGenome project: Sequencing life for the future of life. *Proceedings of the National Academy of Sciences, 115*(17), 4325–4333. https://doi.org/10.1073/pnas.1720115115

Maclurcan, D. C. (2010). Southern roles in global nanotechnology innovation: Perspectives from Thailand and Australia. In S. E. Cozzens & J. Wetmore (Eds.), *Nanotechnology and the challenges of equity, equality and development* (pp. 349–378). Springer. https://doi.org/10.1007/978-90-481-9615-9_21

Mazzucato, M. (2018). Mission-oriented innovation policies: Challenges and opportunities. *Industrial and Corporate Change, 27*(5), 803–815. https://doi.org/10.1093/icc/dty034

McCartney, A. M., Anderson, J., Liggins, L., Hudson, M. L., Anderson, M. Z., TeAika, B., Geary, J., Cook-Deegan, R., Patel, H. R., & Phillippy, A. M. (2022). Balancing openness with indigenous data sovereignty: An opportunity to leave no one behind in the journey to sequence all of life. *Proceedings of the National Academy of Sciences, 119*(4), Article 4. https://doi.org/10.1073/pnas.2115860119

Michiels, F., Feiter, U., Paquin-Jaloux, S., Jungmann, D., Braun, A., Sayoc, M. A. P., Armengol, R., Wyss, M., & David, B. (2021). Facing the harsh reality of access and benefit sharing (ABS) legislation: An industry perspective. *Sustainability, 14*(1), 277. https://doi.org/10.3390/su14010277

Mormina, M. (2019). Science, technology and innovation as social goods for development: Rethinking research capacity building from Sen's capabilities approach. *Science and Engineering Ethics, 25*(3), 671–692. https://doi.org/10.1007/s11948-018-0037-1

Neumann, D., Borisenko, A. V., Coddington, J. A., Häuser, C. L., Butler, C. R., Casino, A., Vogel, J. C., Haszprunar, G., & Giere, P. (2018). Global biodiversity research tied up by juridical interpretations of access and benefit sharing. *Organisms Diversity & Evolution, 18*(1), 1–12. https://doi.org/10.1007/s13127-017-0347-1

OECD (2020). *Enhanced access to publicly funded data for science, technology and innovation*. Available at: https://www.oecd.org/sti/enhanced-access-to-publicly-funded-data-for-science-technology-and-innovation-947717bc-en.htm. Accessed June 21, 2024.

Parker, J., & Crona, B. (2012). On being all things to all people: Boundary organizations and the contemporary research university. *Social Studies of Science, 42*(2), 262–289. https://doi.org/10.1177/0306312711435833

Prathapan, K. D., et al. (2018). When the cure kills—CBD limits biodiversity research. *Science, 360*(6396), 1405–1406. https://doi.org/10.1126/science.aat9844

Pritchard, R., Sauls, L. A., Oldekop, J. A., Kiwango, W. A., & Brockington, D. (2022). Data justice and biodiversity conservation. *Conservation Biology, 36*(5). https://doi.org/10.1111/cobi.13919

Reidpath, D. D., & Allotey, P. (2019). The problem of 'trickle-down science' from the global north to the global south. *BMJ Globalization and Health, 4*(4). https://doi.org/10.1136/bmjgh-2019-001719

Scholz, A. H., Freitag, J., Lyal, C. H. C., Sara, R., Cepeda, M. L., Cancio, I., Sett, S., Hufton, A. L., Abebaw, Y., Bansal, K., Benbouza, H., Boga, H. I., Brisse, S., Bruford, M. W., Clissold, H., Cochrane, G., Coddington, J. A., Deletoille, A.-C., García-Cardona, F., et al. (2022). Multilateral benefit-sharing from digital sequence information will support both science and biodiversity conservation. *Nature Communications, 13*(1) Article 1. https://doi.org/10.1038/s41467-022-28594-0

Scholz, A. H., Lange, M., Habekost, P., Oldham, P., Cancio, I., Cochrane, G., & Freitag, J. (2021). Myth-busting the provider-user relationship for digital sequence information. *GigaScience, 10*(12) Article 12. https://doi.org/10.1093/gigascience/giab085

Secretariat of the Convention on Biological Diversity. (2010). *Nagoya protocol on access to genetic resources and the fair and equitable sharing of benefits arising from their utilization.* Nagoya: United Nations.

Secretariat of the Convention on Biological Diversity. (2022a). Recommendation adopted by the Working Group on the Post-2020 Global Biodiversity Framework: Digital sequence information on genetic resources. CBD/WG2020/REC/3/2, 29 March 2022, Geneva, Switzerland. Available at: https://www.cbd.int/doc/recommendations/wg2020-03/wg2020-03-rec-02-en.pdf. Accessed June 21, 2024.

Secretariat of the Convention on Biological Diversity. (2022b). Decision adopted by the Conference of the Parties to the Convention on Biological Diversity 15/9: Digital sequence information on genetic resources. CBD/COP/DEC/15/9, 19 December 2022, Montreal, Canada. Available at: https://www.cbd.int/doc/decisions/cop-15/cop-15-dec-09-en.pdf Accessed June 21, 2024.

Secretariat of the Convention on Biological Diversity. (2023). Ad Hoc Open-ended Working Group on Benefit-sharing from the Use of Digital Sequence Information on Genetic Resources: Synthesis of views pursuant to decision 15/9: Note by the Secretariat. CBD/WGDSI/1/2/Add.1, 4 October 2023. Available at: https://www.cbd.int/doc/c/9c66/554e/54f525a088e9f0cddd85c645/wgdsi-01-02-add1-en.pdf. Accessed June 21, 2024.

Shen, X., & Williams, R. (2005). A critique of China's utilitarian view of science and technology. *Science, Technology and Society, 10*(2), 197–223. https://doi.org/10.1177/097172180501000202

Sherkow, J. S., et al. (2022). Ethical, legal, and social issues in the earth biogenome project. *Proceedings of the National Academy of Sciences of The United States of America, 119*(4). https://doi.org/10.1073/pnas.2115859119

Sirakaya, A. (2021). Is the Nagoya protocol designed to conserve biodiversity? *Plants, People, Planet, 4*. https://doi.org/10.1002/ppp3.10221

Stevens, H. (2018). Globalizing genomics: The origins of the international nucleotide sequence database collaboration. *Journal of the History of Biology, 51*(4), 657–691. https://doi.org/10.1007/s10739-017-9490-y

Stirling, A. (2010). Keep it complex. *Nature, 468*(7327), 1029–1031. https://doi.org/10.1038/4681029a

Timmers, P. (2022). The technological construction of sovereignty. In H. Werthner, E. Prem, E. A. Lee, & C. Ghezzi (Eds.), *Perspectives on digital humanism* (pp. 213–218). Springer. https://doi.org/10.1007/978-3-030-86144-5_28

Tsosie, R. (2019). Tribal data governance and informational privacy: Constructing 'Indigenous Data Sovereignty'. *Montana Law Review, 80*(2) Available at: https://scholarship.law.umt.edu/mlr/vol80/iss2/4/. Accessed June 21, 2024

UNESCO, 2023. *Open solutions: Free and open source software, open data*. Available at: https://www.unesco.org/en/open-solutions/open-data. Accessed June 21, 2024.

United Nations Conference on Environment and Development (UNCED). (1992). Convention on biological diversity. Rio de Janeiro: United Nations.

Van Vooren, B. (2022). Outcome from COP 15: A new global biodiversity fund paid for by life sciences companies that "Use Digital Sequence Information on Genetic Resources". *Inside EU Life Sciences*. Available at: https://www.insideeulifesciences.com/2022/12/23/outcome-from-cop-15-a-new-global-fund-paid-for-by-life-sciences-companies-that-use-digital-sequence-information-on-genetic-resources/ Accessed June 21, 2024

Von Scherenberg, F., Hellmeier, M., & Otto, B. (2024). Data sovereignty in information systems. *Electronic Markets, 34*(1), 15. https://doi.org/10.1007/s12525-024-00693-4

Woodhouse, E., & Sarewitz, D. (2007). Science policies for reducing societal inequities. *Science and Public Policy, 34*(2), 139–150. https://doi.org/10.3152/030234207X195158

Wynberg, R. (2023). Biopiracy: Crying wolf or a lever for equity and conservation? *Research Policy, 52*(2). https://doi.org/10.1016/j.respol.2022.104674

Korthals, M. & De Jonge, B. (2009). 'Two different ethical notions of benefit sharing of genetic resources and their implications for global development' New Genetics and Society, 28 (1), pp. 87-95. DOI: 10.1080/14636770802670308

Kate, K.T., & Laird, S.A. (1999). The Commercial Use of Biodiversity: Access to Genetic Resources and Benefit-Sharing (1st ed.). Routledge. https://doi.org/10.4324/9780429341540

Marden, E., et al. (2021). 'Sharing and reporting benefits from biodiversity research' Molecular Ecology, 30 (5), pp. 1103-1107. DOI: https://doi.org/10.1111/mec.15702

Espinosa, C. (2014). The Advocacy of the Previously Inconceivable: A Discourse Analysis of the Universal Declaration of the Rights of Mother Earth at Rio+20. The Journal of Environment & Development, 23(4), 391-416. https://doi.org/10.1177/1070496514536049

Stilgoe, J., Owen, R. & Macnaghten, P. (2013). 'Developing a framework for responsible innovation' Research Policy, 42 (9), pp. 1568-1580. DOI: https://doi.org/10.1016/j.respol.2013.05.008 Available at: https://www.sciencedirect.com/science/article/pii/S0048733313000930.

**Bob Kreiken** BSc Public Administration, LL.B. Bachelor of Laws (Leiden University), MSc Forest and Nature Conservation (Wageningen University and Research). 2022 Intern Dutch delegation to the Convention on Biological Diversity, Ministry of Agriculture, Fishery, Food Security and Nature (the Hague), the Netherlands. 2022–present, PhD researcher Department of Values, Technology and Innovation, Ethics and Philosophy of Technology Section, Delft University of Technology. Research areas: Biodiversity, digitalization, international relations, ethics, science and technology policy.

**Adam McCarthy** BSc Cell Biology, Master of Research, Management, PhD candidate in Science, Technology and Innovation Policy, Manchester Institute of Innovation Research (2021–2024). Research areas: Responsible innovation, emerging technology governance, biotechnology.

# Chapter 9
# Resource Exploitation, Transformation for Sustainability, and the Technosolutionism Critique

**Frieder Bögner**

**Abstract** Issues of global justice are ethical concerns for decision-making procedures in wealthy countries of the globe. Many arrangements and technological developments can influence just relations with other regions that are less prosperous. The paper points towards resource exploitation as a common cause for global injustices. Many of the resources extracted in less industrialised and wealthy regions are pertinent for a social transformation for sustainability, e.g. rare earths mined in order to provide electric mobility for a sustainable transition. The paper highlights that sometimes technological strategies in decision-making settings that are relevant to a social transformation for sustainability in wealthy regions have serious discursive effects which have not gained enough attention in the debate. In particular, it is argued that the objection of technosolutionism has not been sufficiently addressed in this regard. This is a critique against societal solution strategies for demanding challenges which would also allow less technologically advanced options. A novel reading of this objection is introduced that allows for a more concrete connection between technological options and global justice issues. Overall, the paper argues for a specific connection between (1) the objection of technosolutionism, (2) social transformation for environmental sustainability, and (3) the practice of resource exploitation. If technological options have a tendency to crowd out other non- or low-technological options in decision-making processes and if global justice problems are more connected to technological options, then there are ethical reasons to object to technosolutionism in this regard.

**Keywords** Global justice · Resource exploitation · Technosolutionism · Collective decision-making · Transformation for sustainability

---

F. Bögner (✉)
RWTH Aachen University, Department of Technology, Society, and Human Factors, Applied Ethics Aachen, Germany
e-mail: frieder.boegner@humtec.rwth-aachen.de



C. Didier et al. (eds.), *Engineering and Value Change*, Philosophy of Engineering and Technology 48, https://doi.org/10.1007/978-3-031-83549-0_9

## 9.1 Introduction

The aim of this chapter is to connect three issues in a productive way. The issues are the objection of technosolutionism, social transformation for environmental sustainability, and the practice of resource exploitation. The objection of technosolutionism is a critique against societal solution strategies for demanding challenges. It is sometimes not transparent which factors actually allow making this objection. I will stay agnostic with regard to the question whether the technosolutionism objection *must* be an ethical objection. Rather, my contention is that the objection of technosolutionism has special significance if it includes reference to global justice problems related to resource exploitation that stem from the impulses towards social transformation. Two features of my argument do not rely on novel contributions, the relation of resource exploitation to global justice and the dependency of innovative technologies on resource exploitation. However, if we can establish a connection between technosolutionism and the push for transformation towards sustainability, this delivers a strong reason to be attentive to strategies involving technological solutions and objections of technosolutionism aimed at these strategies. Hence, the chapter develops one ethically informed way on how to criticise technosolutionism.

The support for this connection will be developed across five sections. The *first* section will introduce a schema for global justice problems. This will be instrumental in linking an objection of technosolutionism to an ethical foundation. A *second* section draws attention to the global justice problems involved in the environmental harms connected to the mining of natural resources. These resources are also important for social transformation towards sustainability, so the *third* section will begin by briefly introducing discourse settings relevant for social transformation. This will narrow the scope of cases to be discussed and here I will introduce the discursive effects of technology to the line of reasoning adopted. This sets the stage for unpacking the objection of technosolutionism, and a novel understanding of this objection in the *fourth* section. The main idea is that technological options may crowd-out other valid alternative options in relevant discourses. It is this reading of the critique, which in the *fifth* section enables the tripartite connection between the drive towards transformation for sustainability and the effects of technological options on decision-making with matters of global justice. If technological options have a tendency to crowd out other non- or low-technological options and if global justice problems are more connected to technological options, then there are ethical reasons to address technosolutionism in this regard.

## 9.2 A Schema for Global Justice Problems

The issue of global justice is a longstanding topic within theoretical reflections on social and political issues and especially in Ethics and Practical Philosophy. Matters of resource exploitation are certainly addressed in these discussions (e.g. Beitz, 1975; Nagel, 2005; Wenar, 2016). Historically, the debate around the appropriate analysis of how inhabitants of one part of the world should relate to inhabitants of other regions has generally revolved around the issues of, as examples, human rights, restitutions, or generally, the demands on socioeconomically affluent countries towards countries that are less wealthy—be this due to natural causes, mismanagement, or historical and present exploitation by the wealthier parts of the globe. What has come into focus recently is the connection between issues of global justice, technology development and production, on the one hand, and certain features of the discourses common in wealthier countries on the other. This contribution examines the above connection and highlights one specific, relevant critique that emerges from this examination: the objection of technosolutionism. This critique objects to technological solution strategies for societal challenges on the grounds that choosing such a strategy is not acceptable since other, social, low- or non-technological approaches are available. I will assume that this is generally a strong objection that should be handled with care. What motivates this chapter is the need for a more concrete reading of this and to highlight the connection to global justice failures which would lend credibility to the objection.

In order to do that, I will assume that issues around global justice are commonly motivated by at least three observations:

1. Activities rooted in one country may sometimes *negatively affect* residents of another country (Brock & Hassoun, 2023).
2. These negative impacts can be *ethically relevant.*
3. There are ethical considerations (rights- *or* value-based) that justify taking these impacts to be a *global justice* problem.

While this schema is certainly not without alternatives, it gives us certain advantages for the topic picked up here. One aspect is that the list may be centred on individual countries and their activities; however, we should not exclude supranational agents, regions, and other collectives of states here. Further, the first observation also allows for (positive) hypothetical variants. That is, if one country or supra-state agent could potentially benefit residents of another country but omits this activity, then this also counts as a negative effect. However, not every negative impact on another country or region is *ethically relevant*. I will rely on this ethical perspective, as I assume that this will ultimately yield a rather innovative and important foundation of technosolutionism objections that, e.g. a political perspective might not necessarily include. That is, global justice problems can of course also be viewed as political problems in the sense that, e.g. global partners have not established just agreements for their economic relations, or that certain partners do not get heard in relevant political arenas. While the legal and political details of just contracts, for example, may in many dimensions be intricate and controversial, I will assume that general ethical relevance of activities negatively affecting other regions is less controversial. It is obvious that, e.g., asylum laws, regulation of the labour market relevant for refugees, and measures of border control in general are ethically relevant, as well as politically. The mere negative effect of, e.g. changes to the labour market policy in one country upon another country is less likely to count as an ethically relevant problem.

This feature is picked up by the third observation: One needs to be able, for example, to point towards *moral rights or values* of the country negatively affected that are infringed by the affecting country. Pointing towards these features will ultimately avoid including too many controversial political topics but allow building the objection as suggested here on potentially more widely acceptable notions. In typical cases, the country or region infringing will also benefit in one way or the other from these ethically questionable practices and this combination justifies it in a particular way to speak of a global justice problem. That is, to take up a recent and common example, one could argue that many countries of the global north, with their highly industrialised and technologically advanced economies, have either taken up more than what their right to a fair share of the atmosphere's ability to absorb greenhouse gas emissions allows, or that the activities of more advanced industries and their emissions work against the basic moral value of human well-being globally, but perhaps especially in more southern regions. These regions and their inhabitants have a moral right to a fair share of the atmosphere and their life conditions are morally valuable, albeit endangered by the manifold effects of climate change which are, and have been historically, caused to a large degree by industrialised countries. Additionally, the latter have benefitted from this in terms of technological progress and material prosperity. Surely, one could make matters more complex here by differentiating between rights of populations to these fair shares and to a more advanced industry. Instead, I will assume that although this might include a prospective right to heavy emissions of novel industries in regions that did not cause the bulk of the global emission in the past, the most relevant injustice here is constituted by the much higher emission rates of industrialised regions coupled with the gained benefits in comparison to the harms and rights violations caused in less industrialised regions. That is, I will rely on the historically greater causal contribution of the industrialised regions. I have pointed towards activities and omissions of wealthy industrialised countries or regions and while I will take up one specific activity in the next section, my main argument will ultimately also draw on omissions by individual and collective agents in these countries.

## 9.3 Global Justice and Resource Exploitation

There is a long history of natural resource extraction by industrialised or technologically more advanced countries from economically and politically less stable and resilient parts of the globe. Mines supplying many of the resources relevant here are based for example in countries like Morocco, South Africa, Zambia, Congo, Indonesia, Chile, or Bolivia. Restricting the discussion to environmental harms, it is especially the mining of raw materials, minerals (like lithium, nickel, cobalt), and rare earths that results in the varied damages that potentially put the environment, workers involved, as well as local residents at risk. Damages include the contamination of soil by chemicals used for extraction, air and water pollution, as well as harms to local ecosystems and biodiversity (Bell, 2018; Velasco-Fernández & Pérez, 2022). If it is companies from the industrialised and technologically advanced countries that are involved in the mining activities or if those countries regulate the contracts and general political frameworks that enable these practices, an ethically justified system would have to deal with the damages caused in the countries of the global south where the mines are located. Apart, e.g., from guideline initiatives for certain industry sectors without regulatory influence (AFI, 2019), there are hardly any just compensations and reparations existing for the damages.

Applying the schema developed in the preceding chapter, we can see that resource exploitation may (1) negatively affect local opportunities and local environmental integrity. (2) It often negatively affects local communities by taking advantage of existing unjust global arrangements, constituting an ethically relevant problem. It may infringe the right of local communities to an intact environment and disregard the value of the living conditions of people directly or indirectly involved. The practices relevant here are exploitative practices, as they rely on unjust background conditions and the damage to ethically relevant features like quality of life, soil integrity, and general environmental conditions is not usually not compensated. In short, (3) exploitation of certain resources may be a *global justice problem* in several dimensions, and I will focus on the environmental aspect here.

At the same time, the relevance of mined resources for the economies of industrialised regions in the current arrangements is high: Raw materials, minerals/rare earths, energy sources, sand, and other resources are needed for a number of products, production facilities, and for economic expansion. For example, it is obvious that energy sources are needed for a vast number of economic activities and minerals as well as rare earths are increasingly used for products that permeate social and economic life. Importantly, it is not only economic activity or expansion per se that is in need of certain raw materials, but there is evidence that a *social transformation for sustainability* aimed for in technologically advanced regions might be no exception. That is, the demand for more sustainable production may be connected to global injustices as grave as the ones mentioned above. Possibly, the extraction of, e.g. fossil fuels so far has caused more environmental damage overall as these extractivist practices have existed for a longer period and it is not only their extraction but their conversion into, e.g. kinetic energy that causes harm due to resultant emissions. However, prospectively, we also need to consider the damages potentially connected to certain resource demands caused by transformative change, even if the environmental harm evoked by this process might for a long time not get to the limits of injustices of more traditional practices of resource extraction.

With reference to this, Ali holds that this "transition to a low-carbon society, [is] a change that will require vast amounts of metals and minerals. […T]he world cannot tackle climate change without an adequate supply of raw materials to manufacture clean technologies" (Ali et al., 2017, p. 367). While this is certainly a contested view in itself (Kallis et al., 2020, p. 4), I will assume it in the following and take as prima facie evidence for it the rising demand for electric vehicles, wind turbines, and smartphones that will prospectively lead to a future explosion in the demand for certain resources. The EU, for example, estimates a 3500% increase in lithium demand for electric mobility by 2050 (Brevini, 2023; EU, 2022). Further, as there is evidence for major environmental damage caused by lithium mining, e.g. due to increased threats to certain animal species (Peltzer et al., 2024), this could have an enormous effect on these species and hence for biodiversity if the estimates mentioned are only roughly correct. The reasons for the estimated increase, to some extent, originates in the introduction of AI technologies, e.g. in the mobility sector as the production of this technology demands further raw materials and minerals. This is an important connective point to be kept in mind. Before we go into that, however, we will next look at the connection between the social transformation for sustainability, nascent technology, and the discursive effects of technology.

## 9.4 Sustainability, Social Transformation, and Emergent Technologies

Technology and technological infrastructure are causes for many global societal challenges that are connected to environmental issues and are thus relevant for sustainability. However, not only are technologies a major cause for emissions, toxic waste, the deterioration of soil quality, and further threats to animal species, they are also an arena for social transformation. That leads to political pressure for transforming, e.g., the energy supply system, waste management, or the value chains of food production systems in wealthy and technologically advanced societies. Moreover, there is a drive in political and economic decision-making for social transformation more generally. There certainly is momentum for social transformation in several societal domains, albeit there being manifold questions in terms of financial resources, regarding the adequate domains to address as well as who should shoulder the main responsibility.

I will restrict myself to political decision-making, discuss how generic illustrations are important for the general task of social transformation, on the one hand, and how discourse features have an impact on the choices that decision-making processes lead to on the other. This will pave the way to introduce the objection of technosolutionism in the next section.

### *9.4.1 Social Transformation and Discourse Settings*

Looking at discourses and their relevance for social transformation first, we should notice (a) that to a large extent, it is the procedures and actions of institutions that are important for social transformation. Policymaking and regulation have *high potential to push wide-ranging transformative measures* and these include, e.g. social and economic policy, educational policy as well as regulation in transportation, mobility, or housing. Notably, these are low- or non-technological options. To take some rather contrasting examples, there are theoretical arguments and initiatives for policy proposals for educational programmes towards sustainability (e.g. Ng, 2019; Zwarthoed, 2015) as well as attempts to incentivise the use of public transport like Germany's "9-Euro" ticket (Bissel, 2023) and major subsidies to enable more sustainable industries, e.g. the USA's "inflation reduction act." To different extents, these measures have the potential to foster social transformation with regard to societal challenges on a large scale.

Further, the institutional settings of collective deliberation relevant here are those that (b) are established *to determine the value of valid options*. Political decision-making especially in democratic systems should adhere to a number of criteria. One of these can be understood as what is sometimes called an "output-oriented" criterion. This refers to the value of the options at hand for solving societal challenges. However, we do not have to determine the value of options solely on the grounds of anticipated consequences of these options after political deliberation and decision-making. For example, in political discourse, the value of an option on how to transform features of the transportation sector may depend on who has voiced it in the process, as well as whose interests it might curtail and how it will help to mitigate emissions. Hence, there are a number of criteria that are relevant. It is the social and political role, the competence, and the epistemic situation of the agents involved in the discourse that is meant to establish the value of specific strategies. In order to address societal problems, participants in regulatory processes need to argue for the value of specific options to regulate towards in order to persuade partners, or oppositional parties.

Finally, this leaves us with the question of which options the decision of the imagined collective is based on. While the aforementioned process should establish the value, to begin with (c) the collective should choose its strategy from a range of *valid options*. For present purposes, and in a simplifying manner, I here take valid options as those that all discourse participants should be able to identify as neither absurd, disadvantageous nor harmful. To make this more specific, e.g. a regulatory body of a ministry concerned with choosing strategies to address climate change issues will have a number of valid options on the table. Among those could be technological solutions that are appropriate as a general means for the respective targets, or rather regulators should be able to recognise this. This does not include the claim that they *in fact* acknowledge this. This is not an obvious view and a challenge to this is that it makes validity dependent on the competences and the epistemic status of the agents involved. If regulators involved fare badly along these features, this will possibly enable a proliferation of options to be seen as valid ones. However, this is of course a problem for any framework of collective decision-making and not specific to the view presented here. There will, presumably, be *valid* options to choose from, but the *value* of each option will be determined in discourse. What this view facilitates is a helpful distinction between the value and the validity of options which will be picked up again in Sect. 9.5.2 below. Next, we will look at a rather specific effect of one type of options: those that include emergent and innovative technologies.

### 9.4.2 Discursive Effects of Technology

Whenever collective decision-making is addressing societal challenges, there is a special interest to have a wide variety of options, to have a fair discussion, and to get to decisions that can be implemented. There may of course be many features of collectives and their decision procedures that negatively impact the choices they make. One should, e.g. try to reduce the impact of agents in the procedure that are either attempting to highlight and emphasise their specific strategies with the help of unacceptable measures like manipulated numbers, or that try to intentionally push against competing view without reasonable justification.

These are relevant features of a single agent or a group of people involved in the procedure that may intentionally distort the overall process. That is, if it is correct, that collective decision-making is, to a certain extent, undertaken in order to establish the value of available options, we should be careful to trust this collective if agents involved display strategies or characteristics like the ones just mentioned. What is problematic about these cases is that the outcome is less trustworthy as, e.g. manipulated numbers might motivate an option that people would favour *if* these numbers applied, while they actually do not. The option might hence distort the process and distract from other views and strategies whose proponents do not display these features. In the end, we fail to establish the value of a specific option if vital information pertaining to the option is not adequately gathered, presented, or valued, but presumably there will be one or several options that the collective decides to put in place. Thus, there is a *discursive effect* involved that leads to an outcome of collective decision-making that is less reliable.

There can be another feature that has similar effects and that sometimes is not attached as clearly to the pertinent concepts and consequences of debates around social transformation for sustainability as it could be. My premise here is that in collective decision-making, and other relevant discourses, *technological options* may sometimes have similar effects without necessarily including an intention of agents to do so.

The effects of technological devices on human decision-making and their relation to the material and social world in general has been researched extensively. Of main importance to my argument are the tendencies of technological devices to influence our decisions and to absorb our attention. Video calls, smartphones, social media, and most recently, chatbots are everyday examples of devices and systems that draw human attention and that in some cases divert from non-technological options to the task at hand. Humans are often drawn to technology, to its abilities, to its assumed benefits in terms of costs, and to the positive effects its use can have on our social environment. My assumption says that technological options in collective decision-making may not only draw our attention but influence the procedure in a way that is, to a certain extent, comparable to the features of single or collective agents mentioned above. These options are not identical, but they can have similar *discursive effects*. That is, although the agents involved might not have intended these consequences, their technological options may influence collective deliberation in a way that distracts from these other options. It is important to the following step of my argument that these effects are specific to technologies and especially to innovative technologies. Anything can of course draw human attention, but emergent technologies may have a special tendency to not only attract attention, but subsequently also appear particularly *valuable* to the respective challenge.

I have highlighted the relevance of collective decision-making for social transformation as well as the potential discursive effects of technology. We can make this more concrete and connect this to a specific objection that has become common in discourses on technology and is of special interest when it comes to AI applications as strategies to address societal challenges like climate change.

## 9.5 The Objection of Technosolutionism

This section introduces the technosolutionism critique which objects to technological solution strategies for disregarding existing non-technological or social approaches. The main idea offered is that this objection should refer to the discursive effects of technology.

### *9.5.1 Societal Challenges and the Promise of Technological Solutions*

The preceding section mentioned the political pressure for transformative action. These impulses take their motivation from a number of interconnected societal problems that the world has to deal with. These are challenges like climate change, diminishing biodiversity, and decreasing soil quality. Sometimes, societal agents refer to technological options as solution strategies to address these challenges. Technologies sometimes promise to offer valid strategies and solutions that are helpful to somewhat desperate situations. There are many examples to this tendency. Take for example the research and the political interest in CDR/CCS (carbon dioxide removal/carbon capture and storage) technologies (Smith et al., 2023; Wettengel, 2023). These are technological strategies to remove carbon emissions from the atmosphere and store it in reservoirs. This has already been seen as a valid alternative to more social and economic strategies for mitigation that would try to reduce consumption of fossil fuels in the private and industry sectors. Climate change, issues of sustainability, and strategies like CDR/CCS make for fitting examples of the more general point. Institutions have a hard time evaluating and agreeing on measures to take. As time is short, some technological options can seem ready solutions to these challenges that appear to be superior to low- or non-technological strategies.

### 9.5.2 The Conceptualisation of Challenges and the Crowding Out Effect

This promise of technological solutions to societal challenges sometimes invites objections of technosolutionism (Sætra, 2023). The objectors usually reject this attitude for its myopic, simplistic, or merely self-interested focus to societal challenges and this may support a rejection to concrete technological strategies in decision-making processes. A place to begin understanding this attitude is Morozov's (2014) treatment of the topic. He writes:

> Recasting all complex social situations either as neatly defined problems with definite, computable solutions or as transparent and self-evident processes that can be easily optimized—if only the right algorithms are in place!—this quest is likely to have unexpected consequences that could eventually cause more damage than the problems they seek to address. […] It's not only that many problems are not suited to the quick-and-easy solutionist tool kit. It's also that what many solutionists presume to be "problems" in need of solving are not problems at all; […T]hese vices are often virtues in disguise. (Morozov, 2014, p. 5, 6)

What Morozov is mainly after here is the way the technological solutions *conceptualise* the challenges they are meant to address. We might, for purposes of this chapter, not go so far as to claim that challenges to be addressed actually should not be considered problems at all, but the way technological strategies shape and conceptualise the understanding of the challenge is important. I will offer a slightly different reading of the attitude of technosolutionism below which relies on three interconnected assumptions.

A first premise of a different model is that we should start to introduce slightly more concrete standards as to when this objection carries some justification. For an objection that may rule out options in social discourse, claiming the opposed strategies are myopic, simplistic, or self-interested seems to lack precision. Further, a way to gain more precision is to at least mention the contexts where this attitude becomes most problematic. The context in question is that of societal discourse and collective decision-making, as it is here that technosolutionism has the biggest impact. Finally, I assume that when analysing the objection of technosolutionism what we should be most interested in is innovative, emergent technology that has the potential to be monetisable on a large scale. With these points in the background, here is my alternative reading of the technosolutionism critique:

> A technosolutionism critique applies, if the suggestion of a technological solution strategy *crowds out other valid alternatives* to technology and is not accompanied by *countermeasures* to this effect.

First of all, this version aims at introducing a more concrete criterion. This is the presence of the *crowding out* effect, which I have presaged above by pointing towards the ability of technologies to absorb our attention in general and influence our decisions more specifically. Technological options may push out non-technological alternatives or other strategies that do not rely as much on innovative, upcoming, and powerful technology, be this CDR/CCS, AI systems, or any other emerging technology with significant economic potential as lucrative products. Technologies structure the way we communicate; however, they also emphasise the strategies they are part of, and deemphasise alternative approaches. Technological options in collective deliberation might have undue influence on other strategies, either those explicitly available, or those that could become available if discourse were not shaped by this effect. Hence, the concept of crowding out does not encompass a purely epistemic feature. It does not only affect what we talk and think about, rather, the criterion for the objection includes that non- or low-technological options are actually chosen more rarely.

To anticipate two objections here, we can first specify this with the help of the means introduced above. Naturally, the effect of crowding out can be a feature of *any* additional option introduced in the decision process. Put simply, a regulatory body, e.g. might decide on which measures to take in order to reduce carbon dioxide in the atmosphere and it might consider building natural carbon sinks, a CDR facility and an AI system that helps reducing emission in the mobility sector. The option of natural carbon sinks might crowd out other options in the debate. However, above I introduced the assumption that innovative technologies and technological options might have a *specific* tendency to crowd out. These technological options may sometimes appear especially valuable and this increases their tendency to crowd out. So there is special reason to be attentive in this regard. Second, one might wonder whether a technological option might not really be appropriate or actually more valuable in the sense of being more promising relative to low-technological options and for the challenge in question. This, however, misconstrues the categories and the processes introduced above. The assumption was that there will be valid options on the table that are generally worth considering at least as potentially estimated by the involved parties. That is, all of the options *are* valid and may not be rejected in a simple manner, or so the assumption says. It is precisely the deliberative process of the collective to determine their more concrete *value*. Hence we cannot object to this by saying that the technological option might in fact be appropriate or (more) valuable. It is the discursive effects in deciding which options are (most) valuable that qualify for the technosolutionism objection.

One should not easily argue against technological solutions on the grounds that they are worthless or carry less value than alternative options. The relevant deliberation processes are usually established to answer the question of which options are most valuable with regards to the challenge at hand. Instead, agents proposing these need to be careful which *effects* the presentation and discussion of, for example, AI systems and other technology systems that are built around powerful chips (as for electric mobility or wind energy) actually have. The reading suggested presents a need for *countermeasures*. That is, in order to rightfully place this objection, one should check for strategies of the opposing party that highlight, support, and underpin low- or non- technologically valid alternatives. Of course, countermeasures cannot be just words, substantial efforts are needed.

## 9.6 Exploitation, Sustainability and Technosolutionism

We can now look at the main argument, which claims to connect the three issues of technosolutionism, social transformation for sustainability, and the practice of resource exploitation in a productive way. The objection of technosolutionism has special significance if it includes reference to global justice problems that stem from the impulse to social transformation. That is, if we employ the reading of the objection introduced in Sect. 9.5.2, this gives us a new approach to address global justice issues in the background. I have restricted the discussion to rather large-scale impact discourses and decision procedures which may refer to technologies like AI systems with the potential for monetisable applications. The objections to damaging and ethically worrying background conditions and global injustices around AI systems are certainly known in the academic debate (e.g. Coeckelbergh, 2021; van Wynsberghe, 2021). As mentioned, AI systems need resources such as rare earths, lithium, or tin (Crawford, 2021, p. 32) and these are by now established views in the debate.

However, what has been highlighted here are the tendencies of technological devices to influence our decision-making and to absorb our attention and that this should be a central concern of the technosolutionism objection. What is disconcerting about this general constellation is the observation that AI technologies get increasingly suggested as options, such as regarding the societal challenges of climate change. While this is certainly not an immediately misguided idea, the hunger for resources and raw materials of these technologies may lead to a sharp rise in the mining of these resources which—as presented above—are a central cause for the global justice issues mentioned.

That is, while climate change and other issues of sustainability can be framed as global justice issues, the strategies taken to address these may include similar problems themselves. Connecting this back to Sect. 9.2, I would like to stress that just as *activities* originating in and primarily benefitting the historically industrialised regions—as is often the case with the extraction of resources—may negatively affect residents of technologically less advanced regions, similar global justice problems can also result from *omissions*. The discussion has shown that omissions with regard to the attention to valid options in collective decision-making are pertinent here. That is, disregarding valid alternative options due to the crowding out effect may be benefitting agents of technologically more advanced regions but negatively affect residents of other parts of the globe.

Overall, the push for more technological solutions might, through the crowding out effect, distract from less technologically advanced alternatives, and due to the resources needed perpetuate global justice problems by accepting exploitative practices. It is clear that many practices of extraction are ethically questionable. What we should add is that the *discursive effects* of certain technologies in the countries of the global north might directly lead to these issues and are part of the problem. The infringement of relevant moral rights as well as actions that offend relevant moral values start with the collective decision-making and discourses among industrialised regions. If resource exploitation comprises *global injustices*, then transformation may lead to unsustainable solutions overall, but most importantly, it is the attitude of *technosolutionism* that can now be connected to global injustice.

## 9.7 Conclusion

There is evidence for productive ways to combine three issues that have been rarely connected so far. The technologies or technological options suggested in decision-making in technologically advanced societies and especially regarding emergent and innovative technologies connected to AI may seriously influence collective decision-making. They may crowd out alternative measures that potentially do not include the environmental issues of technological solutions. To conclude, (1) the objection of technosolutionism can gain further from more concrete conceptions, e.g. the effect of crowding out, on the one hand, and the connection to global justice issues on the other. Moreover, (2) the push for social transformation leads to more technological strategies involved in decision-making. In these processes, the effects of emergent technologies are particularly wide-ranging. Hence, the objection of technosolutionism should be put forward with care and should, when applicable, be connected to the environmental costs these technologies and their development and purveyance entail. If technological options have a tendency to crowd out other non- or low-technological options and if global justice problems are more connected to these technological options, then there are ethical reasons to object to technosolutionism in this regard.[1]

[1] Work on this chapter has been generously funded by the BMBF (Federal Ministry of Education and Research, Germany), *Cluster for Future NeuroSys*, Funding Number 03ZU2106EA. I would like to thank two anonymous reviewers as well as Niël Conradie for valuable comments on earlier versions of this chapter.

## References


AFI/Accountability Framework Initiative. (2019). *Operational guidance on environmental restoration and compensation*. Retrieved from http://tinyurl.com/4snpvs96

Ali, S., Giurco, D., Arndt, N., et al. (2017). Mineral supply for sustainable development requires resource governance. *Nature, 543*, 367–372. https://doi.org/10.1038/nature21359

Beitz, C. (1975). Justice and international relations. *Philosophy and Public Affairs, 4*(4), 360–389.

Bell, R. G. (2018). Protecting the environment during and after resource extraction. In T. Addison & A. Roe (Eds.), *Extractive industries: The management of resources as a driver of sustainable development* (online ed., pp. 318–334). OUP. https://doi.org/10.1093/oso/9780198817369.003.0016

Bissel, M. (2023, August). A public transport ticket that moved a country: Assessing the value of the German 9-Euro-Ticket as a socio-technical experiment. *Findings*. https://doi.org/10.32866/001c.84645

Brevini, B. (2023). Artificial intelligence, artificial solutions placing the climate emergency at the center of AI developments. In H. S. Sætra (Ed.), *Technology and sustainable development. The promise and pitfalls of techno-solutionism* (pp. 23–32). Routledge.

Brock, G., & Hassoun, N. (2023). Global justice. In E. N. Zalta (Ed.), *The Stanford Encyclopedia of philosophy*. Retrieved from https://plato.stanford.edu/archives/fall2023/entries/justice-global/

Coeckelbergh, M. (2021). AI for climate: Freedom, justice, and other ethical and political challenges. *AI Ethics, 1*, 67–72. https://doi.org/10.1007/s43681-020-00007-2

Crawford, K. (2021). *Atlas of AI. Power, politics, and the planetary costs of artificial intelligence*. Yale University Press.

EU. (2022). *European Commission: Communication from the Commission to the European Parliament and the Council "Strategic foresight report twinning the green and digital transitions in the new geopolitical context"*. Retrieved from https://op.europa.eu/en/publication-detail/-/publication/d04b2f02-f86b-11ec-b94a-01aa75ed71a1/language-en

Kallis, G., Paulson, S., D'Alisa, G., & Demaria, F. (2020). *The case for degrowth*. Polity Press.

Morozov, E. (2014). *To save everything, click here. The folly of technological solutionism*. Public Affairs.

Nagel, T. (2005). The problem of global justice. *Philosophy & Public Affairs, 33*(2), 113–147. Retrieved from http://www.jstor.org/stable/3558011

Ng, A. W. (2019). Sustainability and education policy. In W. L. Filho (Ed.), *Encyclopedia of sustainability in higher education*. Springer. https://doi.org/10.1007/978-3-319-63951-2_482-1

Peltzer, P. M., Cuzziol Boccioni, A. P., Attademo, A. M., Simoniello, M. F., Lener, G., & Lajmanovich, R. C. (2024). Ecotoxicological characterization of lithium as a "Timebomb" in aquatic systems: Tadpoles of the South American Toad *Rhinella arenarum* (Hensel, 1867) as model organisms. *Toxics, 12*(3), 176. https://doi.org/10.3390/toxics12030176

Sætra, H. S. (2023). Introduction. The promise and pitfalls of techno-solutionism. In H. S. Sætra (Ed.), *Technology and sustainable development. The promise and pitfalls of techno-solutionism* (pp. 1–9). Routledge. Retrieved from https://www.taylorfrancis.com/chapters/oa-edit/10.1201/9781003325086-1/introduction-henrik-skaug-s%C3%A6tra

Smith, S. M., Geden, O., Nemet, G., et al. (2023). *The state of carbon dioxide removal* (1st edn). Retrieved from https://www.stateofcdr.org

van Wynsberghe, A. (2021). Sustainable AI: AI *for* sustainability and the sustainability *of* AI. *AI Ethics, 1*, 213–218. Retrieved from https://link.springer.com/article/10.1007/s43681-021-00043-6

Velasco-Fernández, R., & Pérez, A. (2022). *Critical minerals in conflict and alternatives for global justice in public procurement: The case of electric vehicles (Report for Medicus Mundi Mediterrània, project OR+ JUST for Global Health: Public Procurement, Jewellery and Responsible Consumption)*. Retrieved from https://www.medicusmundi.cat/wp-content/uploads/2023/05/ENG_ExecutiveSummary.pdf

Wenar, L. (2016). *Blood oil: Tyrants, violence and the rules that run the world.* Oxford University Press.

Wettengel, J. (2023). Quest for climate neutrality puts CCS back on the table in Germany. *CLEW Clean Energy Wire.* Retrieved from https://www.cleanenergywire.org/factsheets/quest-climate-neutrality-puts-ccs-back-table-germany#four

Zwarthoed, D. (2015). Creating frugal citizens: The liberal egalitarian case for teaching frugality. *Theory and Research in Education, 13*(*3*), 286–307. https://doi.org/10.1177/1477878515606620

**Frieder Bögner** BA English Studies and Philosophy (Cologne), MA Philosophy (Münster), PhD Philosophy (Osnabrück 2023). PostDoc in Applied Ethics, RWTH Aachen University. Author of Der Konflikt zwischen legitimer Autorität und der Autonomie von Personen [The Conflict between Legitimate Authority and the Autonomy of Persons] (mentis publishers 2023). Selected research areas: Ethics and Political Philosophy, personal autonomy, legitimate authority, philosophy and ethics of technology, climate ethics.

# Chapter 10
# The Justice Assumptions of Energy Storage Experts: A Reflective Workshop

**Nynke van Uffelen**

**Abstract** Normative assumptions often remain implicit and undebated in technology development. This is unsurprising because many normative ideas seem universal yet are actually particular standards of dominant social groups. As norms pose as universal, they deny legitimacy to alternatives, ignore moral plurality and eliminate the potential for a more just society. Energy technologies, in particular, have been decoupled from ethical reflection for a long time. Although energy justice scholarship aims to fill this gap, its researchers generally focus on the (un)just effects of already developed energy systems. This chapter illustrates how energy justice can be leveraged in the innovation phase to identify implicit and seemingly universal normative assumptions, unlocking a deeply critical potential beyond merely evaluating the consequences of energy technologies. Four workshops were held with energy storage experts to study their energy justice assumptions, whether they acknowledge moral plurality, and how they deal with normative uncertainty. The results show that experts have—and can articulate—specific conceptions of justice about the scale, time, subjects, and principles of justice in relation to energy storage technologies, such as liberal nationalism, anthropocentrism, and utilitarianism. Yet, other defensible normative assumptions circulate in philosophical debates. Some, but not all, normative uncertainties were acknowledged by the participants. When they were, they resolved normative uncertainties by resorting to utilitarian reasonings, dismissing moral questions as subjective, and relying on existing institutional frameworks. The findings stress the importance of a public debate on energy justice, fostering value change in the long run.

**Keywords** Energy justice · Energy storage · Energy transition · Normative uncertainty · Value change

---

N. van Uffelen (✉)
Delft University of Technology, Delft, The Netherlands
e-mail: n.vanuffelen@tudelft.nl



C. Didier et al. (eds.), *Engineering and Value Change*, Philosophy of Engineering and Technology 48, https://doi.org/10.1007/978-3-031-83549-0_10

## 10.1 Introduction

Technologies embody moral values, and as such, they are not neutral (Klenk, 2021; Sovacool et al., 2016; Sovacool & Dworkin, 2014; Winner, 2017). Technologies and their properties are shaped by people—engineers, designers—who have normative assumptions, such as values, visions, and ideas about justice and the good life. Although intentions and normative ideas do not fully determine the use or the (un) just consequences of technologies (Klenk, 2021), they steer innovation and development, and as such, they can engender (in)justice (Karwat, 2019). Normative assumptions play a pivotal role in design decisions, in acquiring funding that prioritises the development of some technologies over others, and in influencing the public debate and the political and scientific agenda (Dignum, 2013; Jacobs & Huldtgren, 2021). Therefore, reflecting on the normative assumptions behind and embedded in technologies is crucial.

Still, normative assumptions often remain implicit and undebated in technology development (Pesch, 2021). This is unsurprising because many normative ideas, such as democracy, freedom, growth, nationalism, and liberalism, seem universally shared and uncontroversial. However, following Iris Marion Young, seemingly universal ideas can often be deconstructed as particular and even Western, as they are the standards of the dominant social groups (Young, 1990). As such ideas pose as universal, they deny legitimacy to alternatives, thus establishing a hegemony, an ideology, or a one-dimensional society (Marcuse, 1964). In doing so, normative uncertainties, which are "situations where there are different partially morally defensible—but incompatible—options or courses of action, or ones in which there is no fully morally defensible option" (Taebi et al., 2020), are denied, and with it the potential for a more just society. Philosophers have a crucial task in identifying the normative assumptions taken for granted, deconstructing them as particular, and questioning them.

Energy technologies, in particular, have been decoupled from ethical reflection for a long time, even though energy innovations and cultural and societal developments are deeply interwoven (Shove & Walker, 2014). As a case in point, it was not until 2013 that researchers stressed the importance of systematically reflecting on justice in relation to energy systems (McCauley et al., 2013). Energy justice frameworks are often considered conceptual, analytical, and decision-making tools to make energy systems more just (Sovacool & Dworkin, 2015; Jenkins et al., 2016). Nevertheless, Sovacool et al. (2021) and Jenkins et al. (2020) separate energy justice from value-sensitive design and responsible research and innovation, arguing that energy justice evaluates the impacts and effects of already developed technologies. Consequently, energy justice research generally focuses on the (un)just effects of energy systems instead of the normative assumptions and values that steer their development.

This chapter explores how energy justice can be leveraged in the innovation phase to identify implicit and seemingly universal normative assumptions, unlocking a deeply critical potential beyond merely evaluating the consequences of energy technologies in terms of justice. To do so, it studies the energy justice assumptions of energy storage experts. Energy storage is chosen as a case because it will be crucial for future renewable energy systems, yet further innovation is required, and because of technological complexities, this generally occurs without philosophical reflection. Four workshops were held with experts on hydrogen production and redox flow batteries from the Netherlands and the United Kingdom to identify their energy justice assumptions. In sum, this chapter studies (a) the energy justice assumptions of energy storage experts; (b) whether these experts acknowledge moral plurality, i.e. the particularity of their own assumptions; and (c) how they deal with or resolve normative uncertainty.

Section 10.2 introduces the energy storage technologies discussed in the workshops, and it explains the workshop design and analysis. Section 10.3 describes the results, in other words, the dominant energy justice assumptions held by the workshop participants. Section 10.4 contrasts these results with other possible justice assumptions, highlighting the normative uncertainties and the ethical questions they trigger. Section 10.5 explores how participants resolve normative uncertainties. Section 10.6 concludes.

## 10.2 Workshops with Energy Storage Experts

### *10.2.1 Energy Storage: Green Hydrogen Production and Redox Flow Batteries*

Realising the energy transition requires a large-scale shift to renewable energy. When most renewable energy sources are intermittent, large-scale energy storage will be vital for a stable and secure net-zero energy system (Gallo et al., 2016; Verzijlbergh et al., 2017). Energy storage solutions that are likely to play an essential role in our energy future are green hydrogen production and redox flow batteries.

Green hydrogen production is crucial for decarbonising industry, long-haul transportation, and seasonal energy storage (Gupta, 2009; Sovacool et al., 2024). Green hydrogen can be produced through electrolysis powered by renewable energy sources. Nevertheless, large-scale implementation faces challenges, including intermittent renewable energy sources, high costs (due to the necessity of precious metals, high electricity prices, and inefficiency due to the technology's relative immaturity), and limited options for storing hydrogen (McDonnell-Worth & MacFarlane, 2018; Griffiths et al., 2021).

Redox flow batteries are a potential solution to the problems intermittent green electricity poses for stabilising the electricity grid, as they can store energy for a relatively long time, from two to 20 h, without using expensive, scarce materials (Hugo et al., 2018). Redox flow batteries are a type of electrochemical energy storage, as they store and release energy through reversible chemical reactions. In theory, there are infinite options in terms of chemistries that can be used. Currently, the most advanced type is the vanadium redox flow battery. However, as vanadium is relatively expensive (Schmidt et al., 2019), researchers are exploring alternative chemistries.

### 10.2.2 *Workshop Design: Measuring Energy Justice Assumptions*

To study the energy justice assumptions of energy storage experts, three workshops on hydrogen production and one on redox flow batteries were held in November 2023. In each workshop, there were 3–6 participants, who were all senior and influential experts and innovators with a technological background and a bird's-eye view of the potential consequences of the technologies under scrutiny. Because the population of such experts is rather small, their job titles and names are not included in this chapter.

Although there are many different types of workshop designs in energy social science research, such as co-creation, backcasting, and futuring workshops, a workshop for identifying and discussing normative assumptions seems lacking, justifying a more explorative approach. The workshop was designed to identify energy justice assumptions, using the revisited energy justice framework by Van Uffelen et al. (2024) as a theoretical framework, as it defines categories of assumptions on which there is normative uncertainty (see Fig. 10.1). The framework discerns four *tenets of justice*, namely, distributive justice (the just distribution of goods, burdens, and benefits), procedural justice (just decision-making procedures), recognition justice (on just relations of recognition, which excludes institutionalised cultural injustices such as devaluing, ignoring, and marginalising social groups), and restorative justice (the just restoration of injustices). Moreover, the framework classifies *justice assumptions*, including the subject (e.g. justice for humans or multispecies justice), scale (e.g. justice in a local, regional, national, continental, or global context), time frame (e.g. justice for the current or also future generations), and principles of justice (e.g. rules or yardsticks that define under what conditions something is just or

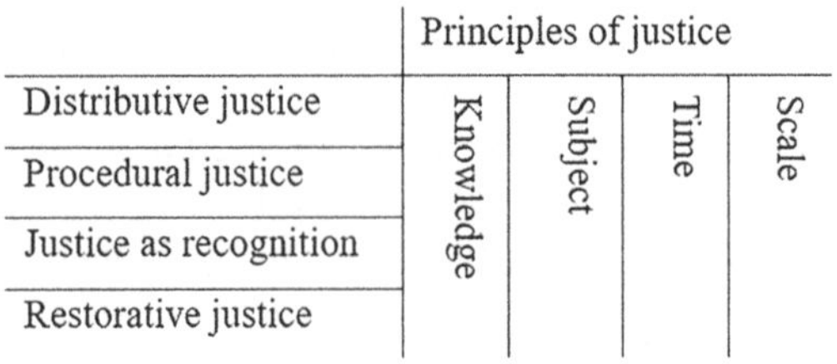


**Fig. 10.1** The energy justice framework (Van Uffelen et al., 2024)

unjust).[1] These categories of assumptions mirror deeply contested philosophical debates on justice (Miller, 2017).

The workshop's design was informed by two scoping interviews with academic experts on hydrogen production and redox flow batteries and a mock workshop with academic colleagues. For practical reasons, the 2-h workshops were held online via Zoom, using the software Miro for the collaborative exercises. The transcripts were pseudonymised and analysed in Atlas.ti.

In the workshops, participants were first asked to write on virtual Post-its what good or bad consequences the technology might have and put them on a matrix (the x-axis being good/bad; the y-axis likely/unlikely). Then, they were asked to pick four consequences—two good, two bad—to dive into during the rest of the workshop. The moderator asked for each Post-it *why* this is a good or bad consequence, in other words, what makes it good or bad. This stimulated the participants to make explicit their normative orientations. Often, further why-questions were required to make these explicit (for example: why is climate change bad?) To dive further into the normative assumptions, participants were asked to articulate for 'who/what' this was a good or bad consequence (which measured both the subject and the scale of justice) and 'when' this would be good or bad (targeting the time component). In the last 20–30 min of the workshop, a stakeholder map was revealed that showed the following stakeholders: ecosystems and animals; people in the Global South; privileged people in the Global North; local people in the Global North; technology innovators and companies; marginalised groups in societies; and future generations. The participants were asked to reflect on what stakeholders they had not yet explicitly mentioned during the workshop, why, and what it would mean to take these stakeholders into account. Moreover, they were asked whether each stakeholder was equally important. This critical reflection exercise encouraged participants to articulate further or question their views on energy justice and challenged participants to imagine alternative conceptions of justice.

The normative assumptions of the participants were analysed through Van Uffelen's energy justice framework. Moreover, attention was paid to whether participants acknowledged alternative conceptions of justice and, if they did so, how they dealt with and resolved normative uncertainty.

Due to the small sample size, the conclusions only pertain to the normative assumptions of these experts, and generalisations are not warranted. However, the implicit normative assumptions taken for granted by the participants shed light on conceptions of justice that are seemingly universalised, and how participants deal with normative uncertainty may reflect tendencies in the larger population. As such, this empirical philosophy study provides critical theorists' claims with empirical grounds and informs philosophical reflections on energy justice.

[1] The fifth category, knowledge, was not part of the workshop design, as this study focuses on normative assumptions, not on epistemic assumptions. However, some epistemic uncertainties were articulated in the workshops, and as such they were part of the analysis.

## 10.3 Dominant Conceptions of Energy Justice

### *10.3.1 The Distributive Paradigm*

In the first part of the workshop, participants were asked to articulate possible 'good' and 'bad' consequences of the energy storage technology at hand and then choose four to elaborate on for the remainder of the workshop. From their interpretations of the chosen Post-its (for an overview, see Table 10.1), it can be derived that most consequences relate to distributive (in)justice. This implies that 'good and bad consequences' were mostly interpreted as an unjust distribution of goods, burdens, and benefits. There were only a few exceptions to this.

First, procedural justice—which is a concern for just decision-making procedures—came up in relation to two Post-its. On the one hand, participants in the second, third, and fourth workshops chose to focus on energy autonomy, independence, or energy security. In all three discussions, autonomy, control, and freedom were central tenets. These relate to procedural justice, as it was generally deemed undesirable that decisions were influenced by foreign and potentially malign states.

> [energy security] makes like a country or a, uh, or a region or whatever less prone to well all kinds of manipulation or blackmail or that kind of thing in a political sense. (workshop three)

**Table 10.1** An overview of the post-its formulated by the participants that were chosen as the focus for the workshops and whether they perceived the Post-it as a 'good' or a 'bad' consequence of the technology

| Workshop | Chosen post-its | Evaluation |
|---|---|---|
| 1—Green hydrogen production | Decarbonisation of hard-to-abate industries/decarbonisation, right production, little leakage | Good |
| | Synergy effects from small to large scale (water, heat, biomass, system integration) | Good |
| | For health, safety, and environment, we need to go through a learning curve | Bad |
| | (New) infrastructure takes time and is expensive | Bad |
| 2—Green hydrogen production | Political/geographical independence energy supply | Good |
| | Decarbonisation | Good |
| | Hydrogen-related safety incidents; safety and image—One huge incident may reduce the current popularity abruptly | Bad |
| | Loss of jobs | Bad |
| 3—Green hydrogen production | Preserve industrial activities | Good |
| | Improved energy security | Good |
| | Higher energy costs | Bad |
| | Perceived risks and safety issues with explosive gas | Bad |
| 4—Redox flow batteries | Decarbonisation of the energy system | Good |
| | Energy autonomy | Good |
| | Increasing costs for citizens | Bad |
| | Lack of regulations | Bad |

On the other hand, the consequence of 'safety incidents and image' was often associated with a lack of public acceptance, which can be read as a procedural justice concern. Nevertheless, public acceptance seemed to be mainly of instrumental importance to the participants because it delays decarbonisation, the energy transition, and the uptake of hydrogen technologies.

Second, restorative justice came up once in workshop four. One participant repeatedly stressed the unequal historical contributions to climate change and the implications for responsibilities towards the Global South. This implies that there is an injustice that ought to be restored.

Third, reflections on recognition justice—which are concerns for institutionalised cultural injustices, such as marginalisation and devaluation (van Uffelen, 2022)—were scarce. In some instances, participants acknowledged that the energy transition has a disproportional impact on people in poverty (workshop one and three) or communities that rely on fossil industries (workshop two). However, without a link to institutionalised cultural injustices, concerns for marginalised actors can be interpreted as distributive justice concerns. To argue that these concerns are instances of misrecognition, some hidden premises would have to be assumed.

### *10.3.2 The Scale of Justice*

Participants were asked to elaborate on 'for whom' the (un)just consequences were unjust. From their answers, the assumed scale of justice can be deduced as each consequence was considered good or bad for a local community, a (supra)national entity, and/or everyone globally. The data shows that participants assume different scales when discussing different topics. Moreover, there were significant tensions between global, local, and (supra)national scales of justice that were sometimes, but not always, made explicit by participants.

Some participants in workshops one and two explicitly dismissed local justice because local claims slow decarbonisation and are thus to be ignored. However, when discussing risks and safety issues, synergy effects, and the loss of jobs, participants stressed local justice. For example, in workshop two, participants stressed that energy infrastructures 'in our backyard' should also benefit communities.

> People don't want to have, let's say, the mining done in their backyard. Whereas we might sit on huge resources that might be very valuable for energy transition. How do you deal with that? And how can you make it indeed also… Valuable for the people who live there? Because that is the main…. (workshop two)

A more frequently adopted scale of justice was that of national or supranational (EU) justice. Participants in the second and third workshops were mostly focused on justice for (countries within) Europe. In the third workshop, 'higher energy costs' were a concern for just transitions within countries, and participants discussed whether 'preserving industrial activities' was desirable or not for the

Netherlands or the UK. Moreover, in workshops two, three, and four, the independence of energy supply was mainly considered good for (countries within) Europe.

However, in all workshops, decarbonisation was discussed in terms of global justice, as it would benefit people globally. In the fourth workshop, both 'increasing costs' and the 'lack of regulations' were generally considered good or bad globally, as they slow down energy transitions. In workshop two, global justice also emerged in the context of energy independence through the notions of 'energy colonialism' and 'fair trade'.

Moreover, in the reflection exercises at the end of the workshops, some participants questioned the assumed (supra)national scale of justice when confronted with the Global South as a stakeholder. Such reflections indicate a global scale of justice. For example, participants started to doubt whether energy autonomy is unequivocally good because, so far, they have only considered this as 'good for Europe'. In workshop three, participants reflected on 'high energy costs' and 'energy security' from the perspective of the Global South.

> I think we have a lot of activities in the Global South where the profit lands in the Global North. And so the people in the Global South may have some jobs from it, but well, they're not really very generously paid and we try to get as much to, well, to our own countries. Yeah. And that's what I would call the wrong way. (workshop three)

Although participants in workshop four started by considering energy autonomy as good for a restricted scale of justice, they gradually started to stress global justice and imply a more elaborate set of responsibilities towards the Global South. They started to explore whether geographical differences between countries are fair or not and what obligations the global north has to the Global South due to the unequal contributions in causing climate change.

> There also has to be a fairness among countries. And this is clearly not there. It is not only that they lack the cash to fund the investment themselves. It's also that they hardly contributed to the mess we're in. (workshop four)

However, there was also push back against a global scale of justice. For example, participants in workshop three argued that we have only limited responsibility towards the Global South.

> You know, we're not doing this [the energy transition] for the benefit it has to the Global South. But I think the mapping out that transition pathway, that's a lot of value, you know, across the world. (workshop three)

### *10.3.3 Subject: Justice for Humans, or Multispecies Justice?*

When discussing the consequences, participants articulated 'for whom' these were good or bad. This not only pertains to the scale of justice but also to the subject of justice. Justice is either *anthropocentric justice*—only for humans—or *multispecies justice*, including animals and/or ecosystems as subjects of justice.

Generally, the participants of all four workshops considered only humans as subjects of justice. This can be drawn from their responses to questions about why decarbonisation is good and from the fact that the impact on animals and the environment was only sporadically mentioned in the first parts of the workshops.[2]

> If we don't do decarbonization, it basically means it goes at the expense of humans, right? So, it's a human perspective on things, yeah? It's human suffering. It's the destruction of human infrastructure and so on, yeah? Yeah. Or a way of living that humans do. (workshop one)

In the second part of the workshops, the participants were confronted with 'animals and ecosystems' as stakeholders. Most participants acknowledged that they have not explicitly discussed 'good or bad' for non-humans.[3] During the reflection exercise, participants from all workshops started to explore animals and ecosystems as subjects of justice. For example, workshop two mentioned the impact of energy infrastructure on birds and sea animals and the possibility (and the legal necessity in permit procedures in the Netherlands) of including this in a design (workshop two). Moreover, workshop two mentioned the possibility of animals going extinct due to climate change.

> We have to be a little bit careful that we don't destroy the world to save the world. (workshop two)

Participants were also asked to reflect on the relative importance of 'animals and ecosystems' as stakeholders. From this discussion, it can be drawn that the overall focus is anthropocentric, and nature is considered of instrumental value to humans.

> I think that, when it comes down to it, we have to make choices, and animals and ecosystems will be less important, if I am being really honest about this. (workshop one)

### *10.3.4 Principles of Justice*

Participants were asked to articulate individually and in plenary why each consequence was considered good or bad. As a result, they formulated values and principles of justice to justify the goodness or badness of the chosen Post-its.

One notable finding was that the goodness or badness of all chosen consequences was determined by whether they contribute to or slow down decarbonisation, except for the consequences related to energy autonomy. In other words, decarbonisation and energy autonomy were perceived as two good consequences that are relatively separate from each other.

---

[2] There is one exception. In the individual exercise in the first workshop, one participant stressed that decarbonisation is good because it avoids harm to the environment. During that workshop, that participant reminded the others twice of the impact of climate change on non-humans.

[3] Workshop two forms an exception; participants claimed they had talked about ecosystems, even though their discussions on decarbonisation have been from an anthropocentric perspective.

Decarbonisation is part of the justification of the goodness or badness of all other chosen Post-its (except energy autonomy), and therefore, the reasons for the goodness of decarbonisation are vital in understanding the normative assumptions of the participants. In the first, third, and fourth workshops, decarbonisation and mitigating climate change were considered good because they reduced costs. The utilitarian principles of cost reduction and economic profit reason that something is good 'because it costs less'. This principle returned in almost all discussions on the chosen Post-its.

> Well, I think if we have no energy transition and we have a lot of climate change, we have a lot of problems, which are a lot worse even than losing your job, I guess, in the longer term. Like, well, food problems, weather problems, big storms, big flooding, droughts, and other things which will, in the end, cost a lot of money. (workshop three)

In the first, second, and fourth workshops, participants measure the goodness of decarbonisation against the principle of maintaining our current living standards. 'Our' Western standard of living was perceived by participants as 'developed', desirable, and exemplary for the rest of the world. This interpretation of a sufficiency principle was often framed as the ultimate justification for why decarbonisation and energy autonomy are good and why other Post-its, such as increased energy costs, are bad. Decarbonisation is also good for health, avoiding mass migration, and creating new jobs.

Next, energy autonomy, independence, and security were signalled as good possible consequences of both hydrogen production and redox flow batteries. Although there was much debate on whether energy autonomy would actually be desirable or feasible, several principles of justice shaped the debates. Returning principles were those of costs (if it costs less, it is good) and maintaining our current living standard. Moreover, participants considered negative freedom and control over 'our' destiny as important principles. These intuitions can be understood as principles of national self-determination.

> There's the ideal situation where, as a nation, that, you know, you have complete control of your energy supply, which is kind of really important. (workshop two)

However, participants acknowledged that dependence might not always be undesirable. In this context, participants in workshop four condemned energy colonialism and raised several fair trade principles. For example, a participant argued that international trade should be "*a two-way traffic.*" Another participant argued that international trade relations should consider causal responsibility for historical damage. Another argued that fair trade is defined by the principle of desert.

> Now you mentioned fairness, yeah okay what is fair? If I work hard I deserve more money. Um so it's uh privileged people in the global north. (workshop four)

### *10.3.5 Time: What About Future Generations?*

Participants were asked 'when' each consequence would be good or bad, in other words, what time frame they had in mind.[4] Generally, participants seemed to draw the line between the short and the long term as around 20 years from now, which can be rounded up to 2050. As such, in most workshops, the concept of 'future generations' did not come up spontaneously. An exception is workshop four, as participants explicitly argued that the interests of future generations should take precedence over current monetary concerns. However, when participants were confronted with 'future generations' as stakeholders during the reflection exercise, they stressed their importance.

Each workshop displayed a tension between the short and the long term.[5] All Post-its from the 'bad consequences' category, plus the 'decarbonisation' and 'energy autonomy' Post-its, were only considered bad in the short term. Moreover, one Post-it ('hydrogen production preserves industrial activities') was considered good in the short term, while its desirability in the long term was questioned. However, the bad consequences were considered sacrifices that needed to be made for the future. To summarise, in general, post-2050 interests were given priority over pre-2050 interests.

> Energy transition is a huge upfront investment, just like previous generations who are making a sewer system or drinking water or the electric grid. To get it started, you need to put a lot of money on the table. And then once it's there, okay, you will expand and maintain it. But for future generations, it's very nice that they already inherited it. (workshop four)

## 10.4 Normative Uncertainties

Experts on hydrogen production and redox flow batteries make specific normative assumptions about justice. However, there were contradictions between several assumptions held by the participants. Moreover, there may be morally defensible alternatives. In other words, energy justice is shrouded in normative uncertainties in relation to the time frame, subject, scale, principles, and tenets of justice.

Normative uncertainty about the subject of justice was hardly acknowledged by the participants. Most participants considered only humans as relevant subjects of justice. Nature and animals were generally seen as instrumental for humans, and this intuition guides decisions on trade-offs between human and non-human interests. This *anthropocentric* conception of justice contrasts with assumptions based

---

[4] At first, this resulted in an epistemic discussion about when a certain technology would be developed and implemented. Yet, for most consequences, the assumed time frame could be deducted.

[5] There was one exception: 'synergy effects' was the only consequence that was considered good both 'now' and in the (far) future (workshop 1).

on *multispecies justice* as defended by some environmental ethics scholars (Celermajer et al., 2021; Tschakert et al., 2021).

There are also vivid philosophical debates on the weight that should be assigned to the interests of future generations. However, normative uncertainty about the assumed temporal modus was largely dismissed in the workshops. Participants acknowledged that certain time frames sometimes clash, as some consequences were conceived as bad in the short term and good in the long term or vice versa. However, negative consequences in the near future—such as high energy prices and safety issues—did not make participants adopt a negative stance towards the technology. Instead, they were tolerated because they were only temporary, and the technology would contribute to a better world in the long term. In other words, techno-optimism and prioritising post-2050 over pre-2050 interests made normative uncertainty about the adopted time frame a non-issue.

> I think the costs will be higher in the short term. In the future, when there is an abundance of renewable energy sources with low marginal cost, the overall cost might decrease for the citizens. (workshop four)

Participants adopted a variety of scales of justice, including local, national, supranational, and global perspectives. In theory, however, different scales of justice can clash. For example, justice within the Netherlands or Europe can give rise to global injustices, while global justice can require sacrifices from nation-states. As such, there is normative uncertainty about whether it can be morally justified to restrict justice concerns to a smaller—local or (supra)national—scale or whether there is a moral obligation to include other countries and continents in our moral circle. In most workshops, normative uncertainty about scale became visible when participants reflected on the Global South as a stakeholder. In this context, notions such as "fair trade," "energy colonialism," "historical responsibility," and "unfair advantage" came up. In these discussions, most—but not all—participants seem to adhere to a form of liberal nationalism, which states that citizens have special relations, and thus special obligations of justice, towards each other that they do not have towards non-citizens (Miller, 1995; Tamir, 1993). Global justice is restricted to fair trade relations and duties of assistance when societies fall below a certain humanitarian threshold (Van Parijs, 2007). People in other countries may benefit from hydrogen production and redox flow batteries, but justice for 'them' is not 'our' primary concern. Only two remarks revealed a more far-reaching commitment to global justice, namely, that the Netherlands would have an 'unfair resource advantage' (which was quickly withdrawn by the participant by saying it was not unfair, but "*competitive*") and that the Global North has a "*historical responsibility*" towards the Global South, both in workshop four.

Participants articulated many different principles of justice. Following Michael Walzer, it is not necessarily problematic to adopt different principles of justice for different situations (Walzer, 1983). For example, although strict equality might be a good principle for distributing love to your children, it seems inappropriate when distributing energy to households because different people have different needs. However, when discussing a good or bad consequence, participants often invoke multiple principles, some of which are contradictory. For example, when discussing energy autonomy, the principles of cost reduction, national self-determination, fair trade as a two-way street, desert, causal responsibility, and sufficiency can clash. These clashes are very much intertwined with the assumed scales of justice. For example, national self-determination might not be the cheapest option, and as such, it may threaten our current living standards. Moreover, cost reduction for 'us' can compete with fair trade or with principles of causal responsibility. Consequently, normative uncertainties about the best course of action are cross-dimensional. For example, the cost reduction principle on a national scale in the short term can constitute trade-offs with concerns for the Global South and future generations.

There is also normative uncertainty about the relations between different tenets of justice. Participants focused on distributive justice, which situates them in the distributive paradigm that "defines social justice as the morally proper distribution of social benefits and burdens" (Young, 1990, p. 16). However, this paradigm has been criticised extensively, in part because it ignores or takes for granted institutionalised cultural injustices—such as racism, ableism, speciesism, ageism, and sexism—that can underlie distributive injustices (Young, 1990). For example, participants generally considered their culture 'developed' and the desirable living standard for the rest of the world, devaluing other views on the good life. By taking this cultural hierarchy as a yardstick for distributive justice, they reproduce an idea that can be considered as misrecognition.

## 10.5 How Normative Uncertainties Were Settled

Despite including a reflection exercise, none of the participants claimed that they changed their minds, only that it made them think a bit more about the social aspects of their technologies. Moreover, there was hardly any explicit disagreement among participants.[6] This raises the question of how they dealt with normative uncertainties on energy justice. Participants seemed to resolve normative uncertainties in three different ways.

First, participants often resorted to utilitarian reasoning and the principle of cost reduction. According to this principle, something is just if it reduces the costs or if it costs less than an alternative. This reduces justice issues to economic sums of gains and losses or morals to markets. Moreover, this type of utilitarian thinking transforms normative uncertainties into *epistemic uncertainties* about the consequences of a certain technology. Techno-optimism colours the probabilities of 'good' outcomes of the technologies, which helped resolve some normative uncertainties. Plausibly, both techno-optimism and utilitarian thinking stem from the scientific and technological background of the participants (Leydens & Lucena, 2018).

[6] This could be because they had just met and that this was a professional context to them.

If this is true, their education has thoroughly influenced how they perceive energy justice.

Second, all participants identified these tensions as political problems and not something experts such as themselves should have to think about. As such, normative uncertainties were sometimes acknowledged and dismissed as political or subjective. For example, distributive justice questions were seen as "*more of a political choice than a technology choice*" (workshop one). According to a participant in workshop three, everyone should have a "*reasonable standard of living. […] But that's more of a political view, I guess.*" In workshop four, a participant argued that "*The needs of the privileged people from the global north might be less important than for example marginalized groups or people in the Global South. Um, yeah, that's just my personal, uh, personal point of view.*" Moreover, their meta-ethical stance seemed to be that of subjectivism or relativism. For example, one participant in workshop four framed normative uncertainty about the importance of future generations as "*a matter of values.*" This finding resonates with early texts on the engineering mindset, such as Gottlieb (1983), who claimed that engineers are generally 'positivists' who strictly separate the domains of knowledge and ethics, the latter being the real of "free and unjustified personal choice" (p. 11).

Third, participants often referred to existing institutional frameworks when faced with normative uncertainty. This point has been made in past research concerning existing codes of engineering ethics that exercise a deontological power (Schmidt, 2014). This study shows that this institutional influence extends beyond ethical guidelines. For example, the well-known milestone of 2050 shaped how participants distinguish between the short and the long term. This hampered the participants in thinking about future generations far beyond 2050. Moreover, the participants were embedded in for-profit business contexts, and it is plausible that this influenced their criteria for good and bad. Similarly, they were tied to (supra) national institutions that regulate the energy transition, such as national or EU governments. Often, the scale of justice matched the scale of their affiliated institution. As stated above, the Post-its were often considered good or bad for a national or European context, which contrasted with articulated intuitions on global injustices. These moral questions stayed out of sight due to the (supra)national institutional context in which the participants operate. In sum, institutional frameworks—including (supra)national state institutions, education, technology, cultural norms, and business—guided, for a large part, how participants perceived justice and how they resolved normative uncertainty.

## 10.6 Conclusion

This study shows, first of all, that energy storage experts could articulate many energy justice assumptions. Generally, participants held a liberal nationalist perspective, which includes some—but limited—responsibilities towards 'other' countries. Most participants considered only humans as subjects of justice, excluding non-humans. The timeframe was largely determined by the 2050 milestone. Moreover, participants mentioned many different principles of justice, including a sufficiency principle (with the Western lifestyle as a yardstick), cost reduction, and national self-determination. Although these insights cannot be generalised, they say something about the 'norms' and what is considered relatively uncontroversial.

This characterisation of participants' most prevalent normative energy justice assumptions also reveals what assumptions participants did not have. Other assumptions—such as multispecies justice, intergenerational justice, and principles of justice—are reasonably defensible, which amounts to normative uncertainty about energy justice. In other words, the perspectives of participants can be deconstructed as particular (Western) conceptions of justice instead of universal, neutral views. On some topics, participants were able to articulate normative uncertainties. However, normative uncertainties often remained unnoticed, and tensions between different assumptions became visible only in data analysis.

Focussing on normative uncertainty stresses the need to include a more elaborate set of assumptions in energy system decisions. Fostering a just energy transition requires explicit reflection on what is just. Currently, reflection is lacking about what assumptions are held and whether these are the most appropriate. In other words, normative assumptions are taken for granted. However, when a broad set of possible assumptions is considered, and the normative uncertainty that comes with it is acknowledged, a different situation arises. Often-held assumptions may be thoroughly critiqued, and less convincing assumptions may be replaced by others. The 'last assumption standing' might be considered the best one for now. Such debates should be organised in the public sphere, and as such, it is not the sole responsibility of engineers or designers but of ethicists, educators, and society as a whole. Such a project of value change might—slowly but steadily—lead to more just energy systems. This study can be seen as an exercise in empirical philosophy to analyse and reflect on energy justice assumptions in energy innovation, which is a first step towards moral progress.

## References


Celermajer, D., Schlosberg, D., Rickards, L., Stewart-Harawira, M., Thaler, M., Tschakert, P., Verlie, B., & Winter, C. (2021). Multispecies justice: Theories, challenges, and a research agenda for environmental politics. *Environmental Politics, 30*(1–2), 119–140. https://doi.org/10.1080/09644016.2020.1827608

Dignum, M. (2013). *The power of large technological visions: The promise of hydrogen energy (1970-2010)*. Technische Universiteit Eindhoven. https://doi.org/10.6100/IR759497

Gallo, A. B., Simões-Moreira, J. R., Costa, H. K. M., Santos, M. M., & dos Santos, E. M. (2016). Energy storage in the energy transition context: A technology review. *Renewable and Sustainable Energy Reviews, 65*, 800–822. https://doi.org/10.1016/j.rser.2016.07.028

Gottlieb, R. S. (1983, December). Engineering and critical communication: The relevance of philosophy to engineering. *IEEE Technology and Society Magazine*.

Griffiths, S., Sovacool, B. K., Kim, J., Bazilian, M., & Uratani, J. M. (2021). Industrial decarbonization via hydrogen: A critical and systematic review of developments, socio-technical systems and policy options. *Energy Research and Social Science, 80*(June), 102208. https://doi.org/10.1016/j.erss.2021.102208

Gupta, R. B. (2009). *Hydrogen fuel: Production, transport, and storage*. CRC Press.

Hugo, Y. A., Kout, W., Sikkema, F., Borneman, Z., & Nijmeijer, K. (2018). Performance mapping of cation exchange membranes for hydrogen-bromine flow batteries for energy storage. *Journal of Membrane Science, 566*(September), 406–414. https://doi.org/10.1016/j.memsci.2018.09.006

Jacobs, N., & Huldtgren, A. (2021). Why value sensitive design needs ethical commitments. *Ethics and Information Technology, 23*(1), 23–26. https://doi.org/10.1007/s10676-018-9467-3

Jenkins, K., McCauley, D., Heffron, R., Stephan, H., & Rehner, R. (2016). Energy justice: A conceptual review. *Energy Research and Social Science, 11*, 174–182. https://doi.org/10.1016/j.erss.2015.10.004

Jenkins, K., Spruit, S., Milchram, C., Höffken, J., & Taebi, B. (2020). Synthesizing value sensitive design, responsible research and innovation, and energy justice: A conceptual review. *Energy Research and Social Science, 69*(January), 101727. https://doi.org/10.1016/j.erss.2020.101727

Karwat, D. M. A. (2019). Engineering for the people: Putting peace, social justice, and environmental protection at the heart of all engineering. In *Frontiers of engineering: Reports on leading-edge engineering from the 2018 symposium* (pp. 1–6). Retrieved from https://www.ncbi.nlm.nih.gov/books/NBK538716/.

Klenk, M. (2021). How do technological artefacts embody moral values? *Philosophy and Technology, 34*(3), 525–544. https://doi.org/10.1007/s13347-020-00401-y

Leydens, J. A., & Lucena, J. C. (2018). *Engineering justice*. Wiley.

Marcuse, H. (1964). *One-dimensional man*. Routledge.

McCauley, D., Heffron, R., Stephan, H., & Jenkins, K. (2013). Advancing energy justice: The triumvirate of tenets. *International Energy Law Review, 32*(3), 107–110.

McDonnell-Worth, C. J., & MacFarlane, D. R. (2018). Progress towards direct hydrogen peroxide fuel cells (DHPFCs) as an energy storage concept. *Australian Journal of Chemistry, 71*(10), 781–788. https://doi.org/10.1071/CH18328

Miller, D. (1995). *On nationality*. Clarendon Press.

Miller, D. (2017). Justice. In *Stanford Encyclopedia of philosophy justice*. Retrieved from https://plato.stanford.edu/archives/fall2021/entries/justice/

Pesch, U. (2021). Imaginaries of innovation: Turning technology development into a public issue. *Science and Public Policy, 0*(0), 1–8.

Schmidt, J. A. (2014). Changing the paradigm for engineering ethics. *Science and Engineering Ethics, 20*(4), 985–1010. https://doi.org/10.1007/s11948-013-9491-y

Schmidt, O., Melchior, S., Hawkes, A., & Staffell, I. (2019). Projecting the future levelized cost of electricity storage technologies. *Joule, 3*(1), 81–100. https://doi.org/10.1016/j.joule.2018.12.008

Shove, E., & Walker, G. (2014). What is energy for? Social practice and energy demand. *Theory, Culture & Society, 31*(5), 41–58. https://doi.org/10.1177/0263276414536746

Sovacool, B. K., & Dworkin, M. H. (2014). *Global energy justice*. Cambridge University Press.

Sovacool, B., & Dworkin, M. (2015). Energy justice: Conceptual insights and practical applications. *Applied Energy, 142*, 435–444. https://doi.org/10.1016/j.apenergy.2015.01.002

Sovacool, B., Heffron, R., McCauley, D., & Goldthau, A. (2016). Energy decisions reframed as justice and ethical concerns. *Nature Energy, 1*. https://doi.org/10.1038/nenergy.2016.24

Sovacool, B., Hess, D. J., & Cantoni, R. (2021). Energy transitions from the cradle to the grave: A meta-theoretical framework integrating responsible innovation, social practices, and energy justice. *Energy Research and Social Science, 75*(March), 102027. https://doi.org/10.1016/j.erss.2021.102027

Sovacool, B. K., Del Rio, D. F., Herman, K., Iskandarova, M., Uratani, J. M., & Griffiths, S. (2024). Reconfiguring European industry for net-zero: A qualitative review of hydrogen and carbon capture utilization and storage benefits and implementation challenges. *Energy and Environmental Science, 17*, 3523–3569. https://doi.org/10.1039/d3ee03270a

Taebi, B., Kwakkel, J. H., & Kermisch, C. (2020). Governing climate risks in the face of normative uncertainties. *Wiley Interdisciplinary Reviews: Climate Change, 11*(5), 1–11. https://doi.org/10.1002/wcc.666

Tamir, Y. (1993). *Liberal nationalism*. Princeton University Press.

Tschakert, P., Schlosberg, D., Celermajer, D., Rickards, L., Winter, C., Thaler, M., Stewart-Harawira, M., & Verlie, B. (2021). Multispecies justice: Climate-just futures with, for and beyond humans. *Wiley Interdisciplinary Reviews: Climate Change, 12*(2), 1–10. https://doi.org/10.1002/wcc.699

van Parijs, P. (2007). International distributive justice. In E. Robert (Ed.), *A companion to contemporary political philosophy* (Vol. 2, pp. 638–652). Wiley. https://doi.org/10.7218/nenpouseijigaku.63.2_331

van Uffelen, N. (2022). Revisiting recognition in energy justice. *Energy Research & Social Science, 92*(August), 102764. https://doi.org/10.1016/j.erss.2022.102764

van Uffelen, N., Taebi, B., & Pesch, U. (2024). Revisiting the energy justice framework: Doing justice to normative uncertainties. *Renewable and Sustainable Energy Reviews, 189*(113974), 1–8. https://doi.org/10.1016/j.rser.2023.113974

Verzijlbergh, R. A., De Vries, L. J., Dijkema, G. P. J., & Herder, P. M. (2017). Institutional challenges caused by the integration of renewable energy sources in the European electricity sector. *Renewable and Sustainable Energy Reviews, 75*(November 2015), 660–667. https://doi.org/10.1016/j.rser.2016.11.039

Walzer, M. (1983). *Spheres of justice: A defense of pluralism and equality*. Basic Books.

Winner, L. (2017). Do artifacts have politics? In *Computer ethics* (pp. 177–192). Routledge. https://doi.org/10.4324/9781315259697-21

Young, I. M. (1990). *Justice and the politics of difference*. Princeton University Press.

**Nynke van Uffelen** BS Philosophy, Master of Philosophy. 2019–2020 Project coordinator ethics and economy, UCSIA (Antwerp), Belgium. Today, Values, Technology and Innovation Department, Delft University of Technology. Research areas: energy justice, environmental justice, energy ethics.

# Chapter 11
# The Role of Epistemic Fairness in Dynamics Models to Support Sustainable Mobility Diffusion

**Camilla Quaresmini, Eugenia Villa, Samira Maghool, Valentina Breschi, Viola Schaffonati, and Mara Tanelli**

**Abstract** Policy frameworks to promote the diffusion of sustainable mobility technologies are increasingly moving toward a data-driven approach. While they investigate incentive policies to foster the widespread diffusion of sustainable mobility technologies leveraging different data sources, these frameworks often fail to include issues of fairness. However, the acceptance of sustainable mobility technologies implies radical changes in the everyday life: besides transport habits, these changes are potentially hampered by socioeconomic individual features, that can raise potential discrimination. Focusing on a data-driven example of a social network of agents, we study how new mobility habits spread based on a combination of personal attitudes and mutual influence. Considering the epistemic dimension of the network, we address a specific type of discrimination, namely, the exclusion of some agents from their epistemic connections due to their own social features. We thus introduce the notion of *epistemic fairness* and consider it in the adoption dynamics. Our approach merges theoretical analysis with technical tools, jointly making them active means to create a preliminary model for exploring and analyzing different scenarios. The ultimate goal is to include elements of fairness in the design of the data-driven models supporting the creation of policy processes aiming at reducing inequalities within the context of sustainable mobility diffusion.

C. Quaresmini (✉) · E. Villa · V. Schaffonati · M. Tanelli (✉)
Dipartimento di Elettronica, Informazione e Bioingegneria, Politecnico di Milano, Milan, Italy
e mail: camilla.quaresmini@polimi.it; eugenia.villa@polimi.it; viola.schiaffonati@polimi.it; mara.tanelli@polimi.it

S. Maghool
Department of Computer Science, University of Milan, Milan, Italy
e-mail: samira.maghool@unimi.it

V. Breschi
Department of Electrical Engineering, Eindhoven University of Technolog, Eindhoven, The Netherlands
e-mail: v.breschi@tue.nl



C. Didier et al. (eds.), *Engineering and Value Change*, Philosophy of Engineering and Technology 48, https://doi.org/10.1007/978-3-031-83549-0_11

**Keywords** Fairness · Epistemic injustice · Opinion dynamics · Optimal control · Sustainable mobility

## 11.1 Introduction

In the study of technology adoption, whose dynamics is often influenced by incentive policies to be designed by governments and policymakers, the notion of fairness plays a fundamental role. In this context, fairness is usually intended as equality, meaning that everybody should get the same amount of resources in terms of incentives. For instance, to encourage the adoption and acceptance of sustainable technologies, especially new mobility solutions like electric vehicles or shared mobility services, economic incentives are traditionally provided through an equality-based approach. These incentives aim to alleviate the disruptive changes in daily life that these technologies might cause, thereby facilitating their broader adoption.[1] While the importance of adopting new mobility technologies is given for granted to promote sustainable mobility (e.g., by Gallo & Marinelli, 2020), the explicit consideration of fairness in the design of the policy framework is not always considered, even if it is essential to redress potential discrimination, as widely highlighted in the literature (Randal et al., 2020; Rode, 2022). Moreover, existing policy strategies often fail to consider fairness in terms of lack of inclusion: several approaches proposed in the literature lead to the consideration of fairness as reduced to distributive criteria, precisely to the classical *equality principle*, i.e., everyone is given the same amount of resources to achieve a given goal.

In this work, we adopt a different approach considering the epistemic dimension of fairness by introducing the notion of credibility. By focusing on a concrete example of a social network of agents, we study how new mobility habits spread based on a combination of personal attitudes and mutual influence. Modeling the interactions between agents by connecting them in a social network allows us to describe the influence of societal dictates and others' credibility on individual choices by leveraging opinion dynamics models for the diffusion of adoption within the network. We build upon our recent works (Villa et al., 2022, 2023) to propose a conceptual shift considering the epistemic component of the network, i.e., the dimension concerning the knowledge and expertise of the agents involved. We do this by reasoning about agents' individual credibility. Credibility is considered here as the actual influencing power attributed by agents to their network connections. Considering the epistemic dimension of the network allows us to highlight a type of discrimination that is not usually addressed in this context, that is the credibility deficit some agents suffer due to their personal socioeconomic features. This

[1] Different approaches in policy-making regarding these issues are listed in Gallo and Marinelli (2020), which proposes a review of the main actions and policies that can be implemented to promote sustainable mobility.

conceptual extension in our modeling framework is crucial for capturing some of the actual effects of social contagion on the diffusion process.

More specifically, we advance the hypothesis that agents' credibility has an effect on the adoption of sustainable mobility solutions in the considered network, being essential in maximizing agents' power of influence. Then we show that if some social groups are not discriminated against and, thus, they are not attributed less credibility than that they would deserve, the spread of adoption is faster, while some inequalities are reduced.

The chapter is structured as follows. In Sect. 11.1, we offer a methodological description of how the social network is built and provide a model for the opinion dynamics that drive the adoption of sustainable mobility solutions. In Sect. 11.2, we review the literature on fairness in models aimed at supporting policy-making for sustainable mobility, introducing the epistemic dimension of fairness in the modeling scenario. In Sect. 11.3, we present and discuss the results of simulations focused on the adoption of sustainable mobility solutions, by looking in particular to the role of the epistemic dimension in influencing adoption. In the last Section, we summarize the results, present some limitations, and lay the foundations for future research.

## 11.2 A Social Network to Model the Diffusion Process

Many studies analyze the features driving the adoption of sustainable mobility solutions at a qualitative level (Liao et al., 2017). On the other hand, studies that take a quantitative direction often neglect social aspects or are narrowly focused at the local level (Chen et al., 2020; Higueras-Castillo et al., 2020), making it challenging to carry out a meaningful cross-national comparison and effective supra-national policy-making strategies (Zhang et al., 2021). We rely on our previous works (Breschi et al., 2023; Villa et al., 2022, 2023) to propose a data-driven approach to identify the main features driving the adoption of sustainable solutions, in particular electric vehicles (EVs) and hybrid-electric vehicles (HEVs), on a large-scale population. Even if sustainable mobility is sometimes thought to be low mobility or car-free (see, e.g., Holden et al., 2020), we assume that EV and HEV are sustainable mobility solutions following (Bradley & Frank, 2009; Kumar & Alok, 2020; Popovich et al., 2021). Here we do not want to discuss whether or not those solutions are the most sustainable ones, but we want to immerse ourselves into the context of governmental policies for the promotion of electric mobility as a means to promote sustainability in general (Biresselioglu et al., 2018; Kazancoglu et al., 2021).

We start by quantitatively characterizing individual inclination toward the adoption of sustainable mobility solutions, including socioeconomic aspects alongside more standard mobility habits. To this end, we exploit public survey data investigating the demographic and social conditions and the mobility habits of a target population to learn classification models capable of interpreting individual attitudes toward the adoption of EV and HEV. This is obtained considering the individual socioeconomic characterization provided by the answers to the survey questions. In particular we analyze the features that can influence the classification outcomes, especially those determining the level of personal predisposition to adopt new mobility solutions. These features define a compact socioeconomic profile that we call *Mobility-DNA*, denoted by $M_v$ and associated to each respondent $v$, which can be easily mapped into the individual inclination toward the adoption of new sustainable mobility solutions (see Villa et al., 2022 for details on how the DNA components are defined).

Differently from individual features, the criteria defining agents' connections are strictly dependent on the application scenario and on the properties of the target population. Nonetheless, in a compact community, it is reasonable to presume that every individual has the potential to engage with others. Conversely, in larger networks encompassing diverse geographic regions, individuals are more likely to interact with a restricted set of peers. We focus here on the case of a community, i.e., a group of people who certainly influence each other. This is modeled via a completely connected graph $G = (V, E)$, where $V$ is the set of the graph's nodes, i.e., agents, $E$ is the set of the graph's edges. Note that we assume that there always exists a direct or indirect path between two nodes. A modeling assumption that we put forward for the network is that its connections are dictated by homophily. This means that agents' similarity is defined on the basis of their socioeconomic profiles, a tendency observed in real-world social networks (Golub & Jackson, 2008).

These modeling choices allow us to combine opinion dynamics models typically employed to study the diffusion of innovations (Acemoglu et al., 2011) with network dynamics, to study how new mobility habits could spread based on a combination of personal attitudes and mutual influence. Specifically, relying on an Irreversible Cascade Model (Granovetter, 1978), we assume that each agent $v$ at time $t$ can be either adopter (state $z_v(t) = 1$ or non-adopter (state $z_v(t) = 0$), remaining an adopter once turned into one. The adoption dynamics is thus described as:

$$z_v(t+1) = \begin{cases} 1, \frac{1}{|N_v|} \sum_{w \in N_v} z_w(t) \geq \rho_v(t), \\ 0, \text{otherwise} \end{cases} \tag{11.1}$$

where $N_v = \{j \in V : (j, v) \in E\}$ is the set of neighbors of the node $v \in V$ and $\rho_v = 1 - \text{Avg}(M_v)$ is the individual resistivity against new mobility solutions, with $\text{Avg}(M_v)$ being the average of the components associated to the Mobility-DNA features. Considering resistivity constant over time, this model allows to study the diffusion of new mobility solutions only triggered by social influence, when no external incentives are allocated.

The model in (11.1) is then augmented by considering the impact of external policies on individual opinions. In particular, we assume that external incentives (e.g., economic nudges) can be allocated to promote large-scale diffusion by reducing personal resistivity in adopting the new mobility solution. Being $\bar{U}$ the resource budget that must be shared among agents, each of them at time $t$ receives an incentive $u_v(t)$ that reduces the individual resistivity according to the following dynamics:

$$\rho_v(t+1) = \rho_v(t) - u_v(t) \tag{11.2}$$

with $\bar{U}$ representing the maximum resource budget at time $t$, i.e., $\sum_v u_v(t) \leq \bar{U}$. Starting from this model, one can leverage (optimal) control techniques (Ciocan & Farias, 2012) to design policies allocating available resources among agents in the target population. This approach does not consider any kind of fairness constraint. In other words, it does not consider the distributive norms in resource allocation to avoid discrimination: in the following section, instead, we explicitly introduce the dimension of fairness in the network.

## 11.3 Fairness in Policy-Making Models for Sustainability

Fairness plays an increasingly crucial role in the context of modeling strategies for policy-making in the specific field of energy sustainability (see Volodzkiene & Streimikiene, 2023 for a systematic review of recent works). Usually, the proposed approaches lead to the consideration of fairness and justice as directly related—and mainly reduced—to classical equality, i.e., individuals are given the same number of resources/opportunities needed to reach the goal. However, there is a lack of clarity on this point in the literature, as sometimes the described approaches explicitly refer to equity (i.e., individuals are given the exact number of resources/opportunities needed to reach the goal) but they ultimately foresee a uniform distribution of incentives across the population. For example, in Savvides (2013), a transport equity planning is proposed to counteract social inequity and exclusion, looking at transport efficiency, costs and individuals' mobility levels, i.e., accessibility levels across race, class, gender, and disability. In Tammaru et al. (2023), an equity-centered model of sustainable urban mobility is proposed, with different levels of resources distribution. In Randal et al. (2020), a conceptual framework of distributive justice is developed, with a set of principles to apply it to transport policy. Here, the equity principle is applied in terms of the *Capabilities Approach*, by considering resources as an input into an individual's capabilities rather than the primary focus. In Rode (2022), different concepts and notions of fairness are presented and discussed. Among others, social fairness is conceived in terms of needs-based equity, that is, the extent to which resource allocation is proportionate to individual needs. Other frameworks are proposed in the context of energy justice (Jenkins et al., 2016), such as recognition and procedural ones. Those approaches consider justice respectively as the correct representation and access to decision-making processes.

While justice and fairness are often intertwined in these discussions, it is important to recognize that they are two distinct concepts, even if this distinction is conceptualized in different ways (see e.g., Goldman & Cropanzano, 2015; Kuppler et al., 2021; Lippert-Rasmussen, 2023). In this work, for the sake of simplicity and following Lackey (2017), we consider justice as understood in terms of fairness.

More specifically, we extend the notion of fairness in an epistemic direction. While emphasis is usually laid on the ethical dimension of fairness (Heidari et al., 2019; Lee et al., 2021), in the sense of the morality of the concepts and the values behind them, the epistemic dimension of fairness has been sparsely considered (Hull, 2023; Nihei, 2022). However, we claim that in a context where agents share information, it is unrealistic to ignore the epistemic dimension of the network, implicitly assuming complete and equal confidence among all the agents. Indeed, while deciding whether to rely upon other agents' ability to adopt a sustainable mobility solution, agents weigh their choices also on some epistemic elements, such as how much agents are reliable and credible in terms of the knowledge they are capable of sharing. Thus, it is crucial to admit that fairness is not only an ethical issue but also an epistemic one. This also implies practical and political effects: if some individuals are epistemically excluded in a systematic way, that is they are deemed as less credible because of some of their social attributes (e.g., gender), this injustice will turn, for instance, into an effective exclusion from the diffusion of sustainable mobility technologies.

In this chapter, we consider the epistemic dimension in terms of credibility. Credibility entitles some agents as knowers while it deauthorizes others (Jones, 2018). Thus our proposal can be considered in line with the recognition justice approach summarized in Jenkins et al. (2016), where injustice is conceptualized in terms of a devaluation of some identities that can be connected to a knowledge deficit.

### 11.3.1 Epistemic Dimension

Our epistemic characterization of the notion of fairness has its philosophical foundations in the work of Miranda Fricker and her conceptualization of *epistemic injustice* (Fricker, 2007). The starting point is the epistemic charge of agents' social features, that have a significant impact on the credibility of the agents with those characteristics. This leads to an identity power that has effects on the overall epistemic system, creating an imbalance in trust relations. According to Fricker, epistemic injustice takes two forms: *testimonial injustice*, i.e., marginalized individuals are less believed because of their identity (e.g., during a trial the judge does not believe the testimony of the defendant because she is a dark-skinned woman, thus attributing her a credibility deficit); and *hermeneutical injustice*, i.e., marginalized individuals are not considered capable of creating concepts or experiences due to a lack of expressions that can define their experience (e.g., before the introduction of the label "sexual harassment" female victims may have had difficulty conceptualizing their experience).

Some attempts to investigate the epistemic dimension of fairness in terms of epistemic injustice have been proposed (Medvecky, 2018). For example, in Kim

(2022), the so-called *epistemic fairness* is defined considering the comparative nature of credibility attribution. This means that relative bias may arise as an excess of credibility attributed to one group which therefore will always remain epistemically superior to another one. Moreover, the criterion of *information unfairness*, which evaluates whether information spreads fairly to all groups in a network, is introduced in Jalali et al. (2020). Formal concepts of epistemic injustice (Bovens & Hartmann, 2004; Marcotte, 2023) have been already applied to a network context (Spiekermann, 2020), revealing *epistemic network injustice*, that is the inability for disadvantaged agents to find individuals with the same interests. The concept has also been linked to sustainability in terms of fair treatment in knowledge-related practices for sustainable development (Cummings et al., 2023).

In our model, epistemic fairness is introduced focusing on credibility, which is at the origin of epistemic injustice (Fricker, 2007). For modeling simplicity, we reduce epistemic injustice to the case of testimonial injustice in trust relations. Here, unfairness arises in terms of epistemic exclusions determined by the identity power of the nodes, because there is a credibility deficit linked to some social features. For example, it could be the case that a dark-skinned woman suggests to a white-skinned man to adopt an electric vehicle, but he does not trust her because of her social features. Furthermore, while being aware that epistemic prejudices can also be positive in terms of credibility excess attribution, we are considering the specific case of credibility deficit as we consider it more relevant to our particular case.

### 11.3.2 A Model for Credibility

In this subsection, we define individual credibility in a formal way to introduce it into the existing model. We are aware that credibility is mainly a relational characteristic, as the attribution of the credibility deficit requires a hearer who attributes it and a speaker who receives this attribution (Fricker, 2007). However, for the sake of simplicity here, we assume credibility as an individual characteristic.

**Definition 1 (Individual Credibility)** *A characteristic $\gamma_v$ of agent $v \in V$ that dictates how much the agent v is trusted in general by any other agent in the network.*

Despite being an individual feature, agents' credibility influences the opinion of each individual in a relational way. Indeed, considering the community network scenario described in Sect. 11.2, we enter credibility in the adoption dynamics dictated by the model in a way such that if an agent is not credible, their actions will not influence those of their peers. Formally, we introduce credibility by weighting the opinions of neighbors dictating the switch of the $v$-th agent's state $z_v(t)$ as follows:

$$z_v(t+1) = \begin{cases} 1, \dfrac{\sum_{j \in N_v} \gamma_j z_j}{\sum_{j \in N_v} \gamma_j} \geq \rho_v, \\ 0, \text{otherwise} \end{cases} \tag{11.3}$$

with $\gamma_v \in [0, 1]$, for all $v \in V$. The state dynamics described above shows that an agent decides to become an adopter (or not) based on the comparison between the ratio of nearby adopters to the total number of nearby agents, weighted by their influence $\gamma_j$, and the individual resistivity, i.e., the personal reluctance to adopt new mobility solutions. That is: if the proportion of the neighbors of agent $v$ who do adopt exceeds $v$'s resistivity and these adopters are sufficiently credible, then $v$ becomes an adopter herself.

According to Fricker, subjects who suffer epistemic injustice are perceived as less credible than they ought to be. This means that the reliability of some agents is undermined by their belonging to some categories. For example, let us assume that in a specific network, all experts are considered reliable. It might be the case that, when some experts are women, they are considered less credible, thus suffering a credibility deficit due to their being women. In other words, we introduce a distinction between reliability and credibility, interpreting the condition of how credible agents should be if there is no prejudice as to their reliability, while the condition of how credible agents actually are as to their credibility.

**Definition 2 (Individual Reliability)** *A characteristic $r_v$ of agent $v \in V$ dictating how much the agent $v$ should be trusted in general by any other agent in $V$, if there was no prejudice.*

While reliability (e.g., how much an agent should be potentially believed in an ideal and unbiased scenario) is the initial condition for the credibility dynamics, credibility (e.g., how much an agent is actually believed because of their social features, and independently of being reliable in terms of their expertise) can evolve over time and actually weights the agents' opinion in real-world scenarios. Thus, in this context, we conceptualize the credibility deficit as a quantity that weakens the agent's reliability. Therefore, we want to bridge the gap between reliability and credibility, making actually credible the nodes that should be credible.

According to the distinction between reliability and credibility, let $\Delta_{r,\gamma}$ represent the credibility deficit, i.e., the gap between reliability and credibility satisfying

$$r_v = \gamma_v + \Delta_{r,\gamma}, \forall v \in V \tag{11.4}$$

Moreover, let $R$ be the subset of reliable agents in $V$. We can now define the concept of epistemic fairness for our network and a given task as follows.

**Definition 3 (Epistemic Fairness)** *A social network G is considered epistemically fair, denoted as EFair(G), iff* $v \in V$ *is provided with an amount of credibility* $\gamma_v$ *s. t.* $\gamma_v$ *is sufficient for v not being wronged in their capacity as a knower, namely* $\gamma_v = r_v, \forall v \in R \subseteq V$.

This means that, for our network to be epistemically fair, reliable agents must reach the credibility value they would have if there was no prejudice $\Delta_{r,\gamma}$.

## 11.4 An Example of Electric Mobility

We now present an example focused on the diffusion of electric mobility in a metropolitan area in Northern Italy. The presented case study is strictly dependent on the data we use, thus implying the contextuality of the approach. This allows us to investigate the impact of epistemic unfairness in models supporting the design of policies promoting the diffusion of EVs, while indicating that our results might not always be generalizable.

By applying the model discussed in Sect. 11.2 to the example, our aim is to investigate the hypothesis that agents' credibility has an impact on the adoption diffusion in the network and has a role in maximizing agents' power of influence. Specifically, we analyze how credibility deficit causes a slowdown in the adoption, thus making the model not only unfair but also less efficient. If the hypothesis is verified, considering credibility in design policies could have a positive impact on the optimal distribution of resources, resulting in savings on incentives distributed to foster the adoption.

### *11.4.1 From Data to Mobility-DNA*

To identify the main socioeconomic features influencing individual attitudes toward electric mobility, we leverage the survey data collected at the European level (Fiorello & Zani, 2015[2]). The survey explores personal information and mobility patterns within a population of more than 26,000 citizens evenly distributed across European countries. The survey's most relevant questions are presented in Table 11.1.

We identified the query "Consider purchasing an EV or HEV" as revealing the attitude regarding the adoption of electric vehicles. The distribution of the target question in the interviewed population is shown in Fig. 11.1.

We focus on individuals with a clear attitude (positive or negative) toward the adoption of electric mobility, discarding all respondents providing other answers.

[2] Data are available at https://doi.org/10.7910/DVN/P82V9X

**Table 11.1** Attributes available from the European survey (Fiorello & Zani, 2015), distinguished based on their type, range of possible values, and encoding

| Attribute | Type | Values | Cat. | Encoding |
|---|---|---|---|---|
| Country | Cat | 28 | Geo | Label |
| Gender | Cat | 2 | Bio | Binary |
| Age | Num | [16,84] | Bio | Ordinal |
| Education | Cat | 4 | Edu | Ordinal |
| Region | Cat | 319 | Geo | Label |
| Profession | Cat | 14 | Prof | Label |
| Work status | Cat | 2 | Prof | Label |
| Household members | Cat | 7 | Fam | Ordinal |
| Income level | Cat | 6 | Fam | Ordinal |
| Location of residence | Cat | 4 | Geo | Ordinal |
| Centre or suburbs | Cat | 2 | Geo | Binary |
| Public transport service | Cat | 4 | Mob | Ordinal |
| Car driving license | Cat | 2 | Mob | Binary |
| Number vehicles in household | Int | [0,10] | Fam | Numerical |
| Consider buying EV or HEV | Cat | 6 | Mob | Label |
| Know what car sharing is | Cat | 3 | Mob | Binary |
| Would subscribe car sharing | Cat | 7 | Mob | Binary |
| Destination most freq trip | Cat | 3 | Mob | Ordinal |
| Frequency most freq trip | Cat | 3 | Mob | Ordinal |
| Freq trip duration [min] | Num | [0,775] | Mob | Ordinal |
| Freq trip distance | Cat | 7 | Mob | Ordinal |
| Concern environmental impacts | Cat | 11 | Env | Ordinal |
| Preference tolls or traffic | Cat | 5 | Mob | Ordinal |

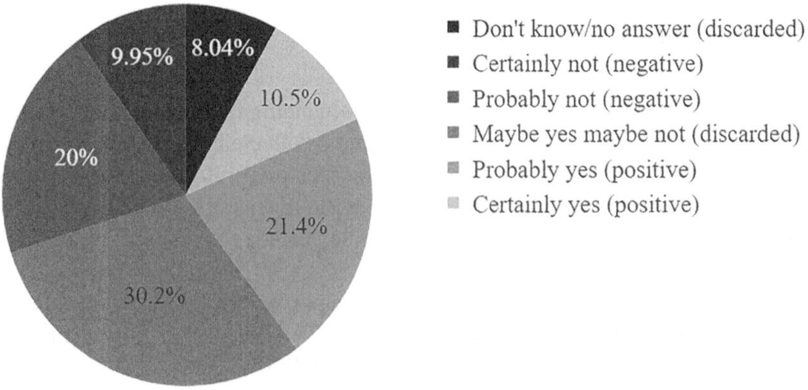


**Fig. 11.1** Pie chart showing the distribution of the target question "Consider purchasing an EV or HEV" in the interviewed population

We train in a supervised fashion a classification model to recognize positive or negative attitudes toward the adoption of EVs based on the available socioeconomic information listed in Table 11.1. The learned model is used as a tool to correlate the main socioeconomic features with the attitude toward the adoption. These are indeed identified by applying ad hoc algorithms, such as Permutation Importance (Altmann et al., 2010), that aim at ranking the available attributes based on their importance in correctly classifying individual attitudes. This allows us to identify seven primary features influencing the target: environmental awareness, income level, age, education, profession, mobility habits, and country. The choice of these features is further sustained by the extensive literature (Coffman et al., 2017; Egbue & Long, 2012).

Based on the answers provided by each respondent to these features we create the *Mobility-DNA*, a representation of the individual predisposition toward innovative and green mobility modes. Each component assumes a value in $[0, 1]$ and represents the empirical likelihood that each agent is positively inclined toward the adoption according to each specific feature. Based on these computations, each agent $v \in V$ is characterized by the Mobility-DNA $M_v$. From this indicator, we can obtain the initial resistivity toward adoption as $\rho_v(0) = 1 - Avg(M_v)$, where $Avg(M_v)$ is the average of the seven components constituting $M_v$ see (Villa et al., 2022 for details).

### 11.4.2 Social Network and Epistemic Dimension

We now construct the social network that represents the mutual influences within the target population: we rely on country, region, and location of residence reported in the survey to select the subset of respondents living in a large city in the region of Lombardy, Italy (Villa et al., 2023). This subset includes 40 individuals that we assume to belong to a compact community in the metropolitan city of Milan. For relatively small communities, we can assume that all agents potentially interact with everyone in the network. This allows us to define the fully connected graph $G = (V, E)$ dictating agents interactions.

Besides personal resistivity $\rho_v(0)$, each agent in the graph is characterized by additional properties related to the epistemic dimension, i.e., reliability and credibility. We assume that in the present scenario, reliability can be represented by the individual expertise level in matters related to EVs. It is therefore defined based on the educational level and the environmental awareness declared by each agent in the survey. Since the allocation of credibility depends on the attribution of reliability, deciding who is reliable is a crucial choice. Note that this choice is strictly guided by the limited information provided by the survey we are considering. Different alternatives could be explored in determining individual reliability, especially when more comprehensive information is available in the data. For instance, reliability could be determined not only by education and environmental awareness but also by individual behaviors and choices related to a particular context. Starting from

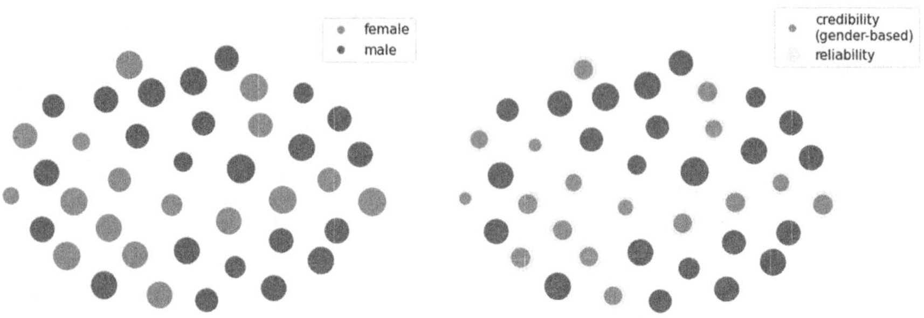


**Fig. 11.2** (**a**) The situation in an ideal scenario. The nodes' power of influence corresponds to their reliability. (**b**) The situation in a real-world scenario. Due to the presence of the credibility deficit, the nodes' power of influence corresponds to their (gender-based) credibility. Social network in the two considered scenarios. Node color denotes gender: light-blue nodes are male, and orange nodes are female. Node size is proportional to the node power of influence, which corresponds to individual reliability in the first case and to gender-based credibility in the second one

reliability, individual credibility is established by a discrimination factor, which is initially defined solely by gender. Therefore, we assume that being part of the set of female agents means being in the marginalized class, that is the class that suffers a credibility deficit (Fig. 11.2).

Note that agents who suffer a credibility deficit are a subset of reliable agents. That is, in the unfair scenario coinciding with the real-world situation, it is shown how effectively the influence power of female experts is reduced due to epistemic discrimination.

We determine the discrimination factor through an intersectional approach (Atrey, 2019) and base our definition on the assumption that in social contexts there exist multiple levels of discrimination that an individual may experience due to the intersection of various aspects of their identity (e.g., age, gender, and income level) that do not operate independently but intersect. In the context of electric mobility, we assume the following forms of discrimination (Potter et al. 2019):

- Gender-based discrimination: female individuals face discrimination.
- Age-based discrimination: young individuals face discrimination.
- Income-based discrimination: individuals with low-income levels face discrimination.

Being part of one or more of these marginalized groups determines an increasing discrimination factor $\phi_v$, ultimately affecting the reliability of the agents, reducing it up to their credibility. We assume that being part of one discriminated group halves the reliability of the agent. While this is a design choice, there are different ways to define how social features influence individual credibility, all without losing the generality of the overall framework. Note that this intersectional assumption makes the individual credibility potentially dynamic, as it depends on transitional features.

By multiplying the discrimination factor $\phi_v$ by the reliability $r_v$, we obtain the credibility $\gamma_v$ of the agent. Namely,

$$\phi_v = 0.5^{\sum_g d_v^g} \tag{11.5a}$$

where $d_v^g$ is a boolean value being 1 if agent $v$ belongs to the discriminated group $g$, thus leading to:

$$\gamma_v = \phi_v \cdot r_v \tag{11.5b}$$

Table 11.2 shows the individual characteristics associated with three different agents in our target population. Gender, age, education, and income columns report binary values denoting whether each agent belongs to the discriminated group for the considered feature. The discrimination factor is calculated by assigning a halving factor for each marginalized group to which the agent belongs.

In Fig. 11.3, node sizes are adjusted to reflect agent reliability in the context of epistemic fairness (Fig. 11.3a) or credibility in the context of epistemic unfairness (Fig. 11.3b).

Comparing Figs. 11.2b and 11.3b, the impact of credibility deficit on the community is higher in the intersectionally unfair scenario. This is because discrimination here operates on different features, all affecting the credibility of the individual.

### 11.4.3 Results and Discussion

The presented framework enables the examination of EV diffusion within the target population by incorporating the epistemic dimension in the form of credibility and considering the effects of credibility deficit. In particular, we focus on comparing two scenarios. The first situation accounts for intersectional discrimination and thus assumes that the social power of each agent is determined by their credibility, limited with respect to their full reliability due to the deficit they suffer from. The second situation idealistically assumes a scenario where epistemic fairness is reached: no discrimination affects the social power of individuals, and hence all agents interact among each other with their full reliability. In both scenarios we simulate a ten-year policy to promote the diffusion of EVs under yearly budget constraints, assuming that the maximum number of resources available at each time step in the simulation is $\bar{U} = 2$, representing the economic budget available to policymakers for designing their incentive plans. The optimal distribution of these incentives

**Table 11.2** Attributes characterizing agents in the network accounting for individual attitude toward the adoption and epistemic properties

| | Resistivity $\rho_v$ (0) | Reliability $r_v$ | Gender | Age | Income | Intersectional $\phi_v$ | Credibility $\gamma_v$ |
|---|---|---|---|---|---|---|---|
| Agent 1 | 0.67 | 0.8 | 0 | 1 | 0 | 0.5 | 0.4 |
| Agent 2 | 0.45 | 0.6 | 1 | 0 | 1 | 0.25 | 0.15 |
| Agent 3 | 0.13 | 0.7 | 1 | 1 | 1 | 0.125 | 0.0875 |

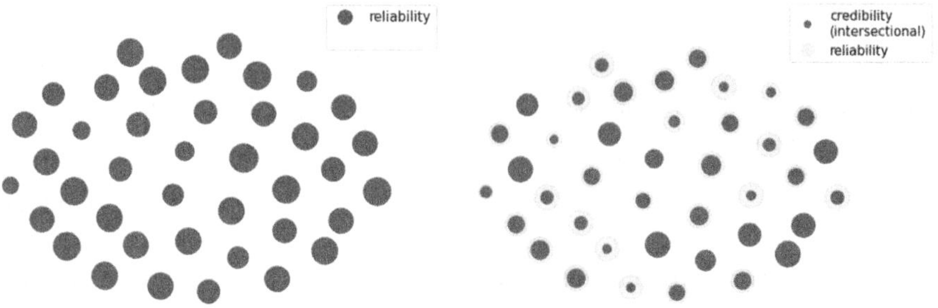


**Fig. 11.3** (**a**) The situation in an ideal scenario. The nodes' power of influence corresponds to their reliability. (**b**) The situation in a real-world scenario. Due to the presence of the credibility deficit, the nodes' power of influence corresponds to their (intersectional) credibility. Social network in the two considered scenarios. Node size is proportional to the node's social power, i.e., individual reliability in the first case and intersectional credibility in the second

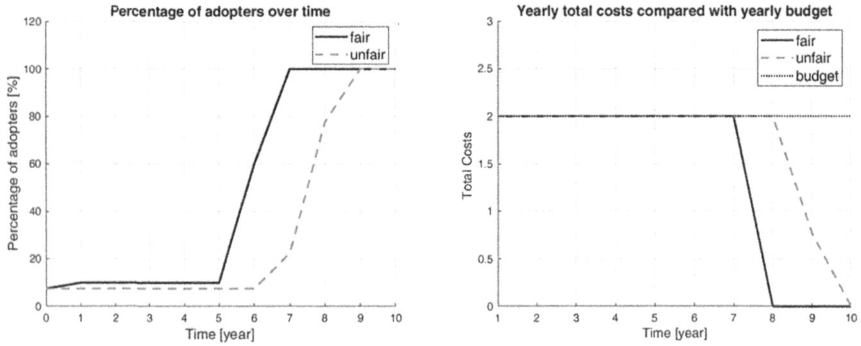


**Fig. 11.4** (**a**) The evolution of the percentage of adopters in the network in the unfair scenario (when credibility deficit is considered) and in the fair scenario (when epistemic fairness is assumed). (**b**) The total amount of resources employed at each time step in the simulation horizon of 10 years. Simulations' results comparing fair and unfair scenarios

among the agents in the target population is determined via an optimal control strategy with a two-fold optimization objective: maximize the diffusion while minimizing policy costs. More details on the choice of the optimization strategy are provided in Villa et al. (2023).

The simulation results for the two considered scenarios are shown in Fig. 11.4.

Figure 11.4a shows the evolution of the percentage of adopters in the network in the unfair scenario (when the credibility deficit is considered) and in the fair scenario (when epistemic fairness is assumed). Similarly, Fig. 11.4b shows the total number of resources employed at each time step in the simulation horizon of 10 years. These basic results intuitively show that epistemic unfairness slows down the diffusion of EVs. Indeed, in the idealistic scenario of epistemic fairness, full-scale adoption is reached at time $t = 7$, while accounting for social discrimination the same result is achieved with a delay of 2 time-steps (2 years based on our simulation assumptions). Furthermore, Fig. 11.4b highlights that in the unfair scenario, higher policy costs are required to foster the adoption. While this result easily follows from our definition of credibility, as a credibility deficit decreases the power of connections, the analysis of this fact is neither obvious nor trivial and changes with the topology of the network. In fact, the credibility deficit impacts both the evolution of the cascade process and the allocation of resources.

These results show a general negative impact of epistemic unfairness in the diffusion of sustainable mobility habits. In the case of social discrimination, the capability of the nodes to influence other agents toward the adoption is reduced. This limits the positive effect of social contagion in the diffusion process, ultimately requiring higher efforts from policymakers to promote the diffusion process. Therefore, fostering policy results in being less effective, when social discrimination exists, requires higher costs and longer times to reach the prefixed diffusion goals. These results highlight the potential benefits of mitigating epistemic unfairness in some social contexts in terms of both the final diffusion of sustainable mobility habits and cost savings.

## 11.5 Conclusion

In this work, we have investigated the impact of credibility deficit on diffusion and costs of adoption of sustainable mobile technologies. In particular, we have analyzed how adoption would be faster if some classes of individuals were not epistemically discriminated against. This work stems from the awareness that data-driven models used to support policy decision-making can have a negative impact on already disadvantaged populations if fairness principles are neglected during the design process. By introducing the notion of credibility, we have aimed at mitigating the epistemic unfairness that might arise in agents' networks (D'Alessandro et al., 2017).

The proposed model is part of a larger work devoted at the integration of philosophy and engineering in devising formal models for exploring different policy strategies to promote the adoption of sustainable mobility solutions. These models not only aim at integrating epistemic fairness, but also at investigating its role in the processes of adoption. In accordance with a well-established tradition (see O'Connor, 2019), we use formal models to explore different (epistemically fair and unfair) scenarios. The advantage of using formal models to explore conceptual possibilities (Mayo-Wilson & Zollman, 2021) is to go beyond empirical investigation, which is fundamental but static, and as such fails to capture any kind of evolution (O'Connor, 2019). We understand that the models we use are partial and based on debatable assumptions, yet we believe they are necessary to explore the unfairness that can arise from the relations we investigate in order to prevent it. We are aware that epistemic fairness, and in particular the non-discriminatory meaning of fairness we have adopted in this chapter, is only a minor part of a wide philosophical discussion. However, we have argued why it is important to extend the concept in an epistemic direction to capture types of discrimination which are usually neglected in the context of adoption dynamics.

This work presents some limitations that will be addressed in future research. First, the role of homophily in the design of the communication network and its being often associated with unfairness (Loveland et al., 2022), as it implies the segregation of social contacts (Currarini & Mengel, 2016). Bias, however, could be also a positive feature toward homophily: if two nodes share a feature, then the bias between them should be smaller than it would be if they didn't share the feature. For instance, two women could not attribute a credibility deficit to each other. In the actual state of the network, however, we consider only the simplified case where the nodes subject to bias also carry it, and have a positive bias in themselves.

Second, the notion of credibility must be extended in a relational way. This solution is not only more accurate from a conceptual point of view but allows us to consider the problem of positive bias in homophily.

Third, the assumption that reliable nodes are the experts, albeit plausible, can be problematic, especially if economic incentives are given: the risk could be of helping those who are already privileged. Furthermore, deciding who is an expert implies a considerable degree of arbitrariness. This questions the decision on which agents are reliable, opening the possibility of other scenarios, such as the case in which reliable nodes are decided randomly. Future work will thus be devoted to reason about the distributive norms governing the fair allocation of the epistemic characteristics in the network (Quaresmini et al., 2023). Other extensions will cover the investigation of more complex models for opinion dynamics (Bass, 1969; Laciana & Rovere, 2011), allowing us to introduce a stochastic component for a more realistic representation of the diffusion process and for reasoning about different similarity measures (see, for example, Maghool et al., 2023).

## References


Acemoglu, D., Ozdaglar, A., & Yildiz, E. (2011). Diffusion of innovations in social networks. In *Paper presented at the 50th IEEE conference on decision and control and European control conference*.

Altmann, A., Toloşi, L., Sander, O., & Lengauer, T. (2010). Permutation importance: A corrected feature importance measure. *Bioinformatics, 26*(10), 1340–1347.

Atrey, S. (2019). *Intersectional discrimination*. Oxford University Press.

Bass, F. M. (1969). A new product growth for model consumer durables. *Management Science, 15*(*5*), 215–227.

Biresselioglu, M. E., Kaplan, M. D., & Yilmaz, B. K. (2018). Electric mobility in Europe: A comprehensive review of motivators and barriers in decision making processes. *Transportation Research Part A: Policy and Practice, 109*, 1–13.

Bovens, L., & Hartmann, S. (2004). *Bayesian epistemology*. Oxford University Press.

Bradley, T. H., & Frank, A. A. (2009). Design, demonstrations and sustainability impact assessments for plug-in hybrid electric vehicles. *Renewable and Sustainable Energy Reviews, 13*(*1*), 115–128.

Breschi, V., Ravazzi, C., Strada, S., Dabbene, F., & Tanelli, M. (2023). Driving electric vehicles' mass adoption: An architecture for the design of human-centric policies to meet climate and societal goals. *Transportation Research Part A: Policy and Practice, 171*, 103651.

Chen, C. F., de Rubens, G. Z., Noel, L., Kester, J., & Sovacool, B. K. (2020). Assessing the socio-demographic, technical, economic and behavioral factors of Nordic electric vehicle adoption and the influence of vehicle-to-grid preferences. *Renewable and Sustainable Energy Reviews, 121*, 109692.

Ciocan, D. F., & Farias, V. (2012). Model predictive control for dynamic resource allocation. *Mathematics of Operations Research, 37(3)*, 501–525.

Coffman, M., Bernstein, P., & Wee, S. (2017). Electric vehicles revisited: A review of factors that affect adoption. *Transport Reviews, 37(1)*, 79–93.

Cummings, S., Dhewa, C., Kemboi, G., & Young, S. (2023). Doing epistemic justice in sustainable development: Applying the philosophical concept of epistemic injustice to the real world. *Sustainable Development, 31(3)*, 1965–1977.

Currarini, S., & Mengel, F. (2016). Identity, homophily and in-group bias. *European Economic Review, 90*, 40–55.

D'Alessandro, B., O'Neil, C., & LaGatta, T. (2017). Conscientious classification: A data scientist's guide to discrimination-aware classification. *Big Data, 5(2)*, 120–134.

Egbue, O., & Long, S. (2012). Barriers to widespread adoption of electric vehicles: An analysis of consumer attitudes and perceptions. *Energy Policy, 48*, 717–729.

Fiorello, D., & Zani, L. (2015). *EU Survey on issues related to transport and mobility*. Publications Office of the European Union.

Fricker, M. (2007). *Epistemic injustice: Power and the ethics of knowing*. Oxford University Press.

Gallo, M., & Marinelli, M. (2020). Sustainable mobility: A review of possible actions and policies. *Sustainability, 12(18)*, 7499.

Goldman, B., & Cropanzano, R. (2015). "Justice" and "fairness" are not the same thing. *Journal of Organizational Behavior, 36*(2), 313–318.

Golub, B., & Jackson, M. O. (2008). *How homophily affects diffusion and learning in networks*. arXiv preprint arXiv:0811.4013.

Granovetter, M. (1978). Threshold models of collective behavior. *American Journal of Sociology, 83(6)*, 1420–1443.

Heidari, H., Loi, M., Gummadi, K. P., & Krause, A. (2019, January). A moral framework for understanding fair ML through economic models of equality of opportunity. In *Proceedings of the conference on fairness, accountability, and transparency* (pp. 181–190).

Higueras-Castillo, E., Molinillo, S., Coca-Stefaniak, J. A., & Liebana-Cabanillas, F. (2020). Potential early adopters of hybrid and electric vehicles in Spain—Towards a customer profile. *Sustainability, 12(11)*, 4345.

Holden, E., Banister, D., Gössling, S., Gilpin, G., & Linnerud, K. (2020). Grand narratives for sustainable mobility: A conceptual review. *Energy Research & Social Science, 65*, 101454.

Hull, G. (2023). Dirty data labeled dirt cheap: Epistemic injustice in machine learning systems. *Ethics and Information Technology, 25(3)*, 38.

Jalali, Z. S., Wang, W., Kim, M., Raghavan, H., & Soundarajan, S. (2020). On the information unfairness of social networks. In *Proceedings of the 2020 SIAM International Conference on Data Mining* (pp. 613–521). Society for Industrial and Applied Mathematics.

Jenkins, K., McCauley, D., Heffron, R., Stephan, H., & Rehner, R. (2016). Energy justice: A conceptual review. *Energy Research & Social Science, 11*, 174–182.

Jones, K. (2018). The politics of credibility. In *A mind of one's own* (pp. 154–176). Routledge.

Kazancoglu, Y., Ozbiltekin-Pala, M., & Ozkan-Ozen, Y. D. (2021). Prediction and evaluation of greenhouse gas emissions for sustainable road transport within Europe. *Sustainable Cities and Society, 70*, 102924.

Kim, J. (2022). Democracy and epistemic fairness: Testimonial justice as a founding principle of aggregative democracy. *The Southern Journal of Philosophy, 60*(2), 173–193.

Kumar, R. R., & Alok, K. (2020). Adoption of electric vehicle: A literature review and prospects for sustainability. *Journal of Cleaner Production, 253*, 119911.

Kuppler, M., Kern, C., Bach, R. L., & Kreuter, F. (2021). *Distributive justice and fairness metrics in automated decision-making: How much overlap is there?* arXiv preprint arXiv:2105.01441.

Laciana, C. E., & Rovere, S. L. (2011). Ising-like agent-based technology diffusion model: Adoption patterns vs. seeding strategies. *Physica A: Statistical Mechanics and Its Applications, 390(6)*, 1139–1149.

Lackey, J. (2017). Norms of credibility. *American Philosophical Quarterly, 54(4)*, 323–337.

Lee, M. S. A., Floridi, L., & Singh, J. (2021). Formalising trade-offs beyond algorithmic fairness: Lessons from ethical philosophy and welfare economics. *AI and Ethics, 1(4)*, 529–544.

Liao, F., Molin, E., & van Wee, B. (2017). Consumer preferences for electric vehicles: A literature review. *Transport Reviews, 37(3)*, 252–275.

Lippert-Rasmussen, K. (2023). Using (un)fair algorithms in an unjust world. *Res Publica, 29(2)*, 283–302.

Loveland, D., Zhu, J., Heimann, M., Fish, B., Schaub, M. T., & Koutra, D. (2022). *On graph neural network fairness in the presence of heterophilous neighborhoods*. arXiv preprint arXiv:2207.04376.

Maghool, S., Casiraghi, E., & Ceravolo, P. (2023, October). Enhancing fairness and accuracy in machine learning through similarity networks. In *International conference on cooperative information systems* (pp. 3–20). Springer Nature Switzerland.

Marcotte, M. (2023). *Using formal epistemology to model epistemic injustice against neurodivergent people* (Doctoral dissertation, The University of Western Ontario (Canada)).

Mayo-Wilson, C., & Zollman, K. J. (2021). The computational philosophy: Simulation as a core philosophical method. *Synthese, 199(1)*, 3647–3673.

Medvecky, F. (2018). Fairness in knowing: Science communication and epistemic justice. *Science and Engineering Ethics, 24*(5), 1393–1408.

Nihei, M. (2022). Epistemic injustice as a philosophical conception for considering fairness and diversity in human-centered AI principles. *Interdisciplinary Information Sciences, 28(1)*, 35–43.

O'Connor, C. (2019). *The origins of unfairness: Social categories and cultural evolution*. Oxford University Press.

Popovich, N. D., Rajagopal, D., Tasar, E., & Phadke, A. (2021). Economic, environmental and grid-resilience benefits of converting diesel trains to battery-electric. *Nature Energy, 6(11)*, 1017–1025.

Potter, L., Zawadzki, M. J., Eccleston, C. P., Cook, J. E., Snipes, S. A., Sliwinski, M. J., & Smyth, J. M. (2019). The intersections of race, gender, age, and socioeconomic status: Implications for reporting discrimination and attributions to discrimination. *Stigma and Health, 4(3)*, 264.

Quaresmini, C., Villa, E., Breschi, V., Schiaffonati, V., & Tanelli, M. (2023). Qualification and quantification of fairness for sustainable mobility policies. In *2nd European Workshop on Algorithmic Fairness, EWAF 2023* (p. 29). CEUR-WS.org.

Randal, E., Shaw, C., Woodward, A., Howden-Chapman, P., Macmillan, A., Hosking, J., Chapman, R., Waa, A. M., & Keall, M. (2020). Fairness in transport policy: A new approach to applying distributive justice theories. *Sustainability, 12(23)*, 10102.

Rode, P. (2022, March 14). *Enabling sufficiency: Towards an actionable concept of fairness in mobility and accessibility. MyFairShare-individual mobility budgets as a foundation for social and ethical carbon reduction*. Working Paper.

Savvides, A. (2013). Equity of mobility in sustainable transportation. *WIT Transactions on Ecology and the Environment, 179*, 1091–1101.

Spiekermann, K. (2020). Epistemic network injustice. *Politics, Philosophy & Economics, 19*(1), 83–101.

Tammaru, T., Sevtsuk, A., & Witlox, F. (2023). Towards an equity-centred model of sustainable mobility: Integrating inequality and segregation challenges in the green mobility transition. *Journal of Transport Geography, 112*, 103686.

Villa, E., Preti, G., Riva, M., Breschi, V., & Tanelli, M. (2022). Sharing-DNA: A data-driven tool to map the attitude towards sharing services across Europe. In *2022 IEEE International Smart Cities Conference (ISC2)* (pp. 1–7). IEEE.

Villa, E., Breschi, V., Ravazzi, C., Dabbene, F., & Tanelli, M. (2023). Fostering the use of sharing mobility solutions via control-oriented policy design. *IFAC-PapersOnLine, 56*(2), 1–6.

Volodzkiene, L., & Streimikiene, D. (2023). Energy inequality indicators: A comprehensive review for exploring ways to reduce inequality. *Energies, 16*(16), 6075.

Zhang, R., Zhang, J., Long, Y., Wu, W., Liu, J., & Jiang, Y. (2021). Long-term implications of electric vehicle penetration in urban decarbonization scenarios: An integrated land use–transport–energy model. *Sustainable Cities and Society, 68*, 102800.

**Camilla Quaresmini** is a PhD student in Science, Technology and Policy for Sustainable Change at Politecnico di Milano. She earned a Degree in Philosophy at Alma Mater Studiorum, Università di Bologna and a Master's Degree in Philosophical Sciences at Università degli Studi di Milano. Her research interests are logic and formal methods to study algorithmic fairness in AI systems.

**Eugenia Villa** is a PhD student in Data Analytics and Decision Sciences at Politecnico di Milano. She received her B.Sc. and M.Sc. in Mathematical Engineering from Politecnico di Milano (Italy) in 2018 and 2020 respectively. During her M.Sc., she focused her studies on modeling and data analysis discussing a thesis on statistical analysis and modeling of human-robot interactions. Her current research work is concerned with the design of diversity-aware, human-centered policy for the promotion of inclusive and sustainable technological solutions exploiting a data-driven and control-oriented approach.

**Samira Maghool** BS Physics, Master of Condensed Matter Physics, PhD Complex Systems (2020). 2023-Today Assistant Professor in Computer Science, University of Milan, Italy. Author of several published papers in Scientific Reports, ACM, IEEE, Springer, Big Data, JASSS, WWW, PLoS One, Complex and Intelligent Systems journal and, participated in multiple international conferences, member of SEcure Service-oriented Architectures Research (SESAR) Lab. Team member of Human Hall project on inclusive Artificial Intelligence. Research areas: Complex networks, Graph Representation Learning and Embedding, Epidemic spreading, Computational Social Science, Explainable Artificial Intelligence, Modelling complex phenomena as agent-based simulations and, FAIR and non-discriminative decision-making process using Machine Learning models. Co-supervising and collaborating in several research projects such as: SMART BEAR European project H2020, MindFoodSHub, UNIMI SEED 2020 Project POPULITE.

**Valentina Breschi** received her B.Sc. in Electronic and Telecommunication Engineering and her M.Sc. in Electrical and Automation Engineering from the University of Florence in 2011 and 2014, respectively. She received her Ph.D. in Control Systems from IMT School for Advanced Studies Lucca in 2018, being a visiting scholar at the University of Michigan From 2018 to 2023, she was with Politecnico di Milano, first as a post-doctoral researcher and then as a junior assistant professor. In 2023, she joined the Department of Electrical Engineering at Eindhoven University of Technology as an Assistant Professor. Her main research interests include data-driven control, system identification, collaborative learning, and human-centered policy design, focusing on mobility systems.

**Viola Schaffonati** (SM'12) is associate professor of Logic and Philosophy of Science at Politecnico di Milano. Her research interests are in the philosophy and ethics of AI and robotics, the epistemology and methodology of experiments. She serves as the Director of CINI National Laboratory on Informatics and Society; member of the steering committee and founder of Polimi META, the study unit of social sciences and humanities for science and technology; Associate Editor of Science and Engineering Ethics; and elected member of the steering committee of the Italian Society for the Logic and Philosophy of Science.

**Mara Tanelli** (SM'12) is a Full Professor of Automatic Control at the Politecnico di Milano, where she is the Rector's delegate for Diversity and Inclusion. Her research activities focus on active control of ground vehicles, smart and inclusive mobility and industrial analytics. She is co-author of more than 200 scientific publications and more than 20 patents in these research areas. Since 2013, she is part of the Conference Editorial Board of the IEEE CSS. From 2020 to 2023 she was Chair of the IEEE-CSS TC of Automotive Controls.

# Part III
# Epistemology

# Chapter 12
# On the Importance of Democratic Debates Regarding Matters of Concern in Value Sensitive Design

**Jérémie Supiot, Philippe Estival, and Mathieu Guillermin**

**Abstract** Ensuring adequate technological developments in our modern societies is a particularly wicked problem. In fact, societal and ecological stakes are often tremendous. They are considered as issues demanding the consideration of extremely various types of information, most of the time highly technical. Although it is the responsibility of engineers to find technical solutions, the ethical and political questions about the evaluation of technologies is supposedly left to politics and citizens. This view comes from the modern distinction between (scientific) facts and (political or moral) values. However, with the development of Value Sensitive Design (VSD) methodologies, answering evaluative questions seems to have become an important part of the work of engineers. One might think that, although engineers are interested in impacts with political and ethical issues at stake, their evaluation involves only matters of fact. We argue that they cover also matters of concern because facts and values are much more entangled than they seem to be at first glance in engineering design. Yet, defining and prioritizing what matters in the development of new (and old) technologies is not uniquely the responsibility of engineers. We need democratic debates such as the ones proposed in the New Humanism in the age of Neuroscience and Artificial Intelligence (NHNAI) project to help delineate matters of concern concerning the future of technologies.

J. Supiot (✉)
Catholic University of Lyon, Lyon, France

LUMSA, Roma, Italia
e-mail: jeremie.supiot@philosophie.ucly.fr

P. Estival
Cartodébat, Saint-Étienne, France
e-mail: estival@cartodebat.com

M. Guillermin
CONFLUENCE: Sciences et Humanités Research Unit (EA 1598), Catholic University Lyon, Lyon, France
e-mail: mguillermin@philosophie.ucly



C. Didier et al. (eds.), *Engineering and Value Change*, Philosophy of Engineering and Technology 48, https://doi.org/10.1007/978-3-031-83549-0_12

**Keywords** Hume's law · Value sensitive design · Engineering ethics · Matters of concern

## 12.1 Introduction

The distinction between facts and values dates back long before modernity, but its most emblematic formulation remains that developed by Hume. Hume's law, set out in his Treatise on Human Nature (1739–40), states that "one cannot get an ought from an is" (Russell, 2021). This principle, which is now widely shared, has long organized philosophy into two distinct branches: the philosophy of what is (ontology, metaphysics, etc.) and the knowledge that relates to it (epistemology, hermeneutics, etc.), and the philosophy of what ought to be (moral philosophy, ethics, politics, justice, etc.). This distinction has long since extended beyond the field of philosophy alone, since it also structures the relationship between techno-scientific activities and political institutions, particularly in Western countries where Hume's influence remains indisputable. While the task of the sciences is to establish facts (what is), it is up to ethics and politics to address the question of values (what ought to be).

Engineers are no exception to the rule. In their work, it seems obvious that they must distinguish between scientifically established facts, and the interests, desires, needs, or values that are their own or those expressed by their clients, employers, or stakeholders. If aerospace engineers make a mistake in their calculations, the rocket may explode in flight, or even never take off at all. If they fail to meet the interests or values of certain stakeholders, this will have no impact on the rocket's smooth operation. Priority is therefore given to analyzing the facts to understand how and how much they can answer to the set of specifications drawn up by clients or employers. When they use a Value-Sensitive Design (VSD) approach (Friedman et al., 2006), engineers include social or ecological values from the outset.

Yet these values are addressed "in the facts." Not only may they need to be established empirically thanks to psychological or social tools (surveys or interviews for instance), but once the values have been established, engineers have to translate them into artifacts' design, criteria, and indicators that will be assessed empirically. For example, if a company or its stakeholders want to know whether a technology complies with ecological values, engineers will perform a Green House Gas (GHG) assessment or evaluate its ecological footprint. In so doing, they are assessing whether facts comply with certain factual indicators whose transgression is forbidden or, at least, unwanted. That such transgression should not be is a value judgment … But here, values are externally defined by stakeholders. In other words, values are not in nature or artifacts (in the things engineers evaluate), but come from society (from humans and their representations).[1] It is up to engineers to establish the link between what must be (what's in their specifications) and what will be (according to the laws of nature). To do so, the neutrality of scientifically established facts ensures that desires are not confused with the order of the world. That's why part of the engineers' job is often to explain to their clients why what they are asking for is impossible.

However, in some cases, it is hard to tell the difference between facts and values. For example, is the development of carbon-free technologies a necessary adaptation to new facts (climate change), or a new value that engineers should implement? Climate change is a complex reality where values (e.g., the protection of future human generations and other species) are intertwined with facts (like those established by the International Panel on Climate Change). Is protecting the environment a question of personal commitment to subjective values, or an objective priority regarding the urgency of the situation described by the IPCC? Are engineers neutral actors that only have to find technical solutions, or do we need their opinion on ethical issues as well?

In this chapter, we will explore some issues related to a too radical distinction between facts and values. We will show that, in the design of technical objects, this distinction can prevent engineers from fully integrating social and environmental issues. We will argue that Latour's conception of facts and values, from which he derives what he calls "matters of concerns" (Latour, 2004, 2014), enables a better integration of ethical concerns in the design of technical objects. Finally, we will present the methodology and tools used in NHNAI project,[2] to show how it may help to collectively establish knowledge about what is valuable using democratic debates. This knowledge should inform the political and technical choices made by public institutions and citizens (including governments and local authorities), as well as companies and employees, particularly engineers.

[1] According to Friedman, "a value refers to what a person or group of people consider important in life" (Friedman et al., 2013, p. 2).

[2] NHNAI (New Humanism in the age of Neurosciences and Artificial Intelligence) project is a capacity-building project coordinated by UR CONFLUENCE: Sciences et Humanités (EA 1598) at UCLy (Lyon Catholic University) under the aegis of the International Federation of Catholic Universities (IFCU). Its overall objective is to empower all relevant stakeholders in the arenas of Neuroscience (NS) and Artificial Intelligence (AI) to address the vast array of ethical issues involved in the ongoing developments in these fields. More information at https://nhnai.org/

## 12.2 Facts and Values: The Genealogy of a Distinction

### *12.2.1 From Hume's Law to Positivism*

The limited space of this chapter does not permit a thorough historical explanation of the current organization of scientific and political institutions and their role in establishing facts and values. Instead, we propose a genealogical analysis of the fact–value distinction, which we believe has its roots in Hume's law. We'll no doubt be taking some shortcuts, and we're aware that the history of ideas can't be summed up in the rough picture we're about to propose. Furthermore, the story we are about to tell is not a particularly original one or a new vision of the history of ideas. However, we hope that telling it will suffice to emphasize what's at stake in engineers' design processes when they have to distinguish between facts and values.

It is first important to recall that the fact–value dichotomy is derived from Hume's more general distinction between what is, i.e., what we experience empirically, and what should be, i.e., what we can induce from experience. According to Hume, we cannot *deduce* universal and necessary laws from the empirical observation of facts. This principle known as "Hume's law" applies of course to moral laws, but also to the laws of physics, and all those laid down by the sciences.

Aware of the problem of induction, Kant proposes in his critiques that natural laws are not primarily based on empirical observations, but rather on the conditions of possibility of experience. For example, time and space are fundamental forms of intuition that shape all phenomena. That is why we can learn something about space independently of any specific empirical observation—a priori knowledge that is therefore necessary, like in geometry for instance. In general, according to Kant, mathematics gives us access to a knowledge which is both a priori and synthetic, i.e., adding to premises, unlike with analytic judgments. In his moral philosophy, a priori knowledge is also necessary to determine what we should do. He calls it categorical imperatives. However, concerning the realm of what *is*, he insists on the fact that our knowledge is limited to phenomena. Because our capacity of reasoning and understanding the world is limited, we can never gain access to the thing-in-itself.

### *12.2.2 The Emergence of Positivism, Pragmatist Philosophy and the Development of the Human Sciences*

A century later, Comtean positivism will interpret this limitation as a positive capacity of approaching reality with the sciences. Positive sciences restrict themselves to what can be empirically ascertained, thereby benefiting from a privileged epistemic status while refusing any form of definitive dogmatism. Very aware of their limits, Comte is convinced that the positive sciences are the foundation of human progress. That is why he founds his moral and political philosophy on a new science that he calls "sociology."[3] Against Kantian rationalism and Platonic idealism, pragmatist, empiricist and utilitarian philosophy emerged in the United Kingdom and the United States as an alternative to the foundationalism of European philosophy. Utilitarians (Bentham, Mill) abandoned the categorical imperatives of morality dictated by reason in favor of a consequentialism rooted in both reason and experience (empirical facts), and pragmatist authors such as Dewey and Rorty proposed an anti-foundationalist vision of philosophy and ethics,[4] from which a renewed vision of democracy based on experimentation emerged. For example, rather than establishing indisputable universal and necessary rights as the foundation of our democracies, Dewey proposes that principles governing public life should be constructed collectively through experimentation and therefore never given *a priori*.[5] Finally, Popper's post-positivism reintroduced the problem of induction in a new fashion with his concept of falsification, and disciplines such as sociology, anthropology, and ethnology moved away from their positivist heritage with the development of cultural studies, post-colonial studies, and gender studies, particularly in the United States. With them grew critiques of moral universalism and the privilege granted to the positive sciences by Comtean positivism concerning knowledge. Through the development of science studies, researchers questioned cultural influences in the psycho-sociological construction of scientific facts, and the validity of the alleged universalism of their veracity.

### *12.2.3 Science Wars*

The attack from some streams of the social sciences on the supposed neutrality of natural science led to critiques from both the human and natural sciences. Criticism of the universality of truth of scientific claims, given its dependency on the cultural and social context in which it has been discovered, rose to the famous "science wars" of the 1990s (Raynaud, 2018). The episode was a reaction to what some authors still do not hesitate to call a "radicalization" (Engel, 2017) of the post-positivist movement initiated by Kuhn and Popper. For some positivist philosophers and scientists, questioning the privileged status of the sciences was tantamount to killing the "goose that lays the golden egg" (Stengers, 2018). However, they did not see that it was not science studies but social and environmental scandals and crises that appeared in many industries in the nineteenth and twentieth centuries which

[3] "According to Comte, this distinction [between natural philosophy and moral philosophy], introduced by the Greeks, is abolished by the existence of sociology" (Bourdeau, 2023).

[4] According to Dewey, "moral theories [...] can be viewed as constructs to solve practical problems" (Festenstein, 2023).

[5] "the merit of even existing 'rudimentary' forms of democracy is that they compel 'recourse to methods of discussion, consultation and persuasion' and in so doing provide the opportunity to improve decisions" (Festenstein, 2023).

escalated public distrust in the reliability of the sciences and science-based technologies (Bensaude-Vincent & Dorthe, 2023).

For some, this questioning of the alleged neutrality of the sciences opened the door to the era of post-truth, which would feed conspiracy theories and various forms of historical and scientific denials; for others, it was an opportunity to rebuild science, opening the way to epistemological pluralism, interdisciplinarity, and to greater democracy in knowledge production. In fact, involving citizens in scientific activities may be the best way to avoid the era of post-truth. Techniques are obviously not neutral because of their social and environmental impacts, and sciences (especially the ones that study important complex problems such as climate change or epidemics) are at the center of political decisions. Not only is scientific knowledge often used to justify political decisions, it also plays an important role in the development of future technologies in engineering design.

## 12.3 Matters of Concern in Engineering Design

### *12.3.1 Designing with Ethics: Deconstructing the Great Divide Between Facts and Values in Engineering Design*

Thanks to Value Sensitive-Design (VSD; see for instance Winkler & Spiekermann, 2021; Gerdes & Frandsen, 2023), near to the tradition of Freeman's stakeholder theory (Freeman, 1984/2015), more and more engineers integrate stakeholders' values in the process of engineering design. In VSD methodologies, goals established on values (e.g., combating climate change) are generally defined in advance in the specifications drawn up with the client (sometimes with larger groups of stakeholders), translated into concrete, falsifiable[6] objectives. Then, their achievement is verified through key performance indicators (KPI) assessment. For example, environmental values of some stakeholders are translated into concrete objectives concerning the protection of the environment (e.g., carbon neutrality) for the design of a new technology. If the objective is achieved, the technology will be more valuable for customers and for society in general: it is a "green" technology. Being "green" amounts to a negative statement that means "not polluting" (or at least "less polluting").

It could appear that such a VSD approach respects the radical distinction between facts and values, that it complies with Hume's law. However, the question may prove far more subtle. If I say that a technology is polluting (or green), it's a value judgment by which I mean that it is a bad technology (or a good one) for the environment. But it is also a matter of fact that I can assess empirically by analyzing negative (or positive) externalities (or their absence). Modern philosophers might

[6] Here we use "falsifiable" in the Popperian sense of the word. An objective is falsifiable if it is "incompatible with possible empirical observations" (Thornton, 2023).

retort that, when determining an ecological footprint for instance, it is not the value of an object that engineers assess, but the *compliance* of facts *with* a value. But that's not what "polluting" means: a technology is polluting independently of the specific values of a given society. So maybe it's just a fact independent of cultural context, and therefore not a value? Not at all! Although the factual description of what is happening is (largely) independent of values upheld by society, reducing its meaning to such a factual description would amount to removing an important part of what "polluting" also means, that is "*bad* for the environment." Terms like "polluting" count as what Putnam calls "thick evaluative terms,"[7] whose evaluative and factual dimensions are irreducibly and irremediably entangled.

Therefore, distance must be taken with a too radical distinction between facts and values, especially in engineering design. For instance, the challenge of eco-design is not to add an ecological value to an object that would originally be neutral—e.g., using biofuels in aviation or developing hydrogen engines to limit carbon impact. There are no such things as neutral technologies: they all have social and environmental impacts (expressed through thick evaluative terms), and they are all designed for a specific use. Furthermore, data and facts produced or used by engineers to design and assess a technology or an artifact are necessarily intertwined with values, interests, needs, and desires. Facts are not neutral in engineering design because, by searching the "best design," engineers rely on values, norms, and social rules. Therefore, evaluative questions are in no way reducible to external problems to be dealt with by outsiders (governments, civil society, ethical committees…) providing factualized constraints for engineers.[8] These evaluative questions, what we might call with Latour "matters of concern" (Latour, 2004, 2014), cannot be removed from the world of engineers. Responding to them or choosing not to—as well as determining how to respond to them—is *at the very core* of engineering design (Stephan, 2015; Brodersen & Pedersen, 2019). That is why ethics by design *is* the very essence of design, which fully assumes itself to be a set of realistic goals (*desseins*) and not just a more or less realistic blueprint (*dessin*).

If a technology is too polluting, it's not just that it doesn't have the ecological bonus that would have made it a better technology. It has been poorly designed regarding its impacts and should be considered as unsuitable for industrial production. For example, it goes without saying that air travel is the fastest and safest form of transport for passengers. However, it is much less safe for all the people and species that are bearing the brunt of climate change. Let's put it bluntly: today's

[7] Ethical concepts are called "thick" when they involve non-evaluative descriptions (Putnam, 2002; Väyrynen, 2021).

[8] Carbon footprints or chemical pollutants are often described as bad externalities. According to Alexandre Monnin, in an interview given to the french media *Horizons publics*, « il ne s'agit pas de simples "externalités négatives," autrement dit, de conséquences malheureuses de nos activités mais de conditions *sine qua non.* » (they are not simple 'negative externalities," i.e., unfortunate consequences of our activities, but *sine qua non* conditions [our translation]) (Monin in Nessi, 2023).

airplanes and all vehicles using fossil fuels are badly designed considering our current scientific knowledge of the climate and its evolution.

### 12.3.2 Scandals and Controversies

In industry—but not only there, scandals are legion: Diesel Gate, the opioid crisis, mad cow disease… Yet, despite their number, they are often considered to be exceptions in industries that remain globally reliable. So, when the question of engineering ethics is raised, it is usually approached from the angle of accountability regarding technical accidents, as in the classical case of the Space Shuttle Challenger or the more recent one of Boeing 737 MAX crashes of Lion Air Flight 610 and Ethiopian Airlines Flight 302. An unfortunate incident has occurred, and the only two questions for engineers are: "Could it have been avoided?" and "How?" But there are far more complicated issues for engineers to tackle than scandals and risk management: controversies.

The good thing about scandals is that they make everyone agree. In the case of technical or scientific controversies, precisely the opposite happens. Not only engineers or politicians cannot find a satisfying solution but the very formulation of the issue at hand is problematic: they are facing what some call wicked problems[9] (Rittel & Webber, 1973; Arpin, 2022). In France, for example, the use of megabasins to meet agricultural water requirements during periods of drought is the subject of much debate among citizens, scientists, and engineers. In such cases, matters of concern are multiple and sometimes contradictory, making it difficult for engineers to know how to proceed.

Once again, one might think that wicked problems are exceptions, requiring a specific methodology. On the contrary, they seem to become more and more numerous, so that they are becoming the rule rather than the exception. Planetary limits and societal needs are becoming the new framework for action for engineers (Raworth, 2018), although they are as difficult to establish as they are important. No technical object is spared. To see this, we need only ask: what technical object requires no non-renewable energy or resources to build? Almost none.

Of course, it would be too easy to lay the blame at the door of engineers who are facing particularly complicated problems, which they should solve even though they sometimes work for companies or governments that don't make this a priority. This is why, rather than making engineers themselves entirely responsible for defining matters of concern, we advocate a more democratic approach to engineering design, in which stakeholders could contribute more directly to the definition of

[9] Even if the "use [of wicked problem] as a buzzword has undermined precise conceptual definition" (Termeer et al.. 2019), its success shows at least an increasing complexity regarding policy science and practice. Although this term is still vague in the scientific literature (Lönngren & van Poeck, 2021), it generally refers to a problem that is difficult to formulate, has no stopping rule (it can never stop being a problem), and which consequently cannot give rise to a single answer.

matters of concern. To this end, we will show how, by organizing debates with concerned citizens (specialists and non-specialists), it is possible to enrich not only VSD practices but engineering design in general.

## 12.4 How to Define Matters of Concern

### *12.4.1 Asking Is Not Enough*

Most existing methodologies of VSD use stakeholders for the definitions of values, and sometimes for their integration in design (Winkler & Spiekermann, 2021). As a result, in most VSD approaches, a value is what is explicitly valued by stakeholders. This comes with three foundational problems for VSD: it supposes that stakeholders' values are transparent, it creates a confusion between values and interests, and it tends to forget the distinction between what is valued and what is valuable.

To collect stakeholders' values, engineers can either ask them directly using surveys and interviews or observe and analyze their behavior or discourses. In both cases, what will become valuable for the engineering project comes from what is valued by some stakeholders, as determined by a statistical generalization from a representative sample. Although it can be seen as a good practice of governance in innovation processes, it has four major limitations. First, stakeholders' values are scarcely as transparent as they seem to be: determining what is valuable for non-humans like animals or forests is tough. Second, representativity is hard to obtain in such diverse groups of individuals where some stakes are sometimes the only thing they share. Third, it is hard to prioritize conflicting values. How to choose between a cheaper but less ecological solution, and a less accessible but more ecological one?

Participatory Design (Holter, 2022) can help address these limits, but it is still affected by the fourth and major one: even with transparent, representative, and collectively prioritized values, there is no guarantee that the new technology respecting them all will actually be valuable for all stakeholders. According to Alrechtslund, "the assumption that a correspondence exists between design intention and use context" is at least problematic (Alrechtslund cited in Gerdes & Frandsen, 2023), even more so in the long term with changing values and the global changes we are facing in the age of the Anthropocene.

To consider long-lasting matters of concern, engineers need to reflect on what will remain valuable for all stakeholders in the long term, i.e., in a perspective of intergenerational and international justice. In this perspective, they cannot respect Hume's law: on the one hand, discussing moral values—what *would now be* good or bad, and on the other, separately establishing scientific facts: what *is* true or false, possible or not. They must discuss openly with concerned citizens what to care (or not care) about and how to take care of (or abandon) it, for all the living beings involved including other species and future generations.[10] This is the only way to design technologies that are as valuable as possible in the age of the Anthropocene (Allard & Monnin, 2020).

VSD is more than a problem of engineering ethics. What is at stake here is the very definition of what engineering design is. Indeed, the development of VSD methodologies is not only about opening some engineering designs to stakeholders' values, but first and foremost about making engineering design a valuable process for every stakeholder, even for the ones that cannot speak, like animals or forests. That is why we argue that engineering design needs a lot more than asking stakeholders what they value the most.

### *12.4.2 Democratic Debates as a Resource for Engineering Design*

If it is hard to know what is valued by stakeholders, it is even harder to evaluate what is valuable for them. Some late Saint-Simonists would defend the belief that we need only technical and scientific skills to do so, because matters of concern could always find a technical answer. For example, if we want to achieve carbon neutrality in 2050, we may have only to develop carbon-neutral technologies. However, according to the historian of technology Jean Baptiste Fressoz, energy transitions never happened in the history of capitalism and there is no reason to think that renewable carbon-free energies will replace coal and oil in the near future (Fressoz, 2024). That is why, the first question is not necessarily "how do we develop or adapt technologies with the right values?" but "should we keep developing, producing and maintaining ill-adapted technologies?" (Monnin, 2023). For example, the development of new technologies like hydrogen-powered airplanes is not the only option to adapt industry to climate change: we could also abandon the production of airplanes and produce something else. Of course, it would not be easy, but the replacement of all existing airplanes with profitable hydrogen-powered ones seems to be an impossible goal to achieve in 25 years from now.

Considering the tremendous difficulties to adapt or abandon "negative commons" (Monnin, 2023) such as polluting industries and technologies, it is important that engineers reflect upon the genuine needs the technologies they design should answer to. Following Jean-Hugues Barthélémy, design should be sensitive first and foremost to unmet needs ("besoins en souffrance," Barthélémy, 2018) before being sensitive to unmet desires and values. In the background, this questioning also means wondering whether what is valued is effectively valuable when applied in the design of a technology. It's important to note that the answer doesn't necessarily

[10] Arguably we may also need to reflect on our own vulnerability (Pelluchon, 2021) regarding our "être-en-dette" (being-indebt [our translation], Barthélémy, 2021) and how we are taken care of (Silberzahn, 2024).

have to be binary, between a proven need and a notorious uselessness. It may be more subtle. For example, we don't necessarily need to abandon aircraft production altogether, but rather to limit its use. Yet, if we are to come up with such an answer, we (citizens) need places and mechanisms to discuss it freely and publicly. This kind of question requires democratic debates about the future of technology, concerning new technologies coming onto the market, but also on existing technologies that can't be maintained as they are.

In this spirit, the project called "New Humanism in the Age of Neurosciences and Artificial Intelligence" (NHNAI) was set up to look at artificial intelligence and neuroscience matters of concern.[11] Face-to-face and remote debates have been and still are organized in nine countries (France, Italy, Kenya, USA, Canada, Chile, Belgium, Portugal, and Taiwan) and are being reported on an online debate platform hosted by CartoDEBAT.[12] Participants can then contribute directly through the online platform. These debates will be analyzed and synthetized with initial public feedback by the end of 2024 and in a white paper expected by 2025. The idea of the project is not to represent all citizens' point of view, but to obtain what Hervé Marchal[13] calls a "significant corpus."

Through the NHNAI project, we do not seek to establish a representative sample of given groups' answers to specific ethical questions concerning artificial intelligence or neuroscience. Rather, we seek out the emergence of rational debates between concerned citizens. Thanks to those debates, we hope to contribute to a process of collective ethical capacity-building of participants, including the researchers that are organizing the debates, on matters of concern about what it means to be human in the time of Neuroscience and Artificial Intelligence.

### *12.4.3 Elaborating Matters of Concern About Neuroscience and Artificial Intelligence*

Artificial Intelligence and Neuroscience are developing fast, raising pressing ethical questions. To what extent should artificial intelligence help humans? Should we enhance our bodies and minds with neurological implants? Such questions cannot be treated as a mere implementation of preexisting societal ideals, values, and goals (equity, immortality, transhumanism...). In the end, it depends on the definition of each term (e.g., "intelligence," "body," or "mind"), which demands a philosophical, ethical, political, technical, and scientific exploration of what they are, what they

[11] See the website: https://nhnai.org/

[12] See the website: https://cartodebat.org/nhnai/

[13] Hervé Marchal is a French sociologist who set up a research project in sociology with a similar approach to understand the effects of COVID 19 on French citizens entitled "Metamorphoses." Refusing the normative approach of sociology based on the figure of *Homo economicus*, Hervé Marchal proposes an alternative sociologic methodology based on the figure of what he calls *homo alterus* (Marchal, 2022).

mean, and how we value them. Facing this huge variety of matters of concern, the NHNAI project aims at contributing to this effort by putting the focus on what it means to be human as a necessary background reflection. However, our goal is neither to build a universal conception of what it means to be human nor to obtain a sociological snapshot of what is valued by participants. Rather, we aim at contributing, through rational discussions, to a collective refinement of participants' understanding of matters of concern regarding neuroscience and artificial intelligence.

The debates organized by the NHNAI are aiming at the development, within a community of argumentation, of a shared understanding of what matters in the definition of our common humanity. However, the goal is not a consensus on a unique straightforward conception of what matters concerning what it is like to be human. It is quite the contrary. We seek to contribute to elaborating a shared recognition of multiple and sometimes conflictual situated matters of concern. In fact, a distinctive characteristic of democratic debates comparing to democratic deliberations is that they do not involve any decision. It is not important that everyone agrees on a unique conclusion. Their goal is only to enrich the participants' understanding of what is at stake. That is why we chose to explore democratic debates instead of democratic deliberation.[14]

### *12.4.4 How to Maximize Rationality in Democratic Debates*

Instead of searching for universal truths, we want to build a tension toward universality in the sense of Habermas (1992/2023) by promoting contributions that are meant to be convincing for any participant. This approach does not come without problems. The first difficulty is to avoid sterile rhetoric conflicts that doesn't advance the debate (as is very common on social networks). That is why the NHNAI online platform, created in partnership with CartoDébat (see https://cartodebat.org/nhnai/), has been designed to increase rationality in the debates by fostering collective exploration and shared acknowledgement of complexities and tensions. The goal is to make participants *feel* their shared obligation regarding the community of argumentation they enter to aim at a universal validity in the arguments they make.[15] In other words, the platform encourages what Habermas calls a discourse ethics (2023).

Once they have registered and selected a topic, participants have access to a short text submitted for discussion (Fig. 12.1 on the left) and the contributions of the other participants (Fig. 12.1 on the right).[16] If they want to contribute, they must select what they want to discuss. It can be an excerpt of the text on the left, or a part of one

[14] On the role of public deliberation in VSD and participatory design, see Borning et al. (2005), or more recently Sistrunk et al. (2024).

[15] About digital online debates, see Le débat numérique, le média d'un renouveau démocratique (The digital debate, a media to renew democracy [our translation], Sallantin et al., 2023).

[16] The full debate is accessible here: https://cartodebat.org/nhnai/db/how-does-ai-challenge-our-understanding-of-the-nature-and-purpose-of-human-education

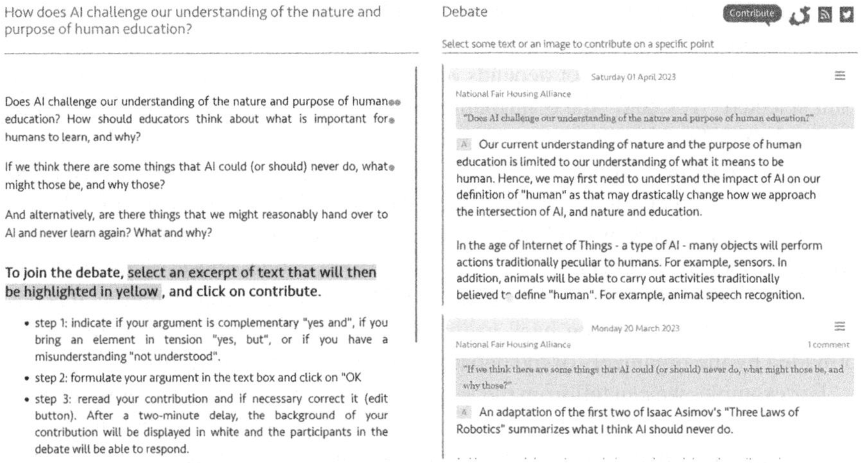


Fig. 12.1 One of the NHNAI's Debate on CartoDébat facilitated by Thor Wasbotten and Brian Patrick Green in Santa Clara University, California, USA

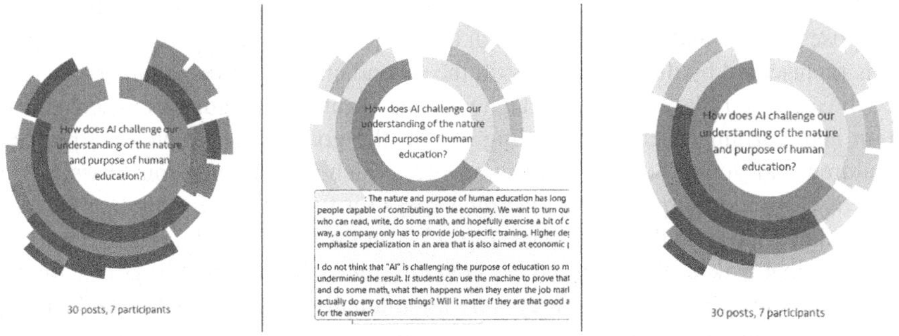


Fig. 12.2 A graphic representation of the same debate (see Fig. 12.1)

of the contributions. This drives participants to specify what they want to react to, instead of giving a general opinion about the text or the contributions.

To support rationality in discussions, we decided to add specific features on the platform. First, when contributing on a given part of the initial text or of other participants' contributions, participants have to choose between three options: either "Yes, and" (I've understood, I agree, and I want to add something that goes in the same direction) or "Yes, but" (I've understood, I agree, but I nevertheless want to add something that is in tension, to point to a conflict), or "I didn't understand" (I can't accept what is said because it makes no sense for me... Maybe you mean this? Would it be ok to reformulate like that?). Dedicated attention is granted to interactions and constructive exchanges between participants who can react precisely to other contributions by selecting a precise extract and by indicating the nature of the reaction. This allows for representing exchanges as exemplified in Fig.12.2.

Figure 12.2 represent a floral-shaped tree of discussions. It has been generated by the platform and is accessible by all the participants. Each contribution is represented by an arc whose color depends on the nature of the contribution: if a participant selected “Yes, and” the contribution will be represented as a green arc, if they selected “Yes, but” it will be red, and if they selected “I don’t understand” it will be blue. The first inner circle (see Fig. 12.2 on the left) represents the participations that answered directly to the initial text provided on the left by debates’ facilitators (see Fig. 12.1). Each new circle represents a level of discussion: contributions of the second circle are answers to the ones of the first circle, those of the third are responses to contributions of the second, etc. Thanks to this representation, it is possible to follow a discussion back to the initial contribution that reacts to the text provided by the organizers of the debates (see Fig. 12.2 on the right).

For example, we can tell from the length of the arc by which it is represented that the first contribution (see Fig. 12.2 in the middle) triggered many contributions. The participant selected the question “Does AI challenge our understanding of the nature and purpose of human education?” and answered that while it has impacts on how we train people and what we expect from our children, it does not challenge the purpose of education (see Fig. 12.2 in the middle). Another contributor disagreed with the first part of his response on the purpose of education, arguing that other purposes existed and that it was difficult to prioritize them (on Fig. 12.2 on the right, it is represented by the large red arc in the second circle). He responded (Fig. 12.2 on the right, third level in green) that it was important to distinguish between education and learning, because according to him the goal of education is to prepare children and young adults for working life, which is different from learning an activity for other purposes (e.g., learning chess for fun). Another participant asked for a clarification on this difference (Fig. 12.2 on the right, fourth level in red): is it the goal that makes education and learning different? He answered (Fig. 12.2 on the right,, fifth level in green) that they were not mutually exclusive, but that education was mostly about graduating to get a job. Here we can see how a specific matter of concern (what is the purpose of education) is getting clearer with each new contribution to the debate.

We can see in this example that rules are not always perfectly respected. For example, the contribution that asked for clarification should have been in blue. Nevertheless, the platform gives to the participants the opportunity to take time to prepare their contributions, which may be more complicated in in-person settings. They can answer one part of another contribution instead of the whole and develop matters of concern in all their complexity instead of polarizing them (searching for simple pros and cons) and reducing them to subjective opinions. Then it is possible to follow each step of a conversation and to monitor how complexities emerge when discussing given matters of concern—here on the role of education.

The aim of the NHNAI debates is first and foremost to offer an enriching discussion experience for persons who feel concerned about societal, anthropological, and ethical issues linked to AI and neuroscience. They are open to all citizens, including engineers who can take part as citizens but also as experts and benefit from such debates as much as debates can benefit from their contributions. In fact, such debates might be a missing step preceding VSD and participatory design toward the democratization of innovation or *désinnovation*[17] processes. Freed from deliberation since no consensus on unique conclusions or solutions is expected from those debates, it is an opportunity for concerned persons to actively and collectively build matters of concern, on the ground of a refined understanding of what is valuable in each context, and a shared awareness of involved tensions, contradictions, and complexities. This preliminary process may allow for enriched and more enlightened subsequent participation of stakeholders in the whole process of engineering design to deliberate on the solutions, using existing VSD methodologies.

## 12.5 Conclusion

Hume's law and its interpretation by the positivist tradition led to the view that to gain more objectivity and stay neutral when designing new technologies, engineers should stick to the establishment of facts (and means) and stay aside from the reflection on values (and ends), merely applying its results in a factualized way. It trapped them in an impossible task. When confronted with matters of concern, they struggled to separate facts from values (attempting to neutrally apply evaluative and political outcomes, to neutrally translate them into design constraints and norms). However, it is quite clear now that engineers are often confronted with thick ethical concepts involving entangled descriptive and evaluative judgments. That is why we need to recognize that Value Sensitive Design cannot be a purely factual work, but also involves (or necessitates) ethical and political reflection. Therefore, VSD should be more democratic with those evaluative and descriptive judgments being based on an open discussion with all concerned citizens (including engineers). That is why we need Value Sensitive Design to integrate democratic debates to improve citizens' judgments (including engineers') on matters of concern.

It is as false as it is dangerous to think that facts could be neutrally discovered, or that technics could be neutrally designed. By separating facts from values, engineers might forget that at the center of every design and every science project there are epistemic and non-epistemic goals, interests, and values (Kitcher, 2001). In this regard, science studies did not open the sciences to gray areas, blurring the difference between knowledge and opinion. It only opened our eyes to the world we are living in (and to our genuine epistemic status).

It is now time for engineers to adapt to this new reality. Because "artifacts have politics" (Winner, 1980), designing an object is a political act which is even more powerful and long-lasting than laws and constitutions. This political act leaves no choice but to follow the rules some people (engineers, CEOs, politicians) ascribed

[17] The French concept of "*désinnovation,*" sometimes called "*innovation par retrait*" (innovation by abandon [our translation], Roby & De Maria, 2022) emerged recently in France. It refers to the voluntary abandonment of certain techniques or technical tools for economic, ecological, or social reasons.

to it, even when values and contexts have changed. If we want a more democratic approach to the development (or the abandonment) of technologies, engineers have a crucial role to play in this transition. From this perspective, developing ethics in engineering design is therefore not a question of integrating values into designs but first of all a political question of governance of innovation and technical democracy about how to best determine matters of concern. VSD therefore needs upstream democratic debates such as the ones proposed by the NHNAI project, to decide on the opportunity for engineering projects and to give them refined ethical guidelines to be further implemented during design processes.

The main issue we have now to tackle is to continue to improve the rationality of democratic debates. We also need to determine how specialists (scientists, engineers, philosophers…) can best help to improve the quality and the rationality of those debates. Their input is useful and indispensable but could also prevent citizens from formulating fruitful questions because they do not correspond to certain scientific standards. We hope to contribute to answering these questions with the work engaged in the NHNAI project.

**Acknowledgments** The authors would like to thank all the reviewers that contributed greatly to improve this chapter, especially William Rose and Diane Mitchelfelder. They are also grateful to Thor Wasbotten and Brian Green for allowing them to use the contributions of the debates they facilitated as a part of the NHNAI project (see https://www.nhnai.org).

## References

Allard, L., & Monnin, A. (2020). Ce que le design a fait à l'Anthropocène, ce que l'Anthropocène fait au design. *Sciences du design, 11*.

Arpin, I. (2022). La recherche scientifique à l'épreuve des « wicked problems » de l'Anthropocène. In *La recherche scientifique à l'épreuve des « wicked problems » de l'Anthropocène* (pp. 2–10). Presses universitaires de Grenoble.

Barthélémy, J.-H. (2018). *La société de l'invention: Pour une architectonique philosophique de l'âge écologique*. Éditions matériologiques.

Barthélémy, J.-H. (2021). *Ego alter: Dialogues pour l'avenir de la Terre*. Éditions matériologiques.

Bensaude-Vincent, B., & Dorthe, G. (2023). *Les sciences dans la mêlée, Pour une culture de la défiance*. Éditions du Seuil.

Bourdeau, M. (2023). Auguste Comte. In E. N. Zalta & U. Nodelman (Eds.), *The Stanford encyclopedia of philosophy (Spring 2023)*. Metaphysics Research Lab, Stanford University. https://plato.stanford.edu/archives/spr2023/entries/comte/

Borning, A., Friedman, B., Davis, J., & Lin, P. (2005). Informing public deliberation: Value sensitive design of indicators for a large-scale urban simulation. In H. Gellersen, K. Schmidt, M. Beaudouin-Lafon, & W. Mackay (Éds.), *ECSCW 2005* (pp. 449–468). Springer.

Brodersen, S., & Pedersen, S. (2019). Navigating matters of concern in participatory design. *Proceedings of the Design Society: International Conference on Engineering Design, 1*(1), 965–974.

Engel, P. (2017). Le relativisme fait pschitt. *Sciences Philosophie*, 3–27.

Festenstein, M. (2023). Dewey's political philosophy. In E. N. Zalta & U. Nodelman (Éds.), The Stanford Encyclopedia of Philosophy (Spring 2023). Metaphysics Research Lab, Stanford University. Retrieved from https://plato.stanford.edu/archives/spr2023/entries/dewey-political/

Freeman, R. E. (1984/2015). *Strategic management: A stakeholder approach*. Cambridge University Press. https://doi.org/10.1017/CBO9781139192675.

Fressoz, J.-B. (2024). *Sans transition*. Éditions du Seuil.

Friedman, B., Kahn, P. H., & Borning, A. (2006). *Value sensitive design and information systems*.

Friedman, B., Kahn, P. H., Borning, A., & Huldtgren, A. (2013). Value sensitive design and information systems. In N. Doorn, D. Schuurbiers, I. van de Poel, & M. E. Gorman (Eds.), *Early engagement and new technologies : Opening up the laboratory* (pp. 55–95). Springer Netherlands. https://doi.org/10.1007/978-94-007-7844-3_4

Gerdes, A., & Frandsen, T. F. (2023). A systematic review of almost three decades of value sensitive design (VSD): What happened to the technical investigations? *Ethics and Information Technology, 25*(2), 26.

Habermas, J., traduction from Hunyadi. (1992/2023). *De l'éthique de la discussion*. Flammarion.

Holter, C. T. (2022). Participatory design: Lessons and directions for responsible research and innovation. *Journal of Responsible Innovation*.

Kitcher, P. (2001). *Science, truth, and democracy*. Oxford University Press.

Latour, B. (2004). *Why has critique run out of steam? From matters of fact to matters of concern*. The University of Chicago.

Latour, B. (2014). What is the style of matters of concern? In N. Gaskill & A. J. Nocek (Éds.), *The Lure of Whitehead* (pp. 92–126). University of Minnesota Press.

Lönngren, J., & van Poeck, K. (2021). Wicked problems: A mapping review of the literature. *International Journal of Sustainable Development & World Ecology, 28*(6), 481–502.

Marchal, H. (2022). Homo alterus. Esquisse d'une anthropologie altérophilique. *Revue du MAUSS, 60*(2), 287–298.

Monnin, A. (2023). Capitalisme de la fermeture et communs négatifs. *Revue du MAUSS, 61*(1), 237–253.

Nessi, J. (2023, May 22). *Alexandre Monnin: «Il faudra renoncer à certaines choses pour maintenir l'habitabilité de la planète»*. Horizons publics. Retrieved from https://www.horizonspublics.fr/environnement/alexandre-monnin-il-faudra-renoncer-certaines-choses-pour-maintenir-lhabitabilite-de

Pelluchon, C. (2021). *Les Lumières à l'âge du vivant*. Éditions du Seuil.

Putnam, H. (2002). *The collapse of the fact/value dichotomy and other essays*. Harvard University Press.

Raworth, K. (2018). *Doughnut economics: Seven ways to think like a 21st-century economist* (1st ed.). Random House Business.

Raynaud, D. (2018). Chapitre 1. La métacontroverse des « Science Wars » (1994-2006). *Sciences Philosophie*, 55–86.

Rittel, H., & Webber, M. (1973). *Dilemmas in a general theory of planning*.

Roby, C., & De Maria, A. (2022). Frédéric Goulet et Dominique Vinck (dir.), Faire sans, faire avec moins. Les nouveaux horizons de l'innovation. Paris, Presse des Mines, 2022. *Quaderni, 107*(3), 137–141.

Russell, G. (2021). How to prove Hume's Law. *Journal of Philosophical Logic, 51*(3), 603–632.

Sallantin, J., Pinet, V., Luzeaux, D., & Jenny, S. (2023). *Le débat numérique : Le média d'un renouveau démocratique*. Spartacus IDH.

Silberzahn, L. (2024). Care, ecology, and the crisis of eco-social reproduction: Politicizing more-than-human care. *Hypatia, 39*. https://doi.org/10.1017/hyp.2024.16

Sistrunk, A., Self, N., Biswas, S., Luther, K., Verdezoto, N., & Ramakrishnan, N. (2024). Redistrict: Online public deliberation support that connects and rebuilds inclusive communities. *Proceedings of the ACM on Human-Computer Interaction, 8*(CSCW1), 1–23.

Stengers, I. (2018). *Another science is possible: A manifesto for slow science*. Polity Press.
Stephan, P. F. (2015). *Designing 'matters of concern' (Latour): A future design challenge?* (pp. 202–226). Birkhäuser.
Termeer, C. J. A. M., Dewulf, A., & Biesbroek, R. (2019). A critical assessment of the wicked problem concept: Relevance and usefulness for policy science and practice. *Policy and Society, 38*(2), 167–179.
Thornton, S. (2023). Karl Popper. In E. N. Zalta, & U. Nodelman (Éds.), *The Stanford Encyclopedia of Philosophy (Winter 2023)*. Metaphysics Research Lab, Stanford University. Retrieved from https://plato.stanford.edu/archives/win2023/entries/popper/
Väyrynen, P. (2021). Thick ethical concepts. In E. N. Zalta (Éd.), *The Stanford Encyclopedia of Philosophy (Spring 2021)*. Metaphysics Research Lab, Stanford University. Retrieved from https://plato.stanford.edu/archives/spr2021/entrieshick-ethical-concepts/
Winkler, T., & Spiekermann, S. (2021). Twenty years of value sensitive design: A review of methodological practices in VSD projects. *Ethics and Information Technology, 23*(1), 17–21.
Winner, L. (1980). Do artifacts have politics? *Daedalus, 109*(1), 121–136.

**Jérémie Supiot** is a PhD student in philosophy at the CONFLUENCE: Sciences et Humanités research unit (EA 1598) of UCLy (Lyon Catholic University) and LUMSA University at Rome. His thesis focuses on ethical empowerment at the interface between society and the university. It is supervised by Mathieu Guillermin (UCLy, Lyon) and Stefano Biancu (LUMSA, ROME), and is part of the Activity 4 of the NHNAI project. His main research topics are epistemology and engineering ethics. He co-organizes the ETIUS seminars at the Université Catholique de Lyon. Co-founder of collectif EPSI (*Études Pluridisciplinaires sur l'Ingénierie*), which organizes the Cultures d'Ingénieurs et Formation seminars.

**Philippe Estival** is a computer scientist co-founder of CartoDEBAT SAS, administrator and animator of debates networks on ecological planning programs and public policies. He designs and develops an educational software, and contributes to Free and Open Source Communities for reproducible research.

**Mathieu Guillermin** is associate professor of philosophy and ethics of science and technology at the CONFLUENCE: Sciences et Humanités research unit (EA 1598) of UCLy (Lyon Catholic University). He holds a PhD in Physics and a PhD in Philosophy. One of his main research interests is the question of the articulations between ethics, scientific research and technological developments. He focuses notably upon philosophical and ethical questions raised by new digital technologies (big data, artificial intelligence, robots …).

# Chapter 13
# Epistemic Achievements of Engineers in Relation to Sociotechnical Systems: From Technological Knowledge to Engineering Understanding

**Vivek Kant and Michael Poznic**

**Abstract** With the growth of modern systems, there is an urgent need to comprehend the epistemic activities of engineers in the design, operation, maintenance, and decommissioning of systems. Typically, these epistemic activities are characterized with the label of technological or engineering knowledge. However, in this article we argue that in complex systems, there is a need for engineering understanding along with engineering knowledge. This paper develops the notion of engineering understanding in relation to existing literature in philosophy of technology as well as philosophy of science and epistemology. It highlights insights from engineers interested in philosophy, as well as from philosophers interested in science and technology, and considers why engineering understanding is vital as a concept and needs further elaboration in the academic literature. We illustrate our claim in the example of troubleshooting problems in transportation systems that levy toll taxes, thereby demonstrating that problems of engineering cannot be handled by discussions of knowledge alone. Hopefully, this engagement with engineering understanding will also help the engineering community in self-reflection and clarification of the epistemic activities of engineers.

**Keywords** Technological knowledge · Engineering epistemology · Engineering understanding · Engineering reflection

---

V. Kant · M. Poznic (✉)
Karlsruhe Institute of Technology (KIT), Institute for Technology Assessment and Systems Analysis (ITAS), KIT, Karlsruhe, Germany
e-mail: vkant@iitk.ac.in; michael.poznic@kit.edu



C. Didier et al. (eds.), *Engineering and Value Change*, Philosophy of Engineering and Technology 48, https://doi.org/10.1007/978-3-031-83549-0_13

## 13.1 Introduction

As the world grows toward the convergence of several technologies such as Nano-, Bio-, Info-, and Cognitive science-based technologies, contemporary systems are becoming increasingly complex to design, operate, maintain, and decommission. Engineers are often developing new knowledge in the form of processes, techniques, and frameworks to address these challenges. However, despite the specialized knowledge developed in individual engineering disciplines, there is also the need to comprehend how various aspects of knowledge coherently merge together in the form of engineering understanding, while dealing with complex sociotechnical systems. Sociotechnical systems are modern-day complex systems defined as those "systems which involve people and technologies interacting with each other at multiple levels involving operations, management, regulations, and governance" (Vermaas et al., 2011, p. 3).

The aim of this article is to propose the concept of engineering understanding as a new theme worthy of investigation for philosophers of engineering and technology, and also as an epistemic concept for engineers, which they can use in a reflective spirit. The key idea is that philosophical investigations of this concept will help the engineering community at a theoretical level, to reflect on and assess the basis of their knowledge and aid in the personal growth and awareness of self-reflective engineers.

Reflecting on engineering concepts, engineering practice and on the philosophical implications of their work is not new for engineers (Florman, 1994; Bucciarelli, 1984; Harlim & Belski, 2013; Blockley, 1992; Mitcham, 2020). A number of engineers as well as philosophers of engineering and technology have addressed the nature of engineering and engineering knowledge under the label of "philosophy and engineering" (e.g., Van de Poel & Goldberg, 2010) or "what engineers know" (Vincenti, 1990; De Vries, 2003). There are three perspectives on the topic of engineering knowledge, or more broadly speaking, on engineering epistemology.[1]

Before the 2000s, the philosophy of engineering was nascent and the primary emphasis was on the larger dimensions of technology which addressed issues of impact of technology on society, along with moral dilemmas and challenges (Brey, 2010; Meijers & Kroes, 2000; Franssen et al., 2016). The literature on the philosophy of engineering knowledge was nascent, as well, and it was primarily addressed in terms of the philosophy of technology.

This dimension of discussion has been supported by the works of sociologists and historians of technology (e.g., Callon, 1980; Hughes, 1987; Hughes & Hughes, 2011; Latour, 1996; Law, 2011; MacKenzie, 1993; Vinck & Blanco, 2003: Williams et al., 2013). Here, researchers provided the emphasis on understanding knowledge as socially construed in terms of the wider community of practitioners.

[1] We use the term 'epistemic achievements' as a general umbrella term to speak about knowledge, understanding, or maybe also other potential epistemic goods that are relevant for engineers.

In addition to these foci of the philosophical and the sociological or historical scholarly literature, engineers themselves have addressed the nature of engineering in a significant manner. A notable emphasis was provided by aeronautics engineer Walter Vincenti (1990) (discussed in detail below), who provided a typology of engineering knowledge based on historical case studies. The discussion on engineering epistemology received a significant emphasis in the early years of the twenty-first century by philosophers and engineers who discussed concepts related to design, failures in products, and systems, among others. This emphasis on engineering epistemology has been strongly related to the approach of the "empirical turn" in the philosophy of technology (Meijers & Kroes, 2000; Franssen et al., 2016).

While the above three represent different perspectives on engineering epistemology in terms of knowledge, epistemic practices, and other epistemic achievements, there is a need for philosophical investigation of engineering. In this line of approach, the key focus is to investigate the concepts employed in engineering analytically in light of the broader issues of knowledge that have been developed as part of the growth of reflections on collective human thought and action. These philosophical discussions are important as they provide an insight into the dissection of engineering knowledge from an analytic standpoint. Thus, they complement the above approaches of sociologists, historians, and engineers themselves.

In this article, we are responding to the tradition of the analytic philosophy of engineering and technology and thereby also to the so-called empirical turn in philosophy of technology. At the same time, we are highlighting the need for a discussion and clarification of the notion of engineering understanding. We argue that this concept is of vital importance when dealing with epistemic practices of engineers in contemporary sociotechnical systems (e.g., cyber-physical systems).[2] These sociotechnical systems require not only singular branches of engineering knowledge such as mechanical or electronics engineering, but also need a considerable comprehension of other domains of knowledge such as information science, management, regulation, and governance. These diverse pieces of knowledge have to be related to each other, and the demanding epistemic achievement of generalized understanding consists in the grasping of these pieces and their interrelationships. Thus, in daily engineering activities, along with engineering knowledge, a generalized engineering understanding is also required.

While the emphasis on engineering understanding is new, the concept of understanding itself is not. It has been discussed by scholars in the German philosophical tradition in the nineteenth and twentieth centuries such as Wilhelm Dilthey or Hans-Georg Gadamer, where the emphasis was placed on the notion of understanding as separate from that of explanation (cf. Stueber, 2012). Typically, in engineering and the natural sciences, the notions of knowledge and explanation have been mostly developed in parallel. In the twenty-first century, Henk De Regt stressed that understanding is a crucial concept for the sciences (De Regt, 2017), but the notion of

[2] Our emphasis on engineering understanding is in tandem with other proposals in comprehending engineering more broadly. For example, Gene Moriarty's (1999) emphasis on "focal engineering" and the dimension of *knowing why* is similar to our focus on *understanding*.

engineering understanding is yet to be extensively studied. We argue that this epistemic category is required in contemporary sociotechnical systems as it is necessary for engineers to provide solutions beyond just using their specialized knowledge (detailed in sect. 13.4). Additionally, to acknowledge the importance of understanding will help engineers in their quest toward a better informed, critical, and self-reflective engineering practice.

The article is divided into four sections. Section 13.2 addresses the various engineering situations where understanding plays a prominent role. Section 13.3 then turns a critical eye toward engineering understanding and its relation to engineering knowledge. Finally, the article concludes with a discussion of one particular example of engineering understanding and its application for engineers in their work practices.

## 13.2 Why Is Engineering Understanding Needed for Complex Sociotechnical Systems?

Complex social technical systems—sociotechnical systems—are classes of systems which involve people and technologies interacting with each other at multiple levels involving operations, management, regulations, and governance (Vermaas et al., 2011). Typical examples of such systems include transportation systems, air traffic control, energy infrastructures, among many others. The key emphasis of the systems is that the people and the technologies have to be comprehended together for the functioning and design of the systems (Vermaas et al., 2011). As a result, the traditional activity of engineering has to be analyzed relative to the activity of experts from a variety of other disciplines. These experts include sociologists, economists, policymakers, along with other experts with whom engineers sustain and develop the functioning of sociotechnical systems (Bijker et al., 2012; Carayon, 2006; Cook, 2000; Hughes, 1988, 1998, 2004; Sommerville & Baxter, 2010; Trist, 1981; Trist et al., 1993; Vermaas et al., 2011; Waterson et al., 2002).

Due to the nature of these complex systems, the traditional approach toward disciplinary engineering knowledge may be not sufficient in order to deal with the tasks of designing, operating, maintaining, or decommissioning the systems. Engineers are oftentimes involved in projects where the limits of their disciplinary knowledge are stretched (Moriarty, 1999). In a number of field studies involving sociotechnical systems, sociologists of engineering and technical knowledge have highlighted that even in the development of technology multiple pathways exist, in which technology is shaped by aspects that go beyond the purely technological aspects (e.g., Bijker et al., 2012).

As a result, the engineers involved in designing, operating, maintaining, or decommissioning sociotechnical systems require a dimension of comprehension that is beyond the disciplinary knowledge of engineering and the individual expertise. Not surprisingly, a major challenge for engineering activities related to sociotechnical systems is to address these systems in terms of a concept beyond engineering knowledge. Typical reflections on engineering knowledge have been made in the form of delineations of what engineers know and how they know it by historians, philosophers, and reflective engineers (see e.g., Vincenti, 1990; Kant & Kerr, 2019). Historians of engineering have highlighted various ways in which engineering knowledge has been shaped and co-developed in society. These include various examples of engineering from countries such as France, Germany, USA, among many other countries. Historians have identified that engineering knowledge has a societal link and has been shaped by the challenges in society in which certain technologies have been developed, which in turn have a reciprocal influence in shaping these societies (see Kant & Kerr, 2019 for a summary).

In all these discussions, the role and creation of engineering knowledge has been exemplified. One major thread has been distinguishing engineering knowledge as differentiated from scientific knowledge (Kant & Kerr, 2019; e.g., Nordmann et al., 2011). Other distinctions have been made in terms of power, corporations, innovations, ideologies, amongst other themes (e.g., Hughes, 1988, 1998, 2004). Similar to historians, philosophers of technology have highlighted the need for studying engineering and its associated knowledge.

While previously, the emphasis on technology as a construct was prominent, the empirical turn in philosophy of technology starting from the 2000s resulted in a shift toward engineering practice, technical artifacts, knowledge, and functions (see, e.g., Meijers & Kroes, 2000; Franssen et al., 2016). In addition to historians and philosophers, reflective engineers have addressed the activity of engineering and engineering knowledge (Mitcham, 1994, 2020; Houkes & Meijers, 2022). Since the aim of this article is to argue for the need of engineering understanding in relation to sociotechnical systems, the next section will delineate the concept further in relation to existing discussions about engineering knowledge.

## 13.3 What Is Engineering Understanding and Its Relation to Engineering Knowledge?

In philosophy, there is a long tradition to discuss epistemological questions and to reflect on the notion of knowledge, in particular. Only recently, understanding as a different but related epistemic state has become the focus of many debates in philosophy of science and epistemology. Some distinctions in the study of knowledge are often taken for granted. These include formulating problems of knowledge as the presence or the absence of knowledge itself. In other words, people either do know a certain fact or they do not. Above that, many scholars aim to describe its usage in the form of sentences. In contrast to knowledge being either present or not, understanding is different, because it comes in degrees. The epistemic agents can understand certain issues better or worse. There is a continuum reaching from minimal understanding over generic and outright understanding toward perfect understanding. A second issue concerns whether understanding has to be true or not. Knowledge that something is the case is said to be "factive." This means that one can only know a proposition in case the proposition is true.[3] Whether understanding is factive or not is debated in philosophy. There are arguments about whether one can understand certain topics or phenomena even if the so-called body of information on which the understanding is based contains false items (e.g., Elgin, 2007, 2017).

The question of what engineers know is a good case in point. After Vincenti's (1990) major contribution, a lively debate emerged in philosophy of technology about different types of knowledge that engineers need in order to successfully finish their projects. The debate takes for granted Gilbert Ryle's (1949) distinction between *knowing how* and *knowing that. Knowing that* received a special attention by philosophers, and a certain consensus developed that this is a fundamental category of knowledge (e.g., Williamson, 2000; Nagel, 2014). This paper treats *knowing that* and *knowing how* as philosophical categories of knowledge. The category of knowing that, also called propositional knowledge, is centrally related to the philosophical concepts of *belief, truth*, and *justification*. Some argue that the conception of knowledge as justified true belief (JTB) can withstand criticisms (Weatherson, 2003) even though another highly influential paper argued already a long time ago that the account of knowledge as justified true belief is not sufficient for knowledge (Gettier, 1963). Forms of *knowing how* to do something in relation to technological knowledge are discussed explicitly by Wybo Houkes (2006), Per Norström (2011), and Meijers and Kroes (2013). In other contributions, "action knowledge" is a term used, which can be interpreted to refer to such a type of practical knowledge or knowledge of how to do something.

One may distinguish practical knowledge from theoretical knowledge, in general. These could be temporarily considered as aspects of knowledge and could also be included in engineering understanding to make it comprehensive in nature. This "comprehensive engineering understanding" would then contain theoretical and practical aspects. The types of knowledge related to propositions could be related to a theoretical aspect of understanding, and the practical knowledge of how to do something could be related to the practical aspect of understanding.[4]

In philosophy of technology, the special character of engineering knowledge is recognized. The tasks of engineers are extremely demanding, and their corresponding knowledge is complex and difficult to achieve. The proponents of the debate about types of knowledge start from the traditional JTB account of knowledge, but they searched for a more adequate theory in order to account for the complex nature of engineering knowledge (De Vries & Meijers, 2013; Meijers & Kroes, 2013).

[3] In philosophy of technology, some authors question whether knowledge always has to be true, though. Some ponder over propositions that are not truth-apt (Meijers & Kroes, 2013), others consider epistemic attitudes beside belief such as acceptance that one might interpret to aim at something else than truth (De Vries & Meijers, 2013).

[4] These relations have been mentioned for ease of exposition of the argument. In fact, we believe that the interrelations between theoretical and practical aspects are much more intricate.

Therefore, they based their proposals on Vincenti (1990), but they developed his ideas further and connected them to the traditional notions of philosophy used in the context of the JTB account that Vincenti did not use. Vincenti proposed six types of knowledge that he called "categories of knowledge": (1) fundamental design concepts; (2) criteria and specifications; (3) theoretical tools; (4) quantitative date; (5) practical considerations; and (6) design instrumentalities. In these six types of knowledge, he did not explicitly distinguish parts that belong to knowing how and parts that belong to knowing that. Vincenti even explicitly refused to use the former notion, "[...] I have resisted the temptation to describe engineering knowledge or some part of it as 'know how'" (1990, p. 261 n.19). However, Vincenti stresses that knowledge includes practical *and* theoretical aspects, in his words: "knowledge of how to perform tasks as well as knowledge of facts" (ibid., p. 13), and he claims that "[k]nowing how shows up as both the knowledge of how to do design and knowledge of how to generate the new knowledge" (ibid.).

Marc De Vries (2003) made a first attempt to interpret Vincenti's categories in terms of philosophical notions. He distinguished four types of knowledge: (1) functional nature knowledge; (2) physical nature knowledge; (3) means-ends knowledge; and (4) action knowledge. These four types of knowledge can be related to the two philosophical categories of knowing that and knowing how. The second type is clearly a form of knowing that, and the fourth type seems to be a form of knowing how or, at least, to be closely related. Whether the first and the third type are also forms of knowing that is debatable. De Vries (2003, p. 119) himself is rather skeptical about the approach of using the JTB account for spelling out what engineers know.[5]

On the positive side, he claims that his types of knowledge map onto Vincenti's categories. De Vries claims that the first type of functional nature knowledge comprises Vincenti's first and fifth category. The second type of physical nature knowledge is claimed to comprise Vincenti's third category (especially the knowledge of scientific laws) and the descriptive part of the fourth category. The third type of means-ends knowledge comprises the second category and the prescriptive part of the fourth one. And, finally, De Vries claims that his type of action knowledge comprises the third category (especially reasoning and use of math) and the sixth one (Table 13.1).

The types or categories of knowledge exemplified in the table are only two examples of proposed sets. Other proposals have been discussed in the literature, and these alternative sets differ more or less from De Vries's contribution (cf. Ropohl, 1997; De Vries & Meijers, 2013; Meijers & Kroes, 2013).

One intermediate conclusion at this point is that engineers are faced with an epistemic challenge. There are several types of knowledge that have to be mastered. Some of these types involve belief or are related to theoretical aspects of knowledge. Others are rather practical or are related to practical aspects of knowledge.

[5] In a later publication by De Vries and Meijers (2013, p. 59 fn.6), it is claimed that "these knowledge types only refer to propositional knowledge. Knowing how is not included in this taxonomy."

**Table 13.1** Mapping from De Vries's types of knowledge to Vincenti's categories of knowledge

| De Vries's types of knowledge | Vincenti's categories of knowledge |
|---|---|
| 1.Functional nature knowledge | 1. Fundamental design concepts<br>5. Practical considerations |
| 2.Physical nature knowledge | 3. Theoretical tools (esp. scientific laws)<br>4. Quantitative date (descriptive) |
| 3.Means-ends knowledge | 2. Criteria and specifications<br>4. Quantitative date (prescriptive) |
| 4. Action knowledge | 3. Theoretical tools (reasoning and math)<br>6. Design instrumentalities |

Discussed types of knowledge that involve practical aspects are, for example, use know-how (Houkes, 2006), or technological know-how from rules of thumb (Norström, 2011). Do the engineers also need to have something like a more demanding knowledge that comprises all of those types? This demanding epistemic achievement would include knowledge of how these different types are related to each other. This epistemic achievement and the achievements, in general, are dependent on integrating the different types of knowledge. Engineers have to grasp a variety of disparate aspects of different branches of knowledge and, more importantly, they also have to integrate them into a system of items that philosophers call differently. A traditional interpretation of these items may be to regard them as beliefs, others may want to call them pieces of information or commitments.

This notion of integration resembles ideas that epistemologists have discussed in the context of the philosophical debate on understanding. Jonathan Kvanvig (2003), for example, regards understanding as the grasping of a body of information. Epistemic agents have to grasp pieces of information and the coherence-making relationships between these pieces when they aim to understand a complex object, a topic, or domain (ibid.). Catherine Elgin builds on Kvanvig's work. She follows him in distinguishing propositional from objectual understanding (cf. 2017, p. 33). For the latter kind of understanding, she requires an account that comprises commitments in reflective equilibrium. The following quote nicely illustrates Elgin's comprehensive approach that also allows practical aspects to play a role in this context:

> At a first approximation, an understanding is an epistemic commitment to a comprehensive, systematically linked body of information that is grounded in fact, is duly responsive to reasons or evidence, and enables nontrivial inference, argument, and perhaps action regarding the topic the information pertains to. (Elgin, 2017, p. 44)

In light of these recent developments in epistemology (and philosophy of science) it is a fruitful idea to think of answering the question of what engineers know rather via such an account of understanding than via an account of knowledge. The engineering knowledge that answers to the epistemic challenge of mastering the complex situation to deal with all these types of knowledge can be perhaps better comprehended as the quest for engineering understanding. If this strategy is adopted, then different types of technological or engineering knowledge become parts of a huge body of information in Kvanvig's (2003) terms. The pieces of this body of information could be seen as pieces of knowledge, but one has to be careful. If one follows the mainstream view in philosophy that knowledge is factive, then one is committed to the view that all of these pieces have to be true. It is, however, debatable whether these pieces are all true. In the following, we will use 'pieces of knowledge' and 'pieces of information' as interchangeable but, thereby, we don't want to imply the strong claim that these pieces are necessarily true. As a result, what competent engineers need on top of just having these pieces of the body of information or body of knowledge is an understanding of how these pieces hang together.

## 13.4 Steps Toward an Engineering Understanding and Its Applicability for Engineers

In our everyday life, complex sociotechnical systems abound; examples include the road transportation system and many other such cases where technologies, people, social entities, rules, and regulations all come together to contribute to the functioning of the system.

A good example where engineering understanding is vitally needed is the repair of an element of a sociotechnical system where the failure was not previously documented or not coherently diagnosed. In such a case, a maintenance crew (technicians and, if required, supervisors who are engineers) will reach the site to make the repair. Typically, two broad categories of faults can be found. First, some faults are straightforward to diagnose and repair. Second, some faults are not so easy to repair and the way of fixing them is not at all clear at the beginning. Maintenance personnel will try several lines of inquiry to localize the fault, and the solution may be reached only after a number of trials. In this second case, the maintenance personnel will try to understand the situation at several levels of abstraction. They may consult a maintenance handbook and look for particular codes or instructions. These could include, "Is an X part working?" "Is the X part receiving current?" "Is the X part giving an output properly?" "If X is working then is there a problem in part Y?" "Is there any dependency on Y due to which the malfunction is displayed?" These and many other such lines of inquiry continue. If the maintainers do not find a reasonable solution in terms of the breakdown of physical components, they move to higher levels of abstraction involving the underlying process, scientific laws, or might even consider the overall purpose of the underlying technology. These several layers of abstraction result in moving away from questions of knowledge found in existing handbooks and manuals toward a broader understanding of the functioning and design of the system as a whole and its intended function in a more comprehensive way that may not have been previously captured.

Problem-solving and sensemaking in sociotechnical systems are similar to the above example. While engineering knowledge is required in most cases, in some other cases, engineering knowledge may not be sufficient to understand operational practices. Understanding in contrast to knowledge aims toward connecting or relating disparate pieces of information in order to provide some sort of grasping of disparate elements, so as to provide coherence for the epistemic agent.

In order to comprehend the value of engineering understanding in sociotechnical systems, let us take the example of a toll booth in a national roadway infrastructural system. Toll booths are established to extract tolls from commuters. These tolls provide the economic support for upkeep and maintenance of existing roadways and road services, along with further development of the road transport system. While the concept of toll booths is not new, in our present world, the influx of automation has connected the act of toll-paying to more complex digital infrastructure that goes beyond the presence of toll-booth attendants and commuters. Thus, it brings about personal and organizational interdependencies that lead to a growth in the complexity of transportation systems. Consider the example of a national electronic toll collection system, known as FAStag, developed in India by the National Payment Corporations India (NCPI) for electronic toll requirements for mobility.[6] This cyber-physical system has several stakeholders, ranging from the customer of the service to the banks and the third-party payment entities that support the functioning of the monetary infrastructure, not to mention the number of employees, law-enforcement agencies, emergency-response agencies, among stakeholder bodies that ensure the seamless functioning of the system. In the FASTag system, the car owners register their vehicles for this service and are provided with a Radio Frequency Identification (RFID) tag. This RFID tag is connected to an electronic payment system, managed through a nationwide network of banks, which allows the customers to upload money into an electronic wallet. This electronic wallet is connected to the RFID tag which can be pasted on the car. When this RFID tag holder car approaches a toll booth then the sensor on the booth detects the tag, deducts the money automatically, and informs the vehicle owner through electronic means. Once the money is deducted, the gate opens letting the vehicle through.

In this example, what would happen if the payment was deducted from the electronic wallet of the car owner but the gate did not open? Is the problem that of a mechanical problem in the gate? Or is it in the electronic payment system? Or is it that the particular tag is old and not functioning reliably? In the case of toll-booth maintenance, certain problems may involve information interchange between the car and the toll-booth sensor as well as the moving mechatronic parts of the toll-booth gate. The answers to these questions have to be coordinated in a way that integrates perspectives of particular disciplines into a generalized epistemic state, which goes beyond the individual types of technological knowledge. Especially, when one needs creative solutions that go way beyond established ways of troubleshooting, then engineering understanding is required. This, however, is not meant to downgrade engineering knowledge. In most cases of everyday operational practices, engineering understanding is required in conjunction with engineering knowledge.

---

[6] For details of the FASTag system: https://www.npci.org.in/what-we-do/netc-fastag/product-overview

## 13.5 Conclusion

In order to comprehend the value of engineering understanding in contemporary sociotechnical systems, we used the example of a toll booth maintenance and repair in a national roadway infrastructural system to illustrate a distinctive kind of engineering epistemic achievement.

Technological knowledge, in general, and different types of technological knowledge, in particular, are crucial for the successful practice of engineering. However, when new or unexpected problems arise, something more than technological knowledge may be required. It is the task of practicing engineers to bring pieces of knowledge together and to find creative solutions that go beyond the once learned pieces of knowledge.

The current article highlights the need for comprehending engineering understanding as separated from engineering knowledge. More work has to be done in order to coherently and systematically develop the concept of engineering understanding. The difference between engineering knowledge and engineering understanding needs to be studied further. The crucial features of understanding and their relationship with engineering knowledge from a comparative viewpoint need to be developed in a systematic way. The aim of this paper has been simply to highlight some fundamental ideas. By addressing these two basic concepts of knowledge and understanding and by considering their interrelations, engineers can develop their own self-reflection as well as aid the engineering community in reasoning about epistemic activities of engineers.

**Acknowledgements** We would like to thank the audience at fPET 2023 in Delft for helpful questions and comments, and we especially would like to acknowledge Carl Mitcham for detailed suggestions to improve this article.

## References

Bijker, W. E., Hughes, T. P., & Pinch, T. J. (Eds.). (2012). *[1987]. The social construction of technological systems: New directions in the sociology and history of technology*. Anniversary edition. MIT Press.

Blockley, D. I. (1992). Engineering from reflective practice. *Research in Engineering Design, 4*(1), 13–22. https://doi.org/10.1007/BF02032389

Brey, P. (2010). Philosophy of technology after the empirical turn. *Techné: Research in Philosophy and Technology, 14*(1), 36–48. https://doi.org/10.5840/techne20101416

Bucciarelli, L. (1984). Reflective practice in engineering design. *Design Studies, 5*(3), 185–190.

Callon, M. (1980). The state and technical innovation: A case study of the electrical vehicle in France. *Research Policy, 9*(4), 358–376.

Carayon, P. (2006). Human factors of complex sociotechnical systems. *Applied Ergonomics, 37*(4), 525–535. https://doi.org/10.1016/j.apergo.2006.04.011

Cook, R. I. (2000). *How complex systems fail (being a short treatise on the nature of failure; how failure is evaluated; how failure is attributed to proximate cause; and the resulting new*

*understanding of patient safety)*. Cognitive technologies Laboratory. https://www.gwern.net/docs/technology/2000-cook.pdf
De Regt, H. W. (2017). *Understanding scientific understanding*. Oxford University Press.
De Vries, M. J. (2003). The nature of technological knowledge: Extending empirically informed studies into what engineers know. *Techné: Research in Philosophy and Technology, 6*(3), 117–130. https://doi.org/10.5840/techne20036318
De Vries, M. J., & Meijers, A. (2013). Beliefs, acceptances and technological knowledge. In M. J. De Vries, S. O. Hansson, & A. Meijers (Eds.), *Norms in technology* (pp. 55–65). Springer.
Elgin, C. Z. (2007). Understanding and the facts. *Philosophical Studies, 132*(1), 33–42. https://doi.org/10.1007/s11098-006-9054-z
Elgin, C. Z. (2017). *True enough*. MIT Press.
Florman, S. C. (1994). *The existential pleasures of engineering*. St. Martin's Press.
Franssen, M., Vermaas, P. E., Kroes, P., & Meijers, A. W. M. (Eds.). (2016). *Philosophy of technology after the empirical turn*. Springer. https://doi.org/10.1007/978-3-319-33717-3
Gettier, E. L. (1963). Is justified true belief knowledge? *Analysis, 23*(6), 121–123. https://doi.org/10.1093/analys/23.6.121
Harlim, J., & Belski, I. (2013). Educating a reflective engineer: Learning from engineering experts. In C. J. Lemckert, G. A. Jenkins, & S. Lang-Lemckert (Eds.), *AAEE2013: Work Integrated Learning-Applying Theory to Practice in Engineering Education, Proceedings of the 24th Annual Conference of the Australasian Association for Engineering Education* (pp. 1–9). Griffith School of Engineering, Griffith University.
Houkes, W. (2006). Knowledge of artefact functions. *Studies in History and Philosophy of Science Part A, 37*(1), 102–113. https://doi.org/10.1016/j.shpsa.2005.12.011
Houkes, W., & Meijers, A. (2022). Engineering knowledge. In S. Vallor (Ed.), *The Oxford handbook of philosophy of technology* (pp. 128–151). Oxford University Press. https://doi.org/10.1093/oxfordhb/9780190851187.013.10
Hughes, T. P. (1987). The evolution of large technological systems. In W. E. Bijker, T. P. Hughes, & T. J. Pinch (Eds.), *The social construction of technological systems: New directions in the sociology and history of technology* (pp. 51–82). MIT Press.
Hughes, T. P. (1988). *Networks of power. Electrification in Western society, 1880–1930*. Johns Hopkins University Press.
Hughes, T. P. (1998). *Rescuing Prometheus: Four monumental projects that changed our world*. Pantheon.
Hughes, T. P. (2004). *American genesis: A century of invention and technological enthusiasm, 1870–1970*. University of Chicago Press.
Hughes, A. C., & Hughes, T. P. (Eds.). (2011). *Systems, experts, and computers: The systems approach in management and engineering, world war II and after*. MIT Press.
Kant, V., & Kerr, E. (2019). Taking stock of engineering epistemology: Multidisciplinary perspectives. *Philosophy & Technology, 32*(4), 685–726. https://doi.org/10.1007/s13347-018-0331-5
Kvanvig, J. L. (2003). *The value of knowledge and the pursuit of understanding*. Cambridge University Press. https://doi.org/10.1017/CBO9780511498909
Latour, B. (1996). *Aramis, or the love of technology*. Harvard University Press.
Law, J. (2011). *Heterogeneous engineering and tinkering*. The Open University.
MacKenzie, D. (1993). *Inventing accuracy: A historical sociology of nuclear missile guidance*. MIT Press.
Meijers, A. W. M., & Kroes, P. (Eds.). (2000). *The empirical turn in the philosophy of technology*. Bingley.
Meijers, A. W. M., & Kroes, P. A. (2013). Extending the scope of the theory of knowledge. In M. J. de Vries, S. O. Hansson, & A. W. M. Meijers (Eds.), *Norms in technology* (pp. 15–34). Springer Netherlands. https://doi.org/10.1007/978-94-007-5243-6_2
Mitcham, C. (1994). *Thinking through technology: The path between engineering and philosophy*. University of Chicago Press.

Mitcham, C. (2020). What is engineering? In D. P. Michelfelder & N. Doorn (Eds.), *The Routledge handbook of the philosophy of engineering* (pp. 11–24). Routledge.
Moriarty, G. (1999). A plea for the practice of focal engineering. *Technology in Society, 21*(4), 439–456. https://doi.org/10.1016/S0160-791X(99)00031-7
Nagel, J. (2014). *Knowledge: A very short introduction*. Oxford University Press.
Nordmann, A., Bensaude-Vincent, B., Loeve, S., & Schwarz, A. (2011). *Science vs. Technoscience: A primer*. Accessed January 3rd, 2024. https://wwwevs.mathematik.tu-darmstadt.de/index.php?evsid=32&evsver=973&evsdir=1013&evsfile=2012-02-07_HandoutNordmann.pdf
Norström, P. (2011). Technological know-how from rules of thumb. *Techné: Research in Philosophy and Technology, 15*(2), 96–109. https://doi.org/10.5840/techne201115210
Ropohl, G. (1997). Knowledge types in technology. In M. J. De Vries & A. Tamir (Eds.), *Shaping concepts of technology: From philosophical perspective to mental images* (pp. 65–72). Springer Netherlands.
Ryle, G. (1949). *The concept of mind*. Hutchinson.
Sommerville, I., & Baxter, G. (2010). Socio-technical systems: From design methods to systems engineering. *Interacting with Computers, 23*(1), 4–17.
Stueber, K. R. (2012). Understanding versus explanation? How to think about the distinction between the human and the natural sciences. *Inquiry, 55*(1), 17–32.
Trist, E. (1981). The evolution of socio-technical systems. *Occasional Paper, 2*.
Trist, E., Murray, H., & Trist, B. (1993). *The social engagement of social science, a Tavistock anthology, Volume 2*. The Socio-Technical Perspective. University of Pennsylvania Press.
Van de Poel, I., & Goldberg, D. E. (Eds.). (2010). *Philosophy and engineering: An emerging agenda*. Springer.
Vermaas, P. E., Kroes, P., Van de Poel, I., Franssen, M., & Houkes, W. (2011). *A philosophy of technology: From technical artefacts to sociotechnical systems*. Morgan & Claypool.
Vincenti, W. G. (1990). *What engineers know and how they know it*. Johns Hopkins University Press.
Vinck, D., & Blanco, É. (Eds.). (2003). *Everyday engineering: An ethnography of design and innovation*. MIT Press.
Waterson, P. E., Older Gray, M. T., & Clegg, C. W. (2002). A sociotechnical method for designing work systems. *Human Factors, 44*(3), 376–391.
Weatherson, B. (2003). What good are counterexamples? *Philosophical Studies, 115*(1), 1–31.
Williams, B., Figueiredo, J., & Trevelyan, J. (2013). *Engineering practice in a global context: Understanding the technical and the social*. CRC Press.
Williamson, T. (2000). *Knowledge and its limits*. Oxford University Press.

**Vivek Kant** is an active researcher in the field of systems design, human factors (cognitive ergonomics/cognitive engineering), sociotechnical systems, and history and philosophy of desigengineering. He heads the Human Factors and Sociotechnical Systems Studios at the Department of Design, Indian Institute of Technology Kanpur.

**Michael Poznic** MA Philosophy (RWTH Aachen University), PhD Philosophy (TU Delft, 2017). 2013–2015 PhD Researcher (TU Delft). 2016–2020 (Postdoctoral) Researcher at Karlsruhe Institute of Technology (KIT). 2021–Today, Senior Researcher at Institute for Technology Assessment and Systems Analysis (KIT). Managing Editor of the journal *Techné: Research in Philosophy and Technology*. Author of several articles in journals such as *Design Science*, *Synthese*, *Erkenntnis*, *Philosophy & Technology*, etc. Research areas: Philosophy of engineering and technology, philosophy of science.

# Chapter 14
# Operators' Experiences with Intelligent Compaction Systems in Road Pavement: A Technological Mediation Approach

**Hans Voordijk, Seirgei Miller, and Faridaddin Vahdatikhaki**

**Abstract** This study delves into the influence of intelligent compaction (IC) systems on the perceptions and actions of roller operators in road pavement practices. These operators occasionally tend to ignore these IC systems or even turn them off while performing their tasks. A major question is why the operators don't always use IC systems. By using the technological mediation approach, operators' reasons for ignoring or even turning IC systems off in road pavement practice are explored. In the technological mediation approach, users and technologies co-shape each other. Based on semi-structured interviews with roller operators, experiences with these systems are investigated. Physical, cognitive, and contextual mediation through IC systems is related to behavior-changing strategies encompassing coercive and persuasive interventions, as well as decisive designs. These interventions are now experienced as conflicting with the individual autonomy and responsibility of the operators for the quality of work. Because IC systems amplify certain facets of reality while reducing others, roller operators must ensure that the output of an IC system is judged in the light of the specific project context. Given their "thrownness" in such a specific context, operators' experience and tacit knowledge remain important for taking action.

**Keywords** Construction practice · Behavior-changing strategies · Technological intentionality · Support systems

H. Voordijk (✉) · S. Miller · F. Vahdatikhaki
University of Twente, Enschede, The Netherlands
e-mail: j.t.voordijk@utwente.nl; s.r.miller@utwente.nl; f.vahdatikhaki@utwente.nl



C. Didier et al. (eds.), *Engineering and Value Change*, Philosophy of Engineering and Technology 48, https://doi.org/10.1007/978-3-031-83549-0_14

## 14.1 Introduction

Road pavement is a complex process influenced by numerous factors. Existing support systems collect both design (e.g., asphalt mix type and pavement design) and real-time (e.g., asphalt temperature and roller pass count) data to help operators better plan (or re-plan) and execute road pavement operations. The cognitive load imposed by existing operator support systems may be reduced by transferring the task of data interpretation and integration to the system as much as possible. Operator support systems can be transformed from descriptive systems, i.e., systems that only provide the information needed for planning asphalt compaction operations, into prescriptive systems, i.e., systems that suggest actionable compaction strategies based on the collected data.

These so-called intelligent compaction (IC) systems continuously record the movement of a roller and prescribe the required level of asphalt compaction and the number of passes required (Hu et al., 2017a). However, it is known from practice that roller operators occasionally tend to ignore these IC systems or even turn them off while performing their tasks. The question is why the operators do not always use the IC systems? To understand the reasons why this is the case the interaction between operators and IC systems will be investigated following the technological mediation approach. In this approach, interactions are not only understood as an encounter between two independent entities, i.e., users and technologies, but also a mechanism through which they co-shape each other (Verbeek, 2008). By using this approach, operators' reasons for ignoring or even turning IC systems off in road pavement practice are explored.

The starting point is Ihde's (1990, 2009) and Verbeek's (2008) analysis of technological and composite intentionality. The frameworks developed by Dorrestijn (2017) and Tromp et al. (2011) provide more detailed insights into the ways technology influences or transforms users' actions and behavior. Next, based on semi-structured interviews with the end users of IC systems, the roller operators, experiences with using these systems are investigated. Subsequently, it is discussed how mediation through IC systems is related to behavior-changing strategies (ibid., 2011). Finally, conclusions are drawn.

## 14.2 Technological Mediation

### *14.2.1 Technological Intentionality*

Ihde reconceptualizes the intentional relation between subject and object in phenomenology by investigating the fundamentally mediated character of this relation by technology (Aagaard, 2017; Ihde, 1990; Rosenberger & Verbeek, 2015). Ihde's post-phenomenology shows that amplification and/or reduction is a basic experiential structure of human–technology relations (Adams & Thompson, 2011): the use of a particular technology causes a person to experience a certain aspect of reality, which is amplified, while simultaneously the experiences of other aspects of reality are reduced.

The hermeneutic human–technology relation involves interpretation, meaning generation, and some kind of reading of a text or graphical representation (Rosenberger & Verbeek, 2015; Turk, 2001). This relation is conceptualized as follows: human → (technology—world) (Ihde, 2009). The technology and the world are like a unified entity, which is represented by (technology—world), in which the technology provides a representation of a specific object or aspect of the world. Users turn their attention to a technology or device in order to read and interpret the representation it provides. The arrow between human and technology indicates this human intentionality.

According to Verbeek (2008), "hermeneutic relations always involve a technologically generated representation of the world." This is the result of a "specific technological directedness at the world: thermometers focus on temperature ... sonograms on how material objects reflect ultrasound" [p. 393]. This technological directedness of the "sensing apparatus" within a technology is designed to capture and process inputs in specific manners: a given technology is oriented toward a specific aspect of reality (Ihde, 1990; Verbeek, 2008; Wiltse, 2014). Adding the "directedness" of a technology to human intentionality results in composite intentionality conceptualized as human → (technology → world) (Verbeek, 2008). This gives rise to a dual intentionality: one from technology toward the world, and another from humans toward the outcomes of this technological intentionality. Technology developers design and/or program a certain intentionality into the technology itself. As a consequence, technologies materially embody human intentions, i.e., they capture a version of reality it is designed or intended to (Redaelli, 2022).

Using intelligent systems and technologies that perceive and respond to the world around them is a clear manifestation of this composite intentionality. An intelligent system not only represents and shapes the world but does so independently, not merely as an extension of the human (Verbeek, 2011). Intelligent systems present emergent properties, which seem to assign them a certain degree of freedom. One could speak of a distribution of agency between humans and technologies. Decisions and actions are not exclusively determined by either humans or technology but by collaborative outcomes involving both, forming a blended form of agency known as "composite agency" (Fritz et al., 2020).

Composite agency means that technological intentionality cannot be reduced to that of humans (Redaelli, 2022). Fritz et al. (2020) talk about a 'triadic agency' between users, designers, and technologies. Composite agency serves as postphenomenology's response to the notion of technology as a neutral tool, highlighting the active role played by technology. However, attributing agency to technologies does not imply determinism. The agency of technology does not reside as an inherent meaning or purpose within the technology; instead, it arises through its use (Verbeek, 2008).

### 14.2.2 Loci and Types of Mediation

The use of digital technologies requires interactions among and alignment with users and technologies. These interactions can be analyzed in terms of locus and type (Verbeek, 2013). For conceptualizing the locus or points where mediation is applied, Verbeek (ibid.) refers to the framework of Dorrestijn (2012). Dorrestijn translates different loci of technological mediation into more specific interactions between human beings and technologies. Dorrestijn categorizes the interactions according to the side from which technologies grasp the human body, i.e., physical, cognitive, or contextual. Tromp et al. (2011) categorize the various types of mediation in terms of their influence (weak versus strong) and their visibility (hidden versus explicit) (see Table 14.1).

This typology is used here to categorize the different physical, cognitive, and contextual influences identified by Dorrestijn in terms of behavior-changing strategies (see Table 14.2).

The first locus of mediation focuses on how an individual physically encounters technologies (Dorrestijn, 2012, 2017). The most direct influences that technologies have on humans are when they steer physical behavior, such as *embodiment* and *coercion*: people acquire skills and learn routines and technologies force people into a certain direction (Latour, 1992). Both embodiment and coercion are forms of strong and explicit physical influences. People are aware of this influence and experience it as a strong external force (Tromp et al., 2011).

The second locus of mediation focuses on how technology cognitively impacts human decision-making (Dorrestijn, 2012, 2017) such as *guidance* and *persuasion*: users are being steered toward an intended or appropriate use or persuaded to change their behavior through technology (Fogg, 2002). Guidance and persuasion can both be apparent but also weak in their influence. Technology can guide behavior by giving signs, but these signs can easily be ignored. Technologies can explicitly encourage people to change their behavior, but their influence can be experienced as weak (Tromp et al., 2011).

The third locus of mediation addresses the contextual influence of the technological environment that is implicit but can be both weak and strong. It relates to 'the material infrastructure that has an impact on our actions and experiences' (Verbeek, 2015). A technology requires infrastructure for its operation. These *background conditions* of a technology can be decisive: strong but implicit. People experience their behavior as externally motivated or regulated but do not see this as a deliberate influence or choice influenced by the designer of a technology (Tromp et al., 2011). Second, *technical determinism* reflects how technical developments may transform human values and needs. The threat of being watched through

**Table 14.1** Types of mediation: force and visibility

| | Hidden | Apparent |
|---|---|---|
| Strong | *Decisive* | *Coercive* |
| Weak | *Seductive* | *Persuasive* |

**Table 14.2** Loci and types of mediation

| | Hidden | Apparent |
|---|---|---|
| Strong | Contextual<br>• *Background conditions* | Physical<br>• *Coercion*<br>• *Embodiment* |
| Weak | Contextual<br>• *Technical determinism* | Cognitive<br>• *Guidance*<br>• *Persuasion* |

technology may generate behavioral changes and could lead to 'anticipatory conformity'. The influence can be seductive: weak and implicit. People are unaware of this influence and experience their behavior as internally motivated (ibid., 2011).

## 14.3 Research Method

The methodological approach employed in this study is built upon two common characteristics of technological mediation studies (Rosenberger & Verbeek, 2015). First, these studies take the role actual technologies play in practice as their starting point and focus on concrete case studies of technologies. Second, Ihde's analysis of human–technology relations serves as a typical heuristic in these case studies (Adams & Thompson, 2011).

### *14.3.1 Intelligent Compaction Systems*

The object of study is IC systems used by roller operators in road pavement practice. An IC system consists of a GPS, accelerometer, temperature sensors, and a display. The function of GPS is to provide the location and the speed of the roller in real time, the number of passes, and the compaction path (Zhang et al., 2019). The accelerometer installed on the roller registers the interaction between the roller and the ground and calculates the ground stiffness (Liu et al., 2016). Infrared temperature sensors located in the front and back of the roller keep the operator informed about the temperature of the pavement (Hu et al., 2017b). In this way, the roller operator knows when he should stop or continue the compaction process.

All data gathered from GPS, accelerometer, and temperature sensors are collected and visualized on a display in the driver's cabin. The temperature contour plot (TCP) depicts the surface temperature conditions of newly paved asphalt. The compaction contour plot (CCP) shows data regarding the achieved compaction. Based on the TCP and the CCP, a compaction priority map (CPM) provides data on the number of passes required (Hu et al., 2017a) (Fig. 14.1).

**Fig. 14.1** Different IC systems in road pavement (https://wwshare.services.wirtgen-group.com/wwShare/=DownloadController? md5 = 933d9c4032 ff5dd57d3b60eb362c3f19)

### *14.3.2 Data Collection and Analysis*

To share experiences with the use of IC systems and to identify reasons why operators ignore or even turn these systems off, 17 semi-structured interviews were conducted with roller operators: ten from the largest, six from second largest, and one from the fifth largest construction company of the Netherlands. These companies were chosen because they recently implemented the newest versions of IC systems: the HAMM Compaction Quality (HCQ), the Smart Doc (SD) and the Trimble Compaction Control (TCC) system. All operators have an intermediate vocational education level. Almost half have more than 20 years of experience, a third between 11 and 20 years and a fifth with a maximum of 10 years of experience. Semi-structured interviews are used to provide the interviewees with a safe environment wherein they feel at ease to reflect on their personal experiences and share them with the researcher, and also because they allow researchers to improvise follow-up questions, if necessary. The length of the interviews was about one hour and were conducted at the construction company or on the construction site.

The interviews have been analyzed using thematic analysis (Braun & Clarke, 2006). The first stage included transcribing the interviews and reading through the interviews' transcriptions. The second stage involved initial coding for interesting content identified in the interviews. Similar input from each interview was clustered under the same code. The third stage aimed to gather all relevant codes under a potential theme by looking for similarities. The next stage was defining and naming themes. The fifth stage was dedicated to refining and defining the themes presented and indicating the story that each theme presents in relation to the major research question.

The interview results were validated internally by the roller operators and externally by validation experts. Internally, the results were summarized and presented

**Table 14.3** Experts involved in focus group validation

| # | *Job title* |
|---|---|
| 1 | Roller operators' trainer at a vocational college |
| 2 | Product manager of digital solutions of an equipment supplier company |
| 3 | Asphalt/equipment manager of a construction company |
| 4 | Head of the central asphalt lab of a construction company |

individually to four roller operators from the first construction company and three roller operators from the second construction company. These roller operators were chosen from those who had been interviewed before. Externally, the results were validated in a focus group discussion in two sessions of two hours each with four road construction experts with management experience (Table 14.3). When the role of the management is mentioned, this is based on the results of these validation sessions. The output of these validation sessions is connected with the interview results.

## 14.4 Interactions Between Operators and IC Systems

Based on the side from which technologies grasp the human body, this study uses the three loci of application of mediation to describe the different types of interaction between IC systems and the roller operators in construction practice.

### *14.4.1 Physical Influences*

#### 14.4.1.1 Lack of Training and Feedback

Interview results have shown that no formal training program has been given to roller operators before using the IC systems in road pavement practice. Alternatively, sporadic explanations are given on subjects such as how to operate the system, how to log in, how to set up connections, and how to insert information about the project. "If something went wrong with the system during my work, I don't know what to do, I have in this case to call and wait for a solution," said one of the interviewees.

Newer employees have not even received any explanation on the HCQ system. These operators have to learn how to use this IC system from their colleagues. As mentioned by an interviewee: "I haven't heard anything from the management about HCQ when I have been employed. I have been told about HCQ by a colleague." Another operator contended: "I've had to figure out how to use the system on my own since there was no manual, explanation or training provided."

Regarding the SD system, operators receive no explanation, training, or manual on its use except some presentations about the system during a visit to the vendor's factory. "Except for a factory visit, we didn't get an explanation on how the SD system works, how we benefit from it, or how to use this information. Training and good explanation are important."

Regarding the TCC system, there is also no proper explanation or training provided. Operators have been provided with a single piece of paper in the roller cabin on how this system should be set up for registration. Most operators feel the IC system is not for them but for the management. "No explanation was given at all, nothing about that. Trimble was put in the roller cabin, you get an A4 paper with it. This is an on-and-off button, you do this with this button. But no one has been with us to say, so you should use it… like this," objected one interviewee.

The lack of training leaves operators with very limited understanding of the interface and advanced functions of the system. On-the-job training is not the only issue. Roller operators are also not trained to use this type of systems during their education. Operators also do not get feedback about their use of IC systems and the data registered, except when something goes wrong. As put by an interviewee: "I think if you receive feedback once in a while about it, for example by sending an overview of such a project, it will be useful because then you have the feeling that you actually do it for something and not just that you have to." Operators recognize that learning from others' experiences may help them advance their own skills. According to the experts involved the focus group validation, the lack of feedback on the use of the IC system is caused by technological difficulties in analyzing the data registered.

#### 14.4.1.2 Mandatory Use and Lack of User Involvement

Operators indicated that while the use of IC systems was only encouraged at the beginning, they later felt pressure to do so. Operators feel compelled to use the system even though there are no explicit management requirements to do so. This feeling is strengthened by the notification operators receive when the IC system is not turned on when they start working. Therefore, operators frequently turn on the IC system but flip the system's screen instead of using the information provided. "You don't do it for yourself. You do it for the client. Because if you don't put the system on, we receive a message from managers to turn it on. The intention is that you have to use the system," argued one of the interviewees. Operators perceive IC systems as tools that are only made for monitoring their work. As noted by an interviewee: "The company wants to be able to see what you have done and that is the most important thing for the company. That's why we have the idea that these systems are for the people in the office and not for us."

Operators are not involved in the decision-making process to choose a particular IC system. "These systems are simply built in our rollers and not discussed with us," said an interviewee. Some older operators mention that they have been asked for feedback on different types of equipment in the past. "If you have to work with it, then you have to work on it. So if you can give your feedback on the system that can make a lot of difference. For example, we went to Germany. I sat down in a new roller. I could not put my feet on the ground. So, I indicated that and they are now arranging something. So imagine if you could do that with an IC system." Roller operators argue that they have been told that these systems are for them while they have not been asked their point of view. "Certainly if we have been asked about our preference between multiple systems, you feel more involved," argued an interviewee.

According to the experts involved in the focus group validation, analysis of the registered data and the obligation of turning on the IC system is not to control but to enhance work efficiency for future projects by learning from mistakes. However, there appears to be a lack of effective communication between operators and management regarding these goals and advantages.

## 14.4.2 *Cognitive Influences*

### 14.4.2.1 Lack of Valuable Decision Support

The support function in IC systems that prescribes the required number of passes to reach optimal compaction is not as appreciated. According to the operators, this prescription is only based on data from laboratory experiments with different asphalt mix types. For them, determining the required number of passes is based on multiple factors that IC systems do not measure, such as weather conditions and the humidity of the foundation. "Nice to get that four roller passes is enough on your screen, but if the asphalt cools down faster because of weather conditions, then you're maybe not done with seven roller passes," raised one interviewee. Other on-site conditions not recognized by IC systems are the color/texture of the ground, the asphalt's quality, and soft patches in the asphalt. Operators use these observations to inform colleagues about the on-site conditions.

IC systems also do also not register compositional alterations in asphalt being compacted, such as changes in the kind or size of the particles and places where roller operators have used the vibration compaction. Moreover, there is doubt about the accuracy of the temperature readings because of the effect of wind or hindrance of the temperature sensor by parts of the roller. Therefore, operators do not solely rely on the IC system but also on their own knowledge and experience to determine the number of roller passes needed to reach the required level of asphalt compaction. Validation specialists recognize the skill of roller operators and their capability to evaluate these conditions.

### 14.4.2.2 Safety Concerns

During compaction, operators have to focus on multiple factors at the same time. For instance, the roller operator has to keep an eye on the gauges in his cabin, and monitor the speed of the asphalt spreader and other rollers and the workers on-site.

They also sometimes ignore the IC system because of safety concerns, e.g., when they work near a bike path. An interviewee expressed concerns about the safety issues caused by the distraction of the IC system by saying "If you for example work on provincial roads with commuters as cyclists, there is something that can happen. So the safety, that's what matters to me." In order to prevent being distracted while working, roller operators say they frequently turn the screen off. They only use the system to see if they miss any locations while compacting the asphalt. "I do still use it when I have rolled a long stretch. Then, I zoom out and check if it's a nice even color. Because then I also know, okay, I've been everywhere. But that's more for myself."

Operators believe they cannot rely only on the IC system because the technology is not mature enough to help them in decision-making during compaction. The major reason is that the IC system does not take all relevant site conditions into account and measurements can be unreliable due to external circumstances. Operators sometimes have to ignore the IC system because of safety concerns. As a result, operators perceive IC systems do not help them save time and increase the quality of the compaction process.

## *14.4.3 Contextual Influence*

### 14.4.3.1 No Interface Between Rollers and Pavers

An IC system is placed on a single roller and there is no interface between or connection with other rollers or pavers. The temperature readings and the speed of the asphalt pavers are not communicated to the operators. They also mention that displaying temperature at one specific location, at the location where the roller moves, is not convenient. Predicting the cooling of the asphalt mat at different timings and locations during the movement of the paver would add value. The interruption of the GPS signal during working in tunnels, between high buildings, or under trees causes loss of equipment tracking on the display and distracts roller operators during their work. The foreign language used on the display also makes the IC system appear complex.

### 14.4.3.2 Being Judged

Roller operators believe the IC system and the registered data are used for the assessment by the management. It happens, for example, that sometimes roller operators make fewer roller passes in some areas resulting in a lower density value. By comparing the density value of a specific spot with the number of roller passes, the management blames the roller operators for that. Roller operators stress the necessity for the management to be open about how it uses the technology as well as the data registered. Validation experts confirmed that lack of clarity and transparency about the application purpose causes distrust among roller operators in the IC systems. While management says it adopts the technology to increase production quality, roller operators believe it is utilized for monitoring and controlling their work.

The use of IC systems is influenced by the lack of training and the mandatory use (physical), no valuable decision support and safety concerns (cognitive), and no interface between roller and paver and the fear of being judged (contextual).

## 14.5 Discussion

The section presents the discussion on how technological selectivities make certain aspects of reality accessible to users of the system through data processing. Additionally, the relationship between behavior-changing strategies and physical, cognitive, and contextual mediation through IC systems is elaborated in this section.

### *14.5.1 Technological Selectivities and the Role of Experience*

When the relation of composite intentionality involves an IC system for road pavement, it can be conceptualized as human → (IC system → road). This equation, in an elementary form, models the connection between what is going into and coming out of the IC system. This connection is based on the built-in technological selectivity of an IC system (Ihde, 2009). This selectivity determines possible differences between the data and image displayed on the screen and on-site compaction conditions as perceived by the roller. These differences are attributed to the so-called epistemic uncertainties, i.e., uncertainties derived from the lack of knowledge of a parameter, phenomenon, or process (Basu, 2017).

In the case of IC systems, epistemic uncertainty occurs either from transforming inputs into outputs within the IC system or from external factors disturbing the monitoring of road pavement operations (Zio & Pedroni, 2012). One specific technological selectivity revolves around the acquisition of data using GPS sensors configured to capture the roller's location and translate them into the images displayed. External factors, such as ground vibration, wind, and hindrance of the sensors by roller parts can also give rise to differences between displayed data and perceived on-site conditions. Consequently, the technological directedness of IC systems does not necessarily result in accurate representations of the world (Verbeek, 2008).

The representations amplify an operator's experience of a certain aspect of reality while restricting the reality representation to the scope and boundaries of a two-dimensional screen (Ihde, 1990). A roller operator could experience "blindness" if he or she relies only on displayed data and prescribed actions by the IC system rather than, in Heideggerian terms, recognizing his or her "thrownness" in a given project context (Dias, 2006). Each road pavement project has unique features with particular idiosyncrasies that require context-sensitive perceptions and actions. As such, a roller operator can be seen as being thrown into a specific project context. The context-sensitivity of actions and interpretations taking place in a road pavement project is based on this "thrownness" (Turk, 2001).

In case these practitioners start to think reflectively, they disconnect themselves from this "thrownness." Roller operators must ensure that the output of an IC system is judged in the light of the specific project context. In terms of Heidegger, roller operators have to be committed primarily to "being-in-the-world" rather than to "being-in-a-model-of-the-world" (Turk, 2001). Given this "thrownness" in a particular project with its particular on-site conditions, operators' experience and tacit knowledge remain important for taking action.

### 14.5.2 Behavior-Changing Strategies

Various intervention and behavior-changing strategies surrounding the use of IC systems in road pavement practice are aligned with the three categories of mediating actions: physical, cognitive, or contextual (see Table 14.2).

First, a coercive intervention related to physical influence is visible because operators are compelled to use IC systems in road pavement practice. Such intervention disrupts the routine behaviors of the operators and is experienced as conflicting with individual autonomy (Tromp et al., 2011). It can be argued that this behavior-changing strategy has not succeeded yet because there is no widespread agreement among operators about the need to use these systems. Operators also do not acquire the necessary skills and expertise to work with the system automatically: the system does not become embodied or 'ready-to-hand'. This gets worse because no feedback is given on the use of the IC systems.

Second, a persuasive intervention can be identified in the use of visualized instructions to influence operators' decision-making (Dorrestijn, 2012). This behavior-changing strategy has also not been successful. The IC system and its compaction priority map that tries to guide the operator in a specific direction is perceived as 'quasi-other' (Hogan & Hornecker, 2011). Operators largely disregard directions provided by these priority maps because the instructions lack context awareness. Operators rely more on their own cognitive abilities and experience. They feel they bear the primary responsibility for the quality of work, resulting in a strong indisposition to delegate decision-making to IC systems.

The contextual intervention is related to a decisive behavior-changing strategy, i.e., IC systems shape the context in which operators operate and determine possible user-technology interactions. The system interoperability of IC systems and possibilities for data analysis shape these interactions. In terms of background conditions, IC systems lack an effective interoperability among different rollers and pavers. This lack of interoperability makes the IC system for the operator less relevant. Registering data makes the tacit dimensions of the skills and knowledge of the operators more explicit and their performance more visible (Ibidunni et al., 2023;

Olaisen & Revang, 2018). This transparency fuels operators distrust in the management since they feel it may use explicit data to punish them in some way or another. This exemplifies technical determinism where technology induces behavioral changes.

## 14.6 Conclusion

This study delved into the influence of IC systems on the perceptions and actions of operators in road pavement practices. The built-in technological selectivity of IC systems plays a pivotal role in mediating perceptions and rendering specific aspects of road pavement operations accessible to its users, the operators. Through the presentation of TCPs, CCPs, and CPMs, an IC system accentuates or amplifies certain facets of reality while diminishing or reducing others. Roller operators could experience "blindness" if they only rely on displayed data and prescribed actions by the IC system. They must ensure that the output of an IC system is judged in the light of the specific project context. Given their "thrownness" in a project with its particular on-site conditions, operators' experience and tacit knowledge remain important for taking action.

The investigation revealed that IC systems in road pavement practice mediate actions physically, cognitively, and contextually. These various forms of mediation actions are connected to behavior-changing strategies used to implement technologies. These strategies encompass coercive and persuasive interventions, as well as decisive designs. A coercive intervention, tied to the physical embodiment of systems, compels operators to adopt IC systems, succeeding only when there is a consensus among operators. Cognitive influences involve interventions where operators are directed and persuaded by IC systems, which act as quasi-others to the agency of operators. Persuasive interventions are most impactful when collective and individual interests align with each other, such as providing incentives for operators to enhance road construction pavement processes through system use. Coercive and cognitive interventions are now experienced as conflicting with the individual autonomy and responsibility of the operators for the quality of work. Contextual influence pertains to the decisive designs of technologies that shape human–technology interactions during use. This influence can even provoke distrust, illustrating technical determinism.

To facilitate the use of IC systems, it is essential to establish trust. Different behavior-changing strategies can be implemented to foster this trust. This requires a shift in training strategies for new operators and building trust in the system by providing more opportunities for exposure and experimentation. Operators could learn from each other by sharing feedback based on the data registered and addressing issues that arise during road pavement construction. Insights into the actual quality of the pavement construction process may persuade operators to take corrective actions. Trust will also increase when operators are involved in the decision-making process to choose a particular IC system. Operators are then more likely to accept influences that increase efficiency and effectiveness through using IC systems.

While this examination presents an innovative theoretical perspective on the interplay between users and digital technologies in road construction, it is subject to certain limitations. The conclusions and analysis are derived from a limited number of interviews, demanding additional research to substantiate external validity. Moreover, a more comprehensive understanding of the scope, use, and evolution of IC systems in road pavement practices is essential for a more nuanced comprehension of user–technology interactions over time. An interesting direction for future research is to study if different behavior-changing strategies to implement and to facilitate the use of IC systems also apply to intelligent systems in other sectors. Another direction for future research is the role of the management and the design of the technology in explaining the operators' reasons for ignoring or even turning IC systems off in road pavement practice. Why did no company provide training to their employee regarding a new system? What were the reasons for the lack of effective communication between operators and management? And will an increasing technological maturity of the IC system facilitate operators to use it? These are many questions for further research.

**Acknowledgments** We would like to thank Wael Asadi for supporting the data collection and analysis.

## References

Aagaard, J. (2017). Introducing postphenomenological research: A brief and selective sketch of phenomenological research methods. *International Journal of Qualitative Studies in Education, 30*(6), 519–533.

Adams, C. A., & Thompson, T. L. (2011). Interviewing objects: Including educational technologies as qualitative research participants. *International Journal of Qualitative Studies in Education, 24*(6), 733–750.

Basu, S. (2017). *Plant hazard analysis and safety instrumentation systems*. Elsevier.

Braun, V., & Clarke, V. (2006). Using thematic analysis in psychology. *Qualitative Research in Psychology, 3*(2), 77–101.

Dias, W. (2006). Heidegger's resonance with engineering: The primacy of practice. *Science and Engineering Ethics, 12*(3), 523–532.

Dorrestijn, S. (2012). Theories and figures of technical mediation. In J. Donovan & W. Gunn (Eds.), *Design and anthropology* (pp. 219–230). Routledge.

Dorrestijn, S. (2017). The product impact tool: The case of the Dutch public transport chip card. In K. Niedderer, S. Clune, & G. Ludden (Eds.), *Design for Behaviour Change* (pp. 26–39). Routledge.

Fogg, B. J. (2002). Persuasive technology: Using computers to change what we think and do. *Ubiquity, 2002*(December), 2.

Fritz, A., Brandt, W., Gimpel, H., & Bayer, S. (2020). Moral agency without responsibility? Analysis of three ethical models of human-computer interaction in times of artificial intelligence (AI). *De Ethica, 6*(1), 3–22.

Hogan, T., & Hornecker, E. (2011). Human-data relations and the lifeworld. In *Proc. 5th Irish Human Computer Interaction Conference*.

Hu, W., Huang, B., Shu, X., & Woods, M. (2017a). Utilising intelligent compaction meter values to evaluate construction quality of asphalt pavement layers. *Road Materials and Pavement Design, 18*(4), 980–991.

Hu, W., Shu, X., Huang, B., & Woods, M. (2017b). Field investigation of intelligent compaction for hot mix asphalt resurfacing. *Frontiers of Structural and Civil Engineering, 11*, 47–55.

Ibidunni, A. S., Agbi, B. D., & Kehinde, B. E. (2023). Interacting effects of tacit knowledge and learning orientation in improving firm performance. *Journal of the Knowledge Economy, 14*(3), 2539–2560.

Ihde, D. (1990). *Technology and the lifeworld: From garden to earth.* Indiana University Press.

Ihde, D. (2009). *Postphenomenology and technoscience: The Peking university lectures.* Suny Press.

Latour, B. (1992). Where are the missing masses? The sociology of a FewMundane artifacts. In W. E. Bijker & J. Law (Eds.), *Shaping technology/building society: Studies in sociotechnical change* (pp. 225–258). MIT Press.

Liu, D., Lin, M., & Li, S. (2016). Real-time quality monitoring and control of highway compaction. *Automation in Construction, 62*, 114–123.

Olaisen, J., & Revang, O. (2018). Exploring the performance of tacit knowledge: How to make ordinary people deliver extraordinary results in teams. *International Journal of Information Management, 43*, 295–304.

Redaelli, R. (2022). Composite intentionality and responsibility for an ethics of artificial intelligence. *Scenari, 17*(2), 159–175.

Rosenberger, R., & Verbeek, P.-P. (2015). A field guide to postphenomenology. In R. Rosenberger & P.-P. Verbeek (Eds.), *Postphenomenological investigations: Essays on human-technology relations* (pp. 9–42). Lexington Books.

Tromp, N., Hekkert, P., & Verbeek, P.-P. (2011). Design for socially responsible behavior: A classification of influence based on intended user experience. *Design Issues, 27*(3), 3–19.

Turk, Ž. (2001). Phenomenologial foundations of conceptual product modelling in architecture, engineering and construction. *Artificial Intelligence in Engineering, 15*(2), 83–92.

Verbeek, P.-P. (2008). Cyborg intentionality: Rethinking the phenomenology of human–technology relations. *Phenomenology and the Cognitive Sciences, 7*(3), 387–395.

Verbeek, P.-P. (2011). *Moralizing technology: Understanding and designing the morality of things.* University of Chicago press.

Verbeek, P.-P. (2013). Technology design as experimental ethics. In S. Van den Burg & T. Swierstra (Eds.), *Ethics on the laboratory floor* (pp. 83–100). Palgrave Macmillan.

Verbeek, P.-P. (2015). COVER STORY beyond interaction: A short introduction to mediation theory. *Interactions, 22*(3), 26–31.

Wiltse, H. (2014). Unpacking digital material mediation. *Techné: Research in Philosophy and Technology, 18*(3), 154–182.

Zhang, Q., Liu, T., Zhang, Z., Huangfu, Z., Li, Q., & An, Z. (2019). Unmanned rolling compaction system for rockfill materials. *Automation in Construction, 100*, 103–117.

Zio, E., & Pedroni, N. (2012). Uncertainty characterization in risk analysis for decision-making practice. In *Number 2012–07, cahiers de la Sécurité Industrielle. Foundation for an Industrial Safety Culture.*

**Hans Voordijk** MA Economics & MA Philosophy, PhD Economics (Maastricht University 1994). 1994–2000 Assistant Professor in Organization & Strategy, Tilburg University, the Netherlands. 2000–Now Associate Professor in Construction Management and Engineering, University of Twente, the Netherlands. 2021–Now Honorary Associate Professor, RMIT University Melbourne, Australia. Research areas: Philosophy of technology and civil engineering, supply chain management, and digital technologies in construction.

**Seirgei Miller** NHD Civil Eng., Master of Science, PhD Civil Engineering (University of Twente, 2010). 2013–Present—Associate Professor in Construction Management and Engineering,

University of Twente, the Netherlands. Research areas: Digital technologies in construction, process control in construction, human capital development in construction, data-driven modelling.

**Faridaddin Vahdatikhaki** BS Civil Eng., Master of Science, PhD Civil Engineering (Concordia University 2015). 2016–Now Associate Professor in Construction Management and Engineering, University of Twente, the Netherlands. Research areas: Digital technologies in construction, simulation and optimization, data-driven modelling, digital twinning, information modeling, virtual reality.

# Chapter 15
# Training Engineers for Sustainability, but Which One? A Discussion of Critical Alternatives to the "Good Anthropocene"

**Hugo Paris, Nicolas Freud, and Caroline Ladage**

**Abstract** This chapter explores the narrative of the 'Good Anthropocene' as one way of depicting the current state of the Earth and how humanity should deal with the social-ecological issues at stake. This narrative is based on an ideal of control and claims that humanity should intensify its attempts to engineer planet Earth under the guidance of engineers and scientists. The Good Anthropocene thus reactivates a long genealogy of what Foucault described as pastoral power, with the figure of a positive authority leading humans toward ecoresponsible behavior. But should scientists really be shepherds? Analysis of this narrative leads us to discuss alternatives to this ecomodernist and elitist approach, and how it allows us to question how students' critical thinking and agency can be developed in the Anthropocene. We then present a perspectivist framework that can help us to reconsider the challenges of sustainability in a way that is more compatible with a democratic approach that takes account of different forms of knowledge and addresses epistemic injustices. Through a brief case study of a lecture on narratives of the Anthropocene, we suggest how this framework can be used to question the required teacher support that would be needed to help them build on students' empathy and understanding of power relations.

**Keywords** Anthropocene · Epistemology · Foucault · Pastoral power · Perspectivism · Neoliberalism · Earth system science

---

H. Paris (✉)
ADEF, Aix-Marseille University, Marseille, France

INSA Lyon, Villeurbanne, France
e-mail: hugo.paris@insa-lyon.fr

N. Freud
INSA Lyon, Villeurbanne, France
e-mail: nicolas.freud@insa-lyon.fr

C. Ladage
ADEF, Université d'Aix-Marseille, Marseille, France
e-mail: caroline.ladage@univ-amu.fr



C. Didier et al. (eds.), *Engineering and Value Change*, Philosophy of Engineering and Technology 48, https://doi.org/10.1007/978-3-031-83549-0_15

## 15.1 Introduction

This chapter revolves around the figure of the "Good Anthropocene" as a way of understanding a narrative that makes sense of significant numbers of scientific data about the current state of the Earth and how humanity is to deal with the socio-environmental issues at stake. A critical analysis of this narrative will lead us to discuss alternatives and allow us to ask how students' critical thinking and agency can be developed in the Anthropocene.

This research is based on a PhD in Educational Science that focuses on engineering teachers, with an interest in how they practice curriculum and course design in the context of the Anthropocene. More specifically, one of the aims of this thesis is to develop and experiment with forms of support for teachers to help them integrate social-ecological issues into the 5-year engineering curriculum at INSA Lyon. From 2019, this French engineering school is implementing important changes to integrate compulsory credits on sustainable development and social responsibility with a significant proportion of new transdisciplinary courses. This means that teachers with different profiles will have to work together in new ways to develop these contents and choose the pedagogical approaches they consider appropriate. This is clearly not an easy task for a number of reasons. One of these may be the "acute" state of knowledge available on social-ecological issues: much of it is not as stabilized as much of what is normally found in an engineering curriculum, which implies uncertainty about the reference knowledge on the subject (Bencze et al., 2020). Decisions must therefore be made. With this "acuteness" comes another dimension with which most teachers are uncomfortable: some of this knowledge appears to be deeply linked to values and political questions. How can this complexity be addressed from the perspective of teachers' different disciplines?

To explore these issues, we will focus on a particular narrative that we encountered during our fieldwork among engineering teachers at INSA Lyon, which has previously been identified as the "Good Anthropocene" discourse, as coined by E. Ellis (Hamilton, 2016). In this contribution, we discuss several problems with this narrative, which can be understood as an ecomodernist position that obscures the political issues at stake in sustainability challenges. Based on Foucault's notion of pastoral power, we propose explanations of how the good Anthropocene inherits the neoliberal world in which it was developed (sect. 15.2). We propose to approach the complexity of sustainability issues from a perspectivist framework, inspired in particular by the work of Viveiros de Castro, and show how this can be useful for making sense of epistemological tensions (sect. 15.3). Finally, we conclude with a discussion of a teaching case study on Anthropocene narratives at INSA Lyon as a possible example of the perspectivist approach, with its advantages and limitations (sect. 15.4).

## 15.2 The "Good Anthropocene": A New Form of Pastoral Power?

### *15.2.1 The Good Anthropocene Narrative and Its Values: Toward Planetary Stewardship*

Let us start with the "Good Anthropocene" narrative. The main idea of this narrative, as expressed by Bonneuil and Fressoz (2016), is that until recently humanity has blindly modified its environment for its own development to the point where it threatens the balance of the Earth system. For Rockström et al. (2009), we would now need to consider different types of limits to the planet's capacity to sustain human activities in order to identify a "safe operating space" for humanity. To this end, they propose a planetary boundaries framework to better understand the systemic impacts of human pressures on Earth system biophysical processes (Steffen et al., 2015). This framework is designed to inform decision-making toward adaptive governance and planetary stewardship.

If we are not to call into question here the great interest of the proposed model for Earth system science research, we will discuss its implications and underlying values in this paper. For Crutzen[1] and Stoermer (2000), it implies that engineers and scientists must "guide mankind toward global, sustainable environmental management" for this new geological epoch — which is disputed by geologists (Witze, 2024). According to Asafu-Adjaye et al. (2015) in their Ecomodernist Manifesto, if we are to control the Earth system, it will be through a new wave of modernization, that would push the efficient use of resources, land and energy management to new levels in order to preserve our natural environment, at least part of it. This means that we will have to continue and accelerate the engineering of the Earth and voluntary direct intervention in climate or biodiversity, which logically leads us to forms of geoengineering. In other words, we would have to move from an uncontrolled Anthropocene to a conscious one, an Anthropocene that we would shape for the "good" of humanity.

Here the sense of "good" is rather minimalist, as it means, in terms of the planetary boundaries model, that humanity will not go extinct because of these ecological issues. This particular ethical stance has received far less criticism than other aspects, such as its naturalistic ontology (in the sense of Descola, 2005) and the particularly anthropocentric view (Dutreuil, 2019). Beau and Larrère (2018) compare the Good Anthropocene to a round-bottomed doll that sometimes recognizes the fragility of human control over the Earth and is horrified by the potential risk of a global catastrophe, but always returns to its original position, i.e., a narrative of an engineered Earth.

[1] Chemistry Nobel Prize in 1995, Paul Crutzen (1933–2021) is a notorious promotor of the notion of "Anthropocene Era". He is one of the co-authors of the remarquable studies led by Johan Rockström and Will Steffen previously mentioned. Crutzen is also known for its plead for geo-engineering as a major response against greenhouse gas induced climate change.

Although many authors, such as Hamilton (2016), have strongly criticized the Promethean dimension of this narrative, Dutreuil (2019) has shown that another stance — the Gaian discourse — can be identified among Earth system scientists: "The continuity of humans with the *living*, chemistry, air and water of the Earth system has been correlated with a shift in metaphors *to consider* our relationship with the planet, from the field of mechanics to the biomedical field: *no more* control of the Earth *spaceship* handled by engineers, *now* the 'care' of Gaia is provided by scientists at the bedside of the Earth system. The theme of caring for the *planet* is thus omnipresent in this literature." For scientists and engineers such as Lovelock or Schellnhuber, the image of an acupuncturist working with tiny needles on the surface of the Earth, or a physician examining the planet, seems a fitting depiction of the new role they should play in the Anthropocene. Thus, Prometheans and Gaians, whether they see themselves as mechanics of spaceship Earth or dedicated surgeons of the planet, might actually be better described as technocrats, i.e., well-informed experts who guide the reshaping of human behavior and build social acceptance (Crutzen, 2002) of already controversial technoscientific "solutions" in the case of geoengineering (Hamilton, 2013), but one might find that "adaptive governance" is already quite problematic. As Dutreuil (2019) highlights, democracy is seen as a burden rather than a benefit, resulting in delays in relation to the environmental emergency. We suggest that the Foucauldian concept of pastoral power might be helpful in discussing the issues with this position, which actually goes well beyond the easily criticized—but effectively defended—advocacy of geoengineering.

### 15.2.2 The Pastoral Power: An Individualized and Benevolent Guidance

Considering the political lineaments of the Good Anthropocene narrative requires us to question the guiding role it gives to engineers and scientists. Michel Foucault's concept of pastoral power (1978, 1979) explores a particular form of power that emerged from early Christianity. It is described as a power over a population and its individuals. The shepherd knows each sheep in his flock and what is good for each of them (individualization), but also the flock itself. The exercise of pastoral power is not that of the shepherd's domination, but of the pastor's total devotion and benevolence, committed to the care of the flock (Büttgen, 2007). Later, according to Foucault, pastoral power was an important framework for the birth of the modern state and political liberalism, with the idea of an active policy that would not be too directive (i.e., biopower).

If Foucault himself was never very specific about how this kind of power would develop with neoliberalism, he did leave several indications that have been followed up by other authors with different interpretations that we cannot go into here[2].

[2] See Verlengia (2019) for more details about the different interpretations of Foucault's indications on the critique of neoliberalism.

Hache (2007) suggests that the neoliberal pastorate retains the idea of governing "hands-free" but is also quite different in that it does not rely on a pastor or a state to directly control the population, but rather tries to make people and private institutions more responsible and integrate new norms and values. Through responsibility, neoliberalism aims to produce a subject capable of governing itself within the framework of a globalized market. It is also a break with classical liberalism because it abandons the distinction between public and private spheres of activity. For Stiegler (2019), in her analysis of the famous Dewey–Lippman debate, the new liberalism[3] described by Walter Lippman conceives human nature from an evolutionary point of view, focusing on how globalization and technology have developed too fast for us. This implies that the population should be helped to catch up with this "cultural *lag*" by experts who are able to understand the world and the needs of the population. However, these needs must be redirected to adapt to the needs of a globalized economy, as it is to Lippman the natural direction of human evolution (the telos) that now constitutes the basic framework of society. Despite its attachment to a "hands-free" conception of public action, Lippman's vision does not fundamentally exclude disciplinary means to produce the required "adaptation" in line with the tension already identified by Foucault in his analysis of classical liberalism. Stiegler (2019) interestingly notes that for Lippman — in the context of the conservation/preservation debate of the 1920s — the (bad) use of natural resources is a good example of why the (irrational) population cannot be left to its own, but must be educated by experts in order to preserve the conditions of possibility of the market economy. In this sense, we can see that there is nothing to prevent adaptive governance inspired by the planetary boundaries framework from playing this role of preserving these conditions.

### *15.2.3 What Becomes the Pastoral Power in the Anthropocene? Scientists as a New Kind of Shepherd?*

We would like to suggest three main connections between this brief account of pastoral power and its applications to the analysis of neoliberalism with the "Good Anthropocene" narrative we described earlier.

First, in the naturalized evolutionary account of human history, we find similarities between the globalization described by Lippman and the progressive human engineering of the Earth as a common telos. Such pseudo-historical narratives are problematic because they construct a simplistic view of very diverse historical,

[3] Stiegler does not claim that contemporary forms of neoliberalism are all inspired by Lippman's political philosophy but indicates that it played a significant role in the structuration of this political current, alongside other sources that may diverge on various aspects (Hayek, Friedman, von Mises, Röpke…).

social, economic and political processes and convey a highly questionable sense of a general, linear and directional movement of progress that prevents discussion of qualitative and quantitative differences in the means and goals of globalization dynamics or environmental engineering. If we admit that there is no such thing as a predefined telos for the evolution of the Earth system or humanity, then the question of purpose and finality remains open. It allows us to question the objectives of the planetary boundaries model: are they boundaries for the viability of the Earth system or boundaries for decision-making? Our point here is that these two dimensions can only overlap and should lead us to consider that, in the case of sustainability, expertise cannot be understood as politically neutral.

Second, if we consider the ethical position of the promoters of the Good Anthropocene on the role of scientists and engineers, which consists of a sincere, selfless and benevolent commitment to humanity, then the resemblance to pastoral power becomes apparent. However, it is not a question of the domination of scientists (even if that is what it means in practice), but of a positive role in manufacturing consent — the expression was first framed by Lipmann — to what is presented as neutral facts. If we said earlier that talking about neutrality is irrelevant on such a scale, Earth System scientists and engineers present themselves as the new experts who must frame, nudge and redefine human behavior to fit the stakes of sustainability: an environmentally responsible behavior. The emphasis on responsibility has already been pointed out as a key aspect of the Anthropocene discourse (cited by Dutreuil, 2018) and must be read as a continuation of its neoliberal framing (Hache, 2007), in line with the idea that human behavior must change at all scales (individual, collective, national, international). If this development is to be led by the individuals themselves, it seems clear to Crutzen (2002) that new pastoral figures will have to emerge among scientists and engineers. This raises new, broader questions about the rupture and continuity of the neoliberal pastorate in the Anthropocene: are there seeds for new techniques of governmentality? What would be the balance between biopower (of which pastoral power is a component) and disciplines in the government of the Anthropocene?

Finally, the idea of a sudden awareness of the environmental consequences of human activity is also historically inaccurate. Such a discourse forgets the various protests (social movements, scientific debates) that have accompanied industrialization from the start. Ignoring these movements helps to perpetuate the idea of an ignorant public (Callon, 1998), incapable of understanding what is at stake, and reinforces the role of experts. For Bonneuil and Fressoz (2016), we need to think our present with the idea that the moderns knowingly destroyed their environment. This leads us to think about modernity in a dialectical way, as a process that produces both knowledge (with a naturalistic ontology) and ignorance (as a desensitization of previous alarms from environmental concerns). But in this perspective, the role of scientists becomes unclear: how do they engage in these activities? What is their concrete role in the production of ignorance? Is this to be understood as an actual destruction of knowledge (e.g., indigenous and traditional knowledge) or as a selective blindness to certain issues and problems? Is it possible to prevent this phenomenon? And how can we address this aspect in a curriculum that aims to produce this technical and scientific 'elite'?

From a critical standpoint, we can ask what space is left for democracy in the Good Anthropocene narrative. Another way of thinking about this problem might be to consider how we can imagine democratic politics of collective self-limitation, which was an important question for the ecological movement in the 1970s (e.g., Illich, 1973). Such politics may not emerge from a global perspective, but rather from the actors themselves (Larrère & Larrère, 2023). In this sense, Dewey's response to Lipmann may be useful to us, as he emphasizes the need for a more liberal state, which should help publics to emerge and draw society's attention to their problems, rather than imposing its own view of what is at stake. But what are the conditions of possibility for an engineer or a scientist to engage in democracy rather than in neoliberal technocracy? In particular, what kind of relations with communities would such a role entail from an epistemological point of view? Our main concern here could be encompassed by the idea of epistemic justice (Frinker, 2007) or global cognitive justice (de Sousa Santos, 2014): how can the dominant Western sciences and technologies be articulated with situated knowledge, local knowledge, minority knowledge? We propose to discuss this problem from a perspectivist approach in order to explore the ontological and epistemological issues.

## 15.3 A Perspective on Perspectivism

### *15.3.1 Perspectivism: Making Realism more Complex*

Perspectivism is a minor current in Western philosophy, represented by authors such as Gottfried Leibniz, Friedrich Nietzsche and Gilles Deleuze. It has recently gained a renewed interest with the work of the Brazilian anthropologist Eduardo Viveiros de Castro on Amazonian perspectivism. Beyond being a conceptual framework for understanding Amazonian cultures, Viveiros de Castro claimed that this framework could be a powerful critical tool against the dominant naturalistic ideology (Dianteill, 2015).

In philosophy, the cornerstone of perspectivism is that everything we know about reality — which exists independently of us — always comes from a particular point of view. The methods used to construct that point of view are important in that matter, and not all different perspectives have equal value. In this sense, perspectivism allows us to distinguish sciences (as rationalized, methodologically explicit, peer-reviewed perspectives) from other forms of point of view. Nietzsche extends the idea of perspectivism by questioning individuality itself, since what we take to be our perceptions are always assembled from different sources, e.g., organs that have a specific way of functioning. As Stiegler (2006) puts it, for Nietzsche, Bildung is based on the idea of a deliberate process of organizing that "carnal chaos" which must lead to the formation of an individual capable of temporarily repelling barbarism — since the chaos will eventually exceed one's capacity to organize it. Every perception is in itself an affection of the perceiving body, which means that perception cannot be reduced to an act of identification of something, but also means experiencing its effect and responding to its presence by adjusting to it (Chamois, 2018).

Viveiros de Castro (2014) uses the example that "what is blood for us, is beer to the jaguar" to explain that perspectivism therefore does not presuppose a Thing in itself that would be drunk by humans and jaguars: it is the "limit" between blood and beer that is fundamental in this example, i.e., the relationship that connects and distinguishes the two real substances from the point of view of the different species. This conception of perspectivism is based on a shared epistemology of an embodied point of view (and a common sense) related to the very matter of bodies, but with a variable ontology (related to different references). Viveiros de Castro suggests that in Amazonian perspectivism, the main question is not the common reference to various perspectives — which would be a crucial aspect for Western naturalism — but the understanding of the relationship between the associated objects. In other words, for a perspectivist, the question is not the difference between facts and values but whether one's perspective might make sense to others. This assumption is consistent with what Chamois (2022) identifies as a fragmentalist position according to Fine's definition (Lipman, 2015), in the sense that it admits the existence of real subjective facts that may be contradictory among themselves.

Perspectivism is inspiring as it can integrate the critique of nature in modernity by authors such as Whitehead or Latour (Debaise, 2018), since it makes it impossible to reify or forget the very means used to construct a perspective: the material conditions of production are essential. It also avoids relying on the idea of a common world that is taken for granted: sharing the same world is only possible if we build a joint understanding, deliberate connections and relations between the different perspectives. This shared understanding is not a naive and peaceful aspiration, but also a very concrete consequence of the Nietzschean will to power of subjects: getting others to share one's perspective allows the subject to gain a form of control over its social and natural environment. In that way, perspectivism allows us to connect epistemology, ontology and politics into a coherent system that can help us rethink "social acceptance" in the context of the Anthropocene.

### 15.3.2 *What Are the Implications of Perspectivism for Engineering Education Regarding the Environmental Issues?*

Chamois (2018) highlights an interesting distinction between the notions of "perspective on" and "perspective of." On the one hand, the idea of embodied perspective makes it impossible to have a "view from nowhere" as coined by T. Nagel, while, on the other, it helps us to rethink the interdisciplinary relations between different sciences. Consequently, these relations should not be seen as additive, in the sense that sociology and psychology are not two pieces of the same puzzle that would be an individual: the articulation between them is more complex and may involve some theoretical and practical contradictions that need to be elucidated.

This idea of a complex relation goes beyond scientific perspectives and also applies to other perspectives, such as indigenous and folk knowledge or experiential knowledge. Here we find again our concern about the particular kind of blindness that positivist sciences can produce when they devalue situated knowledge produced outside their fields. In this sense, we argue that perspectivism can be a coherent paradigm for re-legitimizing these forms of knowledge in relation to academic teaching. Compared to relativism, perspectivism does not imply that individuals are "locked into a sensory sphere," but rather the idea that the richness of the world always exceeds the perspective one can produce of it, and the definite possibility that other ways of relating to the world exist and are relevant. That said, this would be quite consistent with the recurring idea that engineers should be able to practice mediation, have a strong social view of their role, and act as mediators who should be more open to diversity (within the profession and in the ways they work with communities) in order to address contemporary challenges, as Tabas et al. (2019) point out in their review of tomorrow's engineers.

On the other hand, the "perspective of" serves as a reminder that engineers are indeed individuals with a particular perception of the social and natural world, which may be related to their social status or the country they come from, what they studied and so on. It also implies that working on problems as an engineer is a transformative experience. Again, engineers are not floating minds in an abstract space, analyzing a problem from a vantage point of nowhere, they are fully engaged in situations. This implies that neutrality is a difficult position to maintain, and this could prevent us from actually engaging in a quest for objectivity and lucidity (Berlan, 2023). Thus, the problem remains open: how can one experience other points of view? If the first step might be to recognize the possibility and value of such points of view, a second step might be forms of serious role-playing games, which have gained increasing interest in recent years (e.g., Halbe et al., 2015). Under what conditions can academic engineering education benefit from practices of collective analysis of power relations such as Theatre of the Oppressed, pedagogies of liberation (Riley, 2003; Riley & Claris, 2009) or popular education techniques?

## 15.4 Developing Future Engineers' Critical Thinking and Agency through Teaching Plural Anthropocene Narratives at INSA Lyon

### *15.4.1 A Brief Detour: French Engineering Schools*

Our defense of perspectivism would be meaningless if we did not go into more detail about the particular conditions in which we are carrying out this reflection, so we need to make some clarifications about French engineering schools. Compared to other countries, the social status of the engineering profession tends to be higher, with a long-established definition essentially linked to the prestigious diplomas awarded by the so-called "*grandes écoles*" (Bouffartigue & Gadéa, 1997), apart from the universities. The social hierarchy of the profession has been linked to the hierarchy of the schools themselves, with the famous Ecole Polytechnique (founded in 1794) being the most prestigious institution in the system and considered a model for other schools, despite its specific characteristics, in particular its historical primary objective of training high-level civil (and military) servants for the state (Belhoste, 2003). In her doctoral thesis, Catherine Roby (2014) identifies a parallel between the social hierarchy of sciences—with distant links to the classification established by Auguste Comte[4]—and the hierarchy of engineering schools. On the one hand, mathematics and physics occupy the highest positions, while chemistry, life sciences or social sciences are considered less prestigious. On the other hand, engineering schools that focus on theoretical knowledge are often considered superior to other institutions where practical and technical knowledge prevails, which partly overlaps with the difference between engineers working as executives or as technical specialists. In addition, we can note that engineering still often relies on a technical framing of problems in the reductionist sense of applying scientific theories without much regard to social and political stakes (Lemaître, 2003).

This context may explain our tendency in France to consider engineers as part of the elite, whether by their academic distinctions or by their positions of power in private companies or public institutions. The importance of the social hierarchy of knowledge in engineering schools is, for us, a crucial feature to explain some of the difficulties of integrating sustainability in their curricula. From this point of view, it is somehow subversive to give more space to life science or social science concepts, even if the rise of the tertiary and service sectors has gradually led engineering schools to prepare their students more effectively for managerial and executive positions. This openness to the social sciences is more pronounced in schools founded in the second half of the twentieth century, including INSA Lyon, where we conducted our field research. However, for this institution, founded in a — productivist — post-war intellectual context concerned with transmitting humanist values to the new elites, the separation between "sciences for engineers" and "humanities" (comprising social sciences, arts, languages and sports) remains strong.[5] For us, this modern legacy is an important explanation for the recurrent tendency toward technological solutionism among the 26 INSA Lyon teachers we interviewed in our field study. Thus, we found that for these teachers, sustainability is often understood as the ("neutral") pursuit of efficiency, i.e., the reduction of energy consumption, of material quantities or of costs, and the search for better

[4] Comte proposed a hierarchy of sciences ranging from the most simple, independent and focused on general phenomena (i.e. in sequence: Mathematics, Astronomy, Physics) to the sciences that are more complex, studying specific and contextual events (continuing the sequence: Chemistry, Biology, Sociology). For more details, see Comte, Auguste (1830–1842), Cours de philosophie positive (6 vols.), Paris: Bachelier and https://atlas-disciplines.unige.ch/Images/Source_images/Index/Comte2.html.

[5] In a way, this distinction fatally carries the idea that sciences such as anthropology or economy may not be "sciences for engineers", or even not (serious) sciences at all…

system optimization. These goals are seen more appropriate for engineers when compared to what they perceive as unrealistic political or 'moralizing' approaches that are not based on (hard) scientific evidence and principles. If other teachers expressed more nuanced or critical positions, this tendency led us to wonder about possible connections with the good Anthropocene narrative and its ecomodernist approach to sustainability.

### *15.4.2 Experiencing the Plurality of Narratives of the Anthropocene*

From a comprehensive standpoint, it is clear that the weight of the modern heritage is not something that can be easily dealt with, a fortiori for scientists and engineers whose identity has historically been constructed on the naturalization of their object and an alliance with the industrial state and capitalism against religious institutions and even the general public in the French context (Latour, 1993; Bensaude-Vincent, 1999). In a situation where engineers are subject to an identity tension between being professionals or managers, an old debate that takes on new meaning in the Anthropocene, it is quite logical that academic engineering teachers insist on the importance of science and technology as a means of differentiating engineers from other experts or executives. However, it is necessary to collectively question this heritage, which means examining the different narratives we draw from it in order to reconsider our perspectives in the Anthropocene.

In 2022–2023, after 3 years of experimentation and collective work, a new module has been introduced for first-year engineering students at INSA Lyon. This 2-credit module[6] is presented as an introduction to several identified stakes and topics of the common learning base,[7] previously defined by a local steering committee composed of about 80 teachers as representatives of the 11 different academic departments of INSA. This approach is the result of an important curricular reform that will be launched in 2019, following a vote by the Board of Directors. The first-year module includes various activities, such as an introduction to the One Health approach through a study case or a training project on life cycle assessment. Each class of students is accompanied by a team of two teachers, one from the "sciences for engineers" and the other from the "humanities." The module consists of a series of two-hour seminars aimed at introducing and discussing the concept of the Anthropocene and the Planetary Boundaries framework through lectures, collective text analysis and short debates. The critique of the Anthropocene has led to a

[6] The first-year module consists of two credits corresponding to 28 hours in class for groups of about 25 students. 30% of the module is given in dual-teaching.

[7] The learning base is structured in six blocks: a. Anthropocene and planetary boundaries; b. climate and energy issues; c. health and biodiversity issues; d. resource issues; e. action levers—this block cover all the previous blocks; f. possible and desirable futures.

proliferation of proposed narratives that highlight various actors and dynamics to explain the contemporary world, as we can see in the ideas of, for example, the Thermocene (highlighting the impact of fossil fuel consumption), the Polemocene (war), the Plantationocene (colonization), the Eurocene (Europe) (Bonneuil & Fressoz, 2016). Bonneuil (2015) identifies four types of narratives of the Anthropocene based on their representation of the relation between nature and society, while Wallenhorst and Theviot (2020) distinguish six political narratives from their polity standpoint and what we translate here as 'degree of Prometheanism'. The aim for students here is not to learn and know all the possible narratives that could be conceived with varying degrees of accuracy and empirical evidence. Rather, it is to grasp the pluralism of interpretations and to be able to identify the underlying values and implications of such narratives. These classifications are useful to facilitate identification and to build empathy through an exercise of decentering from the self to understand different perspectives.

Although this lecture raises many pedagogical and didactic questions that would need to be properly assessed, we have identified two main risks that could jeopardize the benefits of perspectivism.

First, we have to consider the power relations that remain largely in favor of naturalistic and positivistic points of view, which means that certain narratives appear much more obvious and plain to the students, but also to the teachers who intervene in this module. Indeed, it seems more difficult to detach oneself from such a naturalistic perspective when practicing sciences where this paradigm seems to be consubstantial (is it possible to do physics without a naturalistic standpoint?) So how can we help these teachers to grasp these other narratives and better understand the sense and importance of other perspectives? This questioning is an important aspect of our dissertation work, where we experiment with ways of supporting teaching teams. In particular, organizing time for teachers' reflexivity seems to us to be an important way of helping them to become aware of these power relations, to develop their attention to the plurality of perspectives and to work better together.

Secondly, if it is important for students to acquire a critical and systemic vision of situations, the development of complex narratives incorporating ever more causal factors and protagonists should not become an end in itself. As Viveiros de Castro pointed out in the case of Amazonian perspectivism, the essential challenge is to ensure that perspectives can "meet" and communicate with each other, i.e., in our situation, that these perspectives lead to collective, democratic and concrete ways of transforming society to address the challenges of sustainability.

## 15.5 Conclusion

The Good Anthropocene narrative relies on an ideal of control that sees environmental problems as a difficult but stimulating step on the road to a fully anthropized and engineered planet. This story shares similarities with other teleological approaches, such as globalization, which led us to wonder whether other continuities could be identified. The Good Anthropocene thus reactivates an already long genealogy of pastoral power, underlining the role of positive authority that scientists and engineers could play in adapting human behavior in an ecoresponsible way. This is a fully political role, which runs the risk of favoring a technocratic model of social organization. Based on this initial critique, we have tried to show that perspectivism could help us to rethink the challenges of sustainability in a way that is more compatible with a democratic approach that takes account of different forms of knowledge. The INSA Lyon example of teaching based on the idea of the plurality of narratives of the Anthropocene led us to identify points of tension that feed our research work on supporting teachers in the transformation of engineering curricula in the context of the Anthropocene.

In conclusion, if we are to develop the agency of engineering students in a democratic perspective, which we believe is an important role for our institutions, we must pay attention to the translational issues involved and strive to prepare students not only to understand, but also to care deeply about the whole relationship between power and knowledge in engineering. We see this as a crucial part of the subtle handiwork of public empowerment that Dewey told us about. We are convinced that engineers can contribute to the intensification of democracy if they dare to denaturalize their knowledge so that it can be subject to democratic discussion with other perspectives, to ecologize their reflection as an embodied perspective influenced to some extent by their context and social position, and to recognize that no technology comes without political questions.

## References


Asafu-Adjaye, J. *et al.* (2015). *An ecomodernist manifesto.* Accessed 19 January 2023, http://www.ecomodernism.org/manifesto

Beau, R., & Larrère, C. (2018). Conclusion. In R. Beau & C. Larrère (Eds.), *Penser l'Anthropocène* (pp. 523–542). Presses de Sciences Po.

Belhoste, B. (2003). *La Formation d'une technocratie.* L'École polytechnique et ses élèves de la Révolution au Second Empire.

Bencze, L., Pouliot, C., Pedretti, E., Simonneaux, L., Simonneaux, J., & Zeidler, D. (2020). SAQ, SSI and STSE education: Defending and extending "science-in-context". *Cultural Studies of Science Education, 15*, 825–851. https://doi.org/10.1007/s11422-019-09962-7

Bensaude-Vincent, B. (1999). *L'opinion publique et la science* (2e [2003: Seuil]). La Découverte.

Berlan, A. (2023). Comment l'idée de neutralité scientifique nous aveugle. *Écologie & politique, 67*, 131–146. https://www.cairn.info/revue-ecologie-et-politique-2023-2-page-131.htm

Bonneuil, C. (2015). *The geological turn: Narratives of the anthropocene.* https://doi.org/10.4324/9781315743424-3.

Bonneuil, C. & Fressoz, J-B. (2016). *L'évènement Anthropocène.* Points. [1st Edition: Seuil, 2013].

Bouffartigue, P., & Gadéa, C. (1997). Les ingénieurs français. Spécificités nationales et dynamiques récentes d'un groupe professionnel. *Revue Française de Sociologie, 38*(2), 301–326. https://doi.org/10.2307/3322935

Büttgen, P. (2007). Théologie politique et pouvoir pastoral. *Annales. Histoire, Sciences Sociales, 62*(5), 1129–1154.

Callon, M. (1998). Des différentes formes de démocratie technique. *Annales des Mines/ Responsabilité & Environnement, 9*, 63–73.

Chamois, C. (2018). Perspectivismes. Sur deux usages de la notion de « perspective » dans la philosophie contemporaine. In E. Alloa & E. During (Eds.), *Choses en soi* (pp. 479–489). Presses Universitaires de France.

Chamois, C. (2022). *Un autre monde possible: Gilles Deleuze face aux perspectivismes contemporains*. Presses universitaires de Rennes.

Crutzen, P., & Stoermer, E. (2000). The Anthropocene. *Global Change Newsletter, 41*, 17–18. http://www.igbp.net/download/18.316f18321323470177580001401/1376383088452/NL41.pdf

Crutzen, P. J. (2002). Geology of mankind. *Nature, 415*(6867), 23. https://doi.org/10.1038/415023a

Debaise, D. (2018). L'univers perspectiviste. Nature et subjectivité dans la métaphysique contemporaine. In *Choses en soi* (pp. 491–503). Presses Universitaires de France.

De Sousa Santos, B. (2014). *Epistemologies of the south. Justice against Epistemicide*. Routledge.

Descola, P. (2005). *Par-delà nature et culture*. Gallimard.

Dianteill, E. (2015). Ontologie et anthropologie. Dix ans de controverse (Brésil, France, États-Unis). *Revue européenne des sciences sociales. European Journal of Social Sciences*, 53–52. https://doi.org/10.4000/ress.3314

Dutreuil, S. (2018). Chapitre 21. L'Anthropocène est-il un concept d'histoire de la Terre? In R. Beau & C. Larrère (dir.) *Penser l'Anthropocène* (p. 355–374). Presses de Sciences Po. https://www.cairn.info/penser-l-anthropocene%2D%2D9782724622102-page-355.htm

Dutreuil, S. (2019). Is the decisive issue in geoengineering debates really one of representation of nature? Gaia against (or with?) Prometheus? *Carbon & Climate Law Review, 13*(2), 94–103. https://doi.org/10.21552/cclr/2019/2/4

Foucault, M. (1978). *Sécurité, territoire, population.* https://freefoucault.eth.link/

Foucault, M. (1979). *Naissance de la biopolitique.* https://freefoucault.eth.link/

Frinker, M. (2007). *Epistemic injustice: Power and the ethics of knowing*. Oxford University Press.

Hache, É. (2007). Is responsibility a tool of neo-liberal governmentality? *Raisons politiques, 28*(4), 49–65.

Halbe, J., Adamowski, J., & Pahl-Wostl, C. (2015). The role of paradigms in engineering practice and education for sustainable development. *Journal of Cleaner Production, 106*, 272–282. https://doi.org/10.1016/j.jclepro.2015.01.093

Hamilton, C. (2013). *Earthmasters: Playing god with the climate*. Allen & Unwin.

Hamilton, C. (2016). The Anthropocene as rupture. *The Anthropocene Review, 3*(2), 93–106. https://doi.org/10.1177/2053019616634741

Illich, I. (1973). *Tools for conviality*. Harper & Row.

Larrère, R., & Larrère, C. (2023). Penser globalement, agir localement: Une maxime pour l'écologie politique ? *Recherches de Science Religieuse, 111*(2), 343–356. https://doi.org/10.3917/rsr.232.0343

Latour, B. (1993). *We have never been modern*. Harvard University Press. [1st French ed: Gallimard, 1991].

Lemaître, D. (2003). *La formation humaine des ingénieurs*. Presses Universitaires de France.

Lipman, M. A. (2015). On Fine's fragmentalism. *Philosophical Studies, 172*(12), 3119–3133. https://doi.org/10.1007/s11098-015-0460-y

Riley, D. (2003). Pedagogies of liberation in an engineering thermodynamics class. *Proceedings of the 2003 American Society for Engineering Education Annual Conference & Exposition.* Session 2692. https://peer.asee.org/pedagogies-of-liberation-in-an-engineering-thermodynamics-class.pdf

Riley, D., & Claris, L. (2009). From persistence to resistance: Pedagogies of liberation for inclusive science and engineering. *International Journal of Gender, Science and Technology, 1*(1), 36–60. https://genderandset.open.ac.uk/index.php/genderandset/article/view/25

Roby, C. (2014). *Place et fonction des SHS dans les Écoles d'ingénieurs en France: État des lieux, enjeux et perspectives épistémiques* [PhD thesis, Rennes 2 University]. https://www.theses.fr/2014REN2004

Rockström, J., Steffen, W., Noone, K., Persson, Å., Chapin, F. S., III, Lambin, E., Lenton, T. M., Scheffer, M., Folke, C., Schellnhuber, H., Nykvist, B., De Wit, C. A., Hughes, T., van der Leeuw, S., Rodhe, H., Sörlin, S., Snyder, P. K., Costanza, R., Svedin, U., Falkenmark, M., Karlberg, L., Corell, R. W., Fabry, V. J., Hansen, J., Walker, B., Liverman, D., Richardson, K., Crutzen, P., & Foley, J. (2009). Planetary boundaries: Exploring the safe operating space for humanity. *Ecology and Society, 14*(2). https://doi.org/10.5751/ES-03180-140232

Steffen, W., Broadgate, W., Deutsch, L., Gaffney, O., & Ludwig, C. (2015). The trajectory of the Anthropocene: The great acceleration. *The Anthropocene Review, 2*(1), 81–98. https://doi.org/10.1177/2053019614564785

Stiegler, B. (2006). Nietzsche et la critique de la Bildung. *Noesis, 10*, 215–233. https://doi.org/10.4000/noesis.582

Stiegler, B. (2019). *Il faut s'adapter*. Gallimard.

Tabas, B., Beagon, U., & Kövesi, K. (2019). Report on the future role of engineers in society and the skills and competences engineering will require. *A-STEP, 2030*. https://www.astep2030.eu/sites/sub_site_astep2030/files/2019-05/report_1_literature_review_a-step_2030_a1_t1_pdf

Verlengia, C. (2019). Les origines d'un projet critique et la question du néolibéralisme : Foucault, une philosophie aux frontières. *Astérion. Philosophie, histoire des idées, pensée politique, 20*. https://doi.org/10.4000/asterion.3991

Viveiros de Castro, E. (2014). *Cannibal metaphysics*. Univocal Publishing.

Wallenhorst, N., & Theviot, A. (2020). Les récits politiques de l'Anthropocène. Articulation du politique en Anthropocène et de l'Anthropocène en politique. *Raisons politiques, 77*(1), 5–34. https://doi.org/10.3917/rai.077.0005

Witze, A. (2024). Geologists reject the Anthropocene as Earth's new epoch—After 15 years of debate. *Nature, 627*, 249–250. https://doi.org/10.1038/d41586-024-00675-8

**Hugo Paris** MA Architecture (ENSA Lyon, 2017) & MSc Civil Engineering and Urbanism (INSA Lyon, 2019), PhD Educational sciences (Aix-Marseille University, 2024). Co-founder of collectif EPSI (*Études Pluridisciplinaires sur l'Ingénierie*) and Associate Editor for the European Journal of Engineering Education. 2024–2025: Post Doc researcher at INSA Lyon, UNESCO Chair Training engineers for the socio-ecological transition. Research areas: Higher education for sustainability, teacher continuous training, curriculum studies, anthropological approach to didactics (TAD).

**Nicolas Freud** PhD in physics of X-ray imaging (2003). Physics teacher (1997–2007), associate professor (2007–2022), full professor (2022–) at the Institut National des Sciences Appliquées de Lyon (INSA Lyon, France). Former Head of the Humanities Centre at INSA (2013–2020). In charge of curriculum development to introduce socio-ecological issues (2020–2024). Director of the socio-ecological transformation since 2023. Holder of the UNESCO chair "Training engineers in issues of the socio-ecological transition" since 2022.

**Caroline Ladage** PhD, Educational sciences, is a full professor at the University of Aix-Marseille and a member of the ADEF research unit. She is the editor-in-chief of the journal Questions Vives, recherches en éducation. Her research focuses on the didactic and pedagogical challenges of societal transitions (mostly ecological and digital) to which education and training must respond and which give rise to many didactic questions. Her research interests lie in the anthropological approach to didactics (TAD), with a particular focus on curriculum studies, workplace learning, and inquiry pedagogy.

# Chapter 16
# Manipulating the Scaffolded Agent

**David Spurrett**

**Abstract** What we count as manipulation, and what means we regard as possible to achieve it, depends partly on our view of how agency and cognition work in the target of manipulation. Mainstream thinking about the mechanisms of manipulation tend to assume cognitive internalism, where cognition is performed by brains that relate to the world through sense and motor control. Here I review the two most common cognitive internalist approaches to thinking about manipulation, in expected utility theory and prospect theory. Then I focus on manipulation of behaviour taking as a starting point the view that human cognition is extensively scaffolded. Cognition is scaffolded when it relies in significant ways on structure and resources external to the skull and skin. When cognition itself extends into the outside world, it can be manipulated by means of external interventions. This isn't merely a speculative possibility. There are real examples of manipulating scaffolded human cognition.

**Keywords** Manipulation · Cognition · Scaffolding · Gambling

## 16.1 Introduction

This chapter concerns the manipulations of human agents. Its disciplinary perspective is the philosophy of cognitive science. The chapter has two main parts. The first (Sects. 16.2 and 16.3) reviews mainstream thinking about the means by which an agent, typically a human one, can be manipulated. The second (Sects. 16.4 and 16.5) develops an argument that cognitively scaffolded users are vulnerable to manipulation in ways that the mainstream views tend not to appreciate. In this introduction I briefly clarify how I'll be approaching the topic of manipulation itself.

D. Spurrett (✉)
University of KwaZulu-Natal, Durban, South Africa
e-mail: spurrett@ukzn.ac.za





C. Didier et al. (eds.), *Engineering and Value Change*, Philosophy of Engineering and Technology 48, https://doi.org/10.1007/978-3-031-83549-0_16

The term 'manipulation' comes from the Latin verb 'manipulare' meaning to grip or to hold in the hands. In English it has two main senses. One indicates physical or mental handling towards some goal or purpose, usually involving an element of skill, but without negative moral connotations. A helpful physiotherapist might, in this sense, manipulate our joints or a trumpet player manipulate the valves on their instrument to produce to perform an arpeggio. The other applies when the target of manipulation is another agent and indicates influence or control that is often regarded as devious or morally wrong, especially when entity being manipulated is a person. Shakespeare's Iago, who works to arrange what his master Othello sees and how he interprets it, driving him to murderous rage, is a famous fictional example. Misleading advertising is an example of the latter. We can recognise the second morally objectionable sense as a special case of the first, when the target of skilful control is another agent, and the one manipulated is being harmed. Some hold that being manipulated at all is a harm irrespective of the consequences, others that the consequences determine whether there is harm. This difference won't matter in what follows.

Philosophical treatments of manipulation by an agent have tended to insist that manipulation requires *intent* (e.g. Baron, 2014) in part to do justice to the thought that manipulation is morally objectionable. But others have sought to explain the ways in which things that aren't themselves agents, such as advertisements, can be manipulative, and not insisted that advertisements themselves have intent (Wood, 2014). A useful survey of the problems with conventional intentional notions of manipulation, and with alternatives, that is focused on the possibility of manipulation by machines is provided by Pepp et al. (2022) and Parner (2022) that engages with the idea of design as an instrument of manipulation.

My treatment here won't attempt to settle questions about intent. My own view is that intentional manipulation is wrong, and a worthy topic for moral thinkers. But such manipulation doesn't exhaust the forms that manipulation can take, because the same tools can be used without intention, and the same effects achieved by other tools. Instead I will approach the question of manipulating as an agent from the perspective of mechanism. I want to ask what it is about the targets of manipulation, when those targets are agents, that makes their manipulation a possibility. I'll focus primarily on human agents here. It may help to contrast this question with a similar one about what makes manipulation in the sense of skilful handling of an inanimate object possible. A screw-top jar of pills can ordinarily be opened by an adult with two hands because such a person is able to co-ordinate gripping the jar and twisting the lid while squeezing it. Such jars cannot readily be opened by small children, and deliberately so, because of the size of the lid, or because children are less able to squeeze and twist at the same time. What is required to manipulate the object is to co-ordinate a set of simultaneous mechanical forces in various directions on different parts of the jar. Humans differ in how well they are able to do this. When the target of manipulation is another agent *as an agent*, manipulation is usually understood as involving some kind of cognitive or affective influence. This means that part of our thinking about the manipulation of agents involves, even if implicitly, a theory of how cognition works in the target agent. Just as the physical properties of the medicine jar determine what is required to mechanically manipulate it, so the cognitive properties of an agent will determine whether or how it could be manipulated. Focusing on the mechanisms of agent manipulation means focusing on cognitive mechanisms.

I contend that most contemporary reflection on the manipulation of human agents has taken one of two perspectives on cognition, drawn from expected utility theory and prospect theory, respectively, as its starting point. Each is the subject of one of the following two sections. They are both perspectives that have been taken very seriously, and informed important empirical and theoretical research. They are alike in their commitment, in standard versions, to 'cognitive internalism'—the view that psychological processes occur inside an agent (typically in their brain) in ways that make the most plausible routes to manipulation matters of arranging the inputs to those internal processes, including inputs that are downstream responses to the agent's actions (like rewards and punishments). In science fiction and other speculative scenarios involving brain surgery or implants, we see another imaginable, even if not plausible, option which involves directly interfering with internal cognitive processes. In these scenarios skilful physical manipulation of the brain as an object prepares the way for controlling an agent. Stories like these might be entertaining in thrillers and comedies, but they seem to have little or no policy relevance as long as we think that cognition is internal, and practically very difficult to interfere with directly.

Here are some recent examples of the tendency to omit consideration of manipulating scaffolded agents. Flayelle et al. (2023) develop a high-level and in many ways detailed review of ways in which potentially addictive online behaviours may be facilitated by technological design features, but don't mention extended cognition or scaffolding. Newall (2019) makes a useful editorial argument about the weaknesses of standard thinking about 'responsible gambling' pointing out how ideal 'informed consumer' that guides much policy thinking is a kind of Humean agent (see Sect. 16.2 below), and arguing that we need to recognise the importance of 'dark nudges'. That is, he correctly identifies the importance of behavioural economics (see Sect. 16.3 below), but is also silent about extended cognition and scaffolding, as are Thaler and Sunstein (2008). Calo (2014) on the other hand provides a substantial study of the notion of market manipulation from the perspective of consumer protection law. Calo emphasises the importance of behavioural economics, and the ways that information harvesting can allow manipulation to be individually targeted, and how bespoke online experiences dilute some of the protections associated with the shared spaces of physical retail stores. This valuable analysis also doesn't mention extended cognition or scaffolding, or address the possibility that user's own cognitive processes may be leaking into the interfaces and tools in question. Müller's (2020) review of the ethical issues relating to artificial intelligence and robotics doesn't mention scaffolding, and neither do any of the contributions to the Coons and Weber (2014) edited volume on manipulation. There is no reference to scaffolding in Susser et al.'s (2019) internet policy-focused review of 'Technology, autonomy, and manipulation'. This list could easily be extended.

## 16.2 Humean Agents and Their Manipulation

David Hume was an influential defender of the view that motivation for action cannot merely involve reason, but also requires emotion. Famously he said that 'reason is, and ought only to be, the slave of the passions' (1975, II.3.3, p 415). In the now standard terminology an agent won't act merely because of what they believe, but may act when a belief and a desire jointly indicate some action. While Hume was particularly concerned to argue that reason couldn't motivate by itself, the point goes in the other direction too. I might strongly desire a drink of cold water, but that won't make me go to the kitchen unless I also believe there is a water jug in the fridge. Motivated actions seem generally to require *both* belief(s) and desire(s). Various forms of 'belief/desire psychology' were dominant theoretical approaches for extended periods in psychology and philosophy, and remain significant in economics and decision theory. I call these views 'Humean' here because of Hume's contribution.

The most idealised forms of these views are found in Expected Utility Theory (EUT) in decision theory and economics, in both of which rationality is at least largely a matter of consistency. A crucial early technical advance is in Ramsey (1931, chapter 7) which shows that an agent whose degrees of belief failed to satisfy the axioms of probability would accept, if offered, a series of bets that guaranteed losing money. This argument is commonly called the 'Dutch book argument', although Ramsey refers to a book made by a 'cunning better'. The point is that belief formation or updating that isn't consistent is provably inefficient. In the case of preferences (the decision theorist's counterpart to desires) one key form of consistency is transitivity. A set of preferences that violates transitivity exhibits cyclicity. And an agent with cyclical preferences will accept a repeating series of deals that will eventually drain it of any resources it has. It will serve as what is sometimes called a 'money pump' to any agent that offers it the right deals in the right order. Scenarios of this type are articulated in Davidson et al. (1955) although they don't use the term 'money pump'. Here too, inefficiency is shown to follow from inconsistency, and then consistency proposed as a criterion for rational agency. There are, of course, sizeable literatures on both kinds of consistency, and technical discussion of requirements besides transitivity. Even so, expected utility theorists share the core idea of an agent with orderly degrees of belief, and well-behaved preferences that selects between options on the basis of their expected utility (utility multiplied by probability).

The most rational versions of a Humean agent, then, are characterised by some set of orderly degrees of belief (or 'credences') and a set of consistent preferences or desires. The natural ways to manipulate such an agent are to attempt to introduce beliefs that serve your goals rather than theirs (including beliefs about the means to reach their goals) or to influence, by addition, removal or re-ordering, their preferences. Much advertising seeks to do this, by reinforcing selected associations with a product, or by making more or less explicit factual claims. These claims often concern competing products with little or no discernible difference. Greenberg

(1961) describes the testimony of Ernest Dichter at a United States Senate subcommittee concerned about consumer deception. Dichter noted that many competing products were of similar quality and hence that competitive advantage depended on which company could create the 'illusion of a desirable difference'. In many countries today there are regulations about the labelling of some kinds of product (especially medicines and foods) mandating accurate description of their contents, placing restrictions on what can be said about their effects, imposing requirements on what must be said about their side effects, and with penalties for violating them. Many things that used to be permitted are no longer legal because of recognition that advertising could and did influence people's beliefs and desires, and that some forms of influence (especially dishonest ones) were morally objectionable.

It has long been widely recognised that however useful the idealised rational Humean agent might be for theoretical purposes, it doesn't provide a particularly good model of the behaviour of real people (or other living things). That is, when instead of trying to specify what an ideally rational (which crucially includes maximally consistent) agent *should* do, we try to measure what living agents *actually* do, their behaviour seems regularly and in patterned ways to violate rationality. This brings us to the second group of mainstream conceptions of agents and their manipulation.

## 16.3 Prospect Agents and Their Manipulation

Much of the data that showed how poorly the behaviour of real agents corresponded to ideal rational agency was adopted or produced by a field called behavioural economics, characterised by the view that empirical investigation should determine the properties of the economic agency of people and other animals. There are, unsurprisingly, methodological and empirical disputes over how good some of the data is, and what it actually shows, which would take us far away from the main argument. In any event, I don't need to insist that some particular version of behavioural economics is best or that any one is strictly correct, because my aim is to sketch how those persuaded that a theory like that is correct will think about manipulation. So I'll take one leading group of approaches as paradigmatic, the various versions of 'Prospect Theory' (Kahneman & Tversky, 1979), treating it as an umbrella term including various additional findings about heuristics and biases that flesh out the behavioural economic picture.

Prospect theory seeks to accommodate and organise a set of the ways in which observed economic behaviour violated Expected Utility Theory. First, prospect theory supposes choice begins with an 'editing phase' where factors about the framing of a choice (that should be irrelevant according to EUT) make a difference to the attractiveness of options, crucially including how the outcomes are characterised in terms of gains and losses. A key early finding here was that which of two strictly identical options people preferred in a hypothetical public health scenario depended on whether one was characterised in terms of 200 people out of 600 being *saved*, or

400 out of 600 *dying* (Tversky & Kahneman, 1981). Second, prospect theory characterised the evaluation after framing as relative to a 'reference point', where gains and losses relative to that point are valued in EUT violating ways. This includes different risk responses in gains and losses relative to the reference point, with loss-aversion (relative to gains), and biased evaluation of high probability events (over-weighted) and low probability ones (under-weighted). One instance of this is the 'endowment effect' where people tend to demand more money to give up an object than they would be willing to pay to acquire it (Kahneman et al., 1991; Thaler, 1980).

We can think of the 'Prospect agent' (that is the agent described by the theory) as being a kind of *warped* version of the idealised Humean agents. Their valuations and responses to evidence exhibit deviations from EUT, and from Bayesian rationality. Some of these warps create opportunities for influence, including manipulation. The promise of 'nudging' is that these influences can be beneficial to those influenced. Thaler and Sunstein (2008, p. 6) characterised a nudge as an intervention that would tend to alter 'people's behavior in a predictable way without forbidding any options or significantly changing their economic incentives. To count as a mere nudge, the intervention must be easy and cheap to avoid'. Thaler and Sunstein call the idealised rational agents that I'm calling 'Humean' the 'Econs', and contrast them to empirically revealed 'Humans'. An example is making being in some prudent or prosocial state such as saving for a pension or being an organ donor the default setting while allowing people to opt out, as opposed to making not being in it the default. This should make no difference to an ideal Humean agent, but can make a difference to the Prospect agent, judging from their reference point and valuing something more because they already have it. Even well-intended nudges can be controversial precisely because they're regarded, by some at least, as manipulative (e.g. Yan & Yates, 2019).

Nudges are at least intended to benefit those who are their targets, but manipulations informed by the properties of the Prospect agent are sometimes extractive. Newall (2017, 2019), for example, points out that bookmakers offer psychologically seductive (and very profitable) betting options on combinations of events that exploit human tendencies to overestimate some probabilities. He explicitly refers to many techniques used by bookmakers as examples of 'dark nudges'. One significant behavioural economic finding is the 'peak end rule' indicating that both experienced and remembered utility (or pain) are disproportionately sensitive to the evaluation of the most extreme point, and the final stage, of an experience rather than duration or total (Kahneman et al., 1993). One of many areas in which assessment of the overall value of an extended experience matters is gambling. A gambler who hasn't run out of money might stop when they feel as though other options have become better. In 2005 the Harrah casino group planned to 'optimize' player value with a system estimating, given historical data, how much a player could lose over how long while still continuing play. They predicted a 'pain point' for that player, and dispatched a 'Luck Ambassador' to give out vouchers (for further gambling, or in-house entertainment) shortly before the pain point (Schüll, 2012, p. 154). Introducing a small 'win', even if not in the form of a cash payout, was supposed to defer the predicted cessation of play. Descendants of that system include directly distributing non-cashable credits to the players at individual machines at times determined by casino analytics. More generally, and leaving gambling aside, Calo (2014) notes that big data presents a further risk, given that it allows retailers and social media companies to try to 'pinpoint profitable anomalies' in individualised ways (2013, p. 1010).

It is time to draw this and the preceding section together. Humean and Prospect views of agency don't exhaust the ways of thinking about cognition as a possible target of manipulation, but they dominate mainstream thinking. The Humean agent is the ideal of much normative thinking in decision theory and economics, and some policy thinking. The Prospect agent is the template for much thinking in advertising and marketing, and an increasing fraction of policy discussion. Both approaches are, furthermore, *valuable*. How people behave is related to what they believe and want, and considerable evidence supports the behavioural economic view of how we respond to evidence and assess prospects. Each of the two shows us part of the truth, and what follows should not be understood as rejecting either, but rather as complementing and extending them.

## 16.4 Scaffolded Agents and Their Manipulation

The Humean and Prospect agent are, as already noted, typically understood in a cognitively internalist way. The traditional conception of cognition in many fields saw cognition as a set of processes that occurred between peripheral processes of perception and action execution, forming the middle stage in a 'sense-think-act' cycle. This internalist is shared by most Humean and Prospect views despite other differences between them, such as whether the agent contains a rational utility function or comparably quirky prospects. I highlight this shared feature to contrast it with an increasingly widely held view that cognition is not 'all in the head' (Clark & Chalmers, 1998, p. 8) but that the body and external world often do significant cognitive work. Defenders of extended cognition, which comes in many forms, share a picture of cognition as looping out into the environment, and of action itself, as well as tools and external media, as sometimes significantly cognitive. There isn't space to do justice to all forms of extended cognition here, including ones focusing on embodied cognition, niche construction and affect, so I'll focus specifically on the idea of scaffolding.

Cognition is scaffolded when it relies in significant ways on structure and resources external to the skull and skin. Human cognition is distinguished in part by the magnitude of our reliance on scaffolding, and by the variety and sophistication of the forms of scaffolding we have developed and transmit culturally. Many of us organise significant parts of our lives by means of self-directed scaffolding like shopping and task lists that help relate our current activities to our considered intentions, or by going out only holding as much money as we were prepared to spend in cash. Hutchins (1995) emphasises how tools and external media, including slide rules and navigation charts, allow significant cognitive processing to be conducted outside the head. He describes many examples of artefacts which transform the computational demands of navigation, including the 'three scale nomogram' which converts multiplication and division operations relating distance, time and speed into drawing a straight line that connects two known values and crosses all three scales (Hutchins, 1995, p. 148). Cases of scaffolding sometimes overlap with what Kirsh and Maglio (1994) called 'epistemic actions' which are physical actions that function to change a cognitive state, in contrast to pragmatic actions which directly advance towards final goals. Manually rotating a jig-saw puzzle piece while moving it around a partially completed puzzle, thereby replacing difficult mental rotation with simple visual comparison to determine possible placement, is an exemplary epistemic action. The two overlap because some of an agent's scaffolding is produced by their own epistemic actions, and because some scaffolding (whoever made it) supports specialised epistemic actions.

Most arguments defending the idea that cognition is scaffolded focus on benign cases, which augment or support the agent attempting some task (e.g. Clark, 2003). But an agent that tends to offload cognitive work onto the environment is potentially vulnerable to manipulation that interferes with the external cognitive factors they depend on.

I noted above how cash money could enable some kinds of self-management. This used to be possible in casinos, when gambling machines had to be fed coins, and chips for table games bought from cashiers. This allowed customers to decide before going into the casino how much they were willing to lose, and only take that much money in with them. Running out of notes and coins could then reliably cue remembering what they have previously thought about their budget. Putting acceptable losses in cash into a pocket or wallet was a kind of epistemic action, setting up a reminder in the event that the money ran out. Even if the customer went on to give into temptation they would need to leave the casino, and undergo a delay, before gambling more. Casinos obviously prefer it if people don't stop gambling when they still have access to money. And the industry has worked—ostensibly in the name of player convenience—to remove barriers between players and their money. Putting ATM machines in casinos, and moving to cashless play where electronic credits were bought with bank cards reduced various kinds of friction and delay. But they didn't stop there. Since the late 1990s the kiosks used at casinos to buy playing credits combine ATM functionality with tools to bypass daily withdrawal limits, draw on multiple accounts and increase or exceed credit limits. In some places it is legal for terminals to access bank cards to be mounted directly on gambling machines, so that gamblers don't even need to stand up to make further dents on their available credit (Schüll, 2012: 72). What I want to emphasise here is that these changes, which have been highly profitable for casinos, can be understood in part as *preventing* simple kinds of self-scaffolding that cash previously permitted.

Some scaffolded manipulation is more direct. Early mechanical slot machines in which players observed parts of rigid spinning reels allowed reliable cognitive offloading (at least on fair machines). This is because the number of symbols on the reels and their spacing and motion could be observed through the viewing window, and gave a reliable sense of the actual odds of a winning outcome. This was possible because the spinning reels were genuine parts of the mechanism that determined the outcome of play. In electronic gambling machines, in contrast, the outcome of play is determined by a random number generator, and the visible animation generated afterwards. This made it possible to break the relationship between the animated or simulated reels and the actual odds. The technology of ‘virtual reel mapping’ allowed the odds in the random number generator to be decided against an arbitrarily large set of possibilities, which were then mapped onto the visible reels, with many non-paying outcomes corresponding to few visible outcomes. (Having a large set of possibilities was useful because lower odds of winning allowed larger jackpots to be offered on still profitable machines.) As well as blocking accurately inferring the odds from inspecting the display, machines with virtual reel mapping present misleadingly many apparent ‘near miss’ outcomes. Machines with design features like this are very effective at getting customers to play for longer, and so lose more money. Part of how they work is by encouraging relating to the visible display as though it showed real physical movements, inviting cognitive offloading. But the scaffolded inferences the machines encourage are systematically inaccurate, and the behaviour they encourage serves the casino at the expense of the customer. (For a longer discussion of scaffolding in gambling technology see Timms & Spurrett, 2023.)

Epistemic action and scaffolding are found in non-human animals, but humans are generally regarded as unusual in how much of their cognitive lives depend on external cognitive infrastructure (e.g. Clark, 2003; Sterelny, 2012) including inhabiting extensively labelled and standardised environments, using maps, diaries, conventions, symbol systems, and more recently mobile computing devices. The fact that humans are so extensively scaffolded should make us more concerned to understand the ways in which this makes us vulnerable to manipulation in ways that wouldn’t arise if cognitive internalism was true.

I noted earlier that some manipulation scenarios in popular imagination seem to depend on fantasy or science fiction premises, where control involves brain implants, or other kinds of brain manipulation. What makes these seem speculative or at least currently incredible is the thought that cognition happens in brains, and that our brains are inside us and relatively difficult to manipulate. But if cognitive processes are themselves leaky, things are less clear. Part of Clark’s point in calling his (2003) book ‘Natural Born Cyborgs’ is precisely that long before digital technology, humans relied on cognitively engineered environments in ways that made them biotechnological hybrids. The highly scaffolded individual who leans on the appearance of a slot machine to judge their chances of winning on the next play can have the cognitive substrates of their appraisals engineered *without* having to get inside their skull.

## 16.5 Conclusion

Having made the point that scaffolding tends to be ignored in treatments of manipulation, it must be noted that there is a growing body of recent work on harms and risks associated with extended cognition. Liao and Huebner (2021) argue that there are 'oppressive things', meaning material artefacts and environments which contribute to oppression, when their effects are biased in the same direction as, and they are causally embedded in, an oppressive system. One of their key examples is the ways that early consumer photography maintained light skin bias because of inattention to images of dark skinned people in developing the technology, and exclusive use of light skinned models in the images used to calibrate and set up developing systems. Slaby (2016) argued that some employers cultivate affective workplace culture where conforming might go 'discernibly against' someone's 'prior orientations', and call such cases of 'hacking' examples of 'mind invasion' if they are 'in the long run detrimental to [the] personal flourishing' of those subjected to it (Slaby, 2016: 2). And Ross (2020) has argued that gambling technology and environments are engineered to exploit natural vulnerabilities in human reward systems. All of these approaches take scaffolding seriously, and recognise distinctive ways in which scaffolded agents are potentially vulnerable. Timms and Spurrett (2023) argue for the possibility and reality of 'hostile scaffolding' which scaffolds the cognition of one agent in order to serve the interests of another. Nguyen (2023) argues that we should recognise some environments as epistemically hostile in ways that complicate our thinking about epistemic vices, and who bears what responsibility when people form some false beliefs. Parner's (2022) account of manipulative design through gamification argues that some design can enable manipulation 'at a distance', and while it doesn't explicitly mention scaffolding could be accommodated by an analysis in terms of scaffolding. The same point goes for the account of 'malicious design' in Conti and Sobiesk (2010).

There are other examples, most of them similarly recent. These aren't counterexamples to the neglect of scaffolding in mainstream thinking about manipulation noted in my introduction, because there is so little overlap between the literatures. Only a few and very recent collections on manipulation include reference to scaffolded agency, such as Jongepier and Klenk (2022). It remains the case that thinking about manipulation in ethics and policy contexts takes cognitive internalism for granted, and is dominated by either Humean or Prospect models of the agent. Given that human cognitive lives are characterised by such extensive reliance on scaffolding and other forms of cognitive engineering of our environments, internalist conceptions of human agency are bound to incompleteness or inaccuracy. To the extent that Clark is correct that 'the local mechanisms of mind [...] are not all in the head' and that cognition 'leaks out into body and world' (2008, xviii), the possibilities for manipulating cognition *itself*, rather than merely arranging inputs to it, will themselves be partly outside skulls.

Accepting this point doesn't mean rejecting the Humean or Prospect views, but rather the cognitive internalism that usually (but not necessarily) accompanies them.

If you reject internalism and recognise cognitively scaffolded human agency, what you'll mostly find is scaffolded Prospect agents. And, in institutional settings scaffolded in particular ways, humans can even sometimes behave like almost Humean agents.

Contemporary manipulators, using large data sets, including intimate and fine-grained details of our interaction with virtual environments, don't need a theory of scaffolding in order to find ways of manipulating scaffolded agents. As Calo (2014) noted, in companies operating large online stores, or social media environments, data can be mined for patterns to target and manipulate individual users and consumers without needing theory to generate hypotheses. But formulating policy and systems of values applicable in the design process will require recognising the reality of scaffolded agency.

## References

Baron, M. (2014). The *Mens Rea* and moral status of manipulation. In C. Coons & M. Weber (Eds.), *Manipulation: Theory and practice* (pp. 98–120). Oxford University Press.

Calo, R. (2014). Digital market manipulation. *The George Washington Law Review, 82*(4), 995–1051.

Clark, A. (2003). *Natural born cyborgs*. Oxford University Press.

Clark, A. (2008). *Supersizing the mind*. Oxford University Press.

Clark, A., & Chalmers, D. (1998). The extended mind. *Analysis, 58*, 7–19.

Conti, G., & Sobiesk, E. (2010). Malicious interface design: Exploiting the user. In *WWW 2010 (conference paper)* (pp. 271–280).

Coons, C., & Weber, M. (Eds.). (2014). *Manipulation: Theory and practice*. Oxford University Press.

Davidson, D., McKinsey, J. C. C., & Suppes, P. (1955). Outlines of a formal theory of value, I. *Philosophy of Science, 22*(2), 140–160.

Flayelle, M., Brevers, D., King, D. L., Maurage, P., Perales, J. C., & Billieuz, J. (2023). A taxonomy of technology design features that promote potentially addictive online behaviours. *Nature Reviews Psychology, 2*, 136–150.

Greenberg, D. S. (1961). Food packaging: Industry, spokesmen differ on rationality in the supermarket. *Science, 134*, 1408–1410.

Hume, D. (1975). *A treatise of human nature*. ed. L.A. Selby-Bigge, rev. P.H. Nidditch. Clarendon Press.

Hutchins, E. (1995). *Cognition in the wild*. MIT Press.

Jongepier, F., & Klenk, M. (Eds.). (2022). *The philosophy of online manipulation*. Routledge.

Kahneman, D., Fredrickson, B. L., Schreiber, C. A., & Redelmeier, D. A. (1993). When more pain is preferred to less: Adding a better end. *Psychological Science, 4*, 401–405.

Kahneman, D., Knetsch, J. D., & Thaler, R. D. (1991). The endowment effect, loss aversion, and status quo bias. *Journal of Economic Perspectives, 51*(1), 193–206.

Kahneman, D., & Tversky, A. (1979). Prospect theory: An analysis of decision under risk. *Econometrica, 47*(2), 263–291.

Kirsh, D., & Maglio, P. (1994). On distinguishing epistemic from pragmatic action. *Cognitive Science, 18*, 513–549.

Liao, S., & Huebner, B. (2021). Oppressive things. *Philosophy & Phenomenological Research, 103*(1), 92–113.

Müller, V. C. (2020). Ethics of artificial intelligence and robotics. In *Stanford encyclopedia of philosophy*. https://plato.stanford.edu/entries/ethics-ai/

Newall, P. W. S. (2017). Behavioral complexity of British gambling advertising. *Addiction Research & Theory, 25*(6), 505–511.
Newall, P. W. S. (2019). Dark nudges in gambling. *Addiction Research & Theory, 27*(2), 65–67.
Nguyen, C. T. (2023). Hostile epistemology. *Social Philosophy Today, 39*, 9–32.
Parner, W. J. (2022). Manipulative design through gamification. In F. Jongepier & M. Klenk (Eds.), *The philosophy of online manipulation* (pp. 216–234). Routledge.
Pepp, J., Sterken, R., McKeever, M., & Michaelson, E. (2022). Manipulative machines. In F. Jongepier & M. Klenk (Eds.), *The philosophy of online manipulation* (pp. 91–107). Routledge.
Ramsey, F. (1931). *The foundations of mathematics and other logical essays*. Harcourt, Brace.
Ross, D. (2020). Addiction is socially engineered exploitation of natural biological vulnerability. *Behavioural Brain Research, 386*.
Schüll, N. D. (2012). *Addiction by design: Machine gambling in Las Vegas*. Princeton University Press.
Slaby, J. (2016). Mind invasion: Situated affectivity and the corporate life hack. *Frontiers in Psychology, 7*, 266.
Sterelny, K. (2012). *The evolved apprentice*. MIT.
Susser, D., Roessler, B., & Nissenbaum, H. (2019). Technology, autonomy, and manipulation. *Internet Policy Review, 8*(2), 1–22.
Thaler, R. (1980). Toward a positive theory of consumer choice. *Journal of Economic Behavior and Organization, 1*, 39–36.
Thaler, R. H., & Sunstein, C. (2008). *Nudge: Improving decisions about health, wealth and happiness*. Yale University Press.
Timms, R., & Spurrett, D. (2023). Hostile scaffolding. *Philosophical Papers, 52*(1), 53–82.
Tversky, A., & Kahneman, D. (1981). The framing of decisions and the psychology of choice. *Science, 211*, 453–458.
Wood, A. (2014). Coercion, manipulation, exploitation. In C. Coons & M. Weber (Eds.), *Manipulation: Theory and practice* (pp. 17–120). Oxford University Press.
Yan, H., & Yates, J. F. (2019). Improving acceptability of nudges: Learning from attitudes towards opt-in and opt-out policies. *Judgment and Decision Making, 14*(1), 26–39.

**David Spurrett** BA in Philosophy and English, Master of Arts, PhD Philosophy (University of Natal, 2000). 1999–2003 Lecturer in Philosophy, University of Natal. Today, Philosophy, University of KwaZulu-Natal. Co-author of Everything Must Go: Metaphysics Naturalized (2007), Midbrain Mutiny (2008). Co-editor of Distributed Cognition and the Will (2007), What is Addiction? (2010). Fellow of UKZN. Research areas: Philosophy of cognitive science, philosophy of biology, evolution of mechanisms of action selection and their situated subversion.

# Chapter 17
# Maintaining Scientific Instruments: Artifacts, Malfunction, and Values

**Alexandra Karakas**

**Abstract** The Hubble Space Telescope (HST) stands as an iconic symbol of scientific exploration, its launch in April 1990 sparking great expectations. However, the initial images were disappointing, revealing blurred starlight and a failure to capture faint sources of light. NASA and optical technicians embarked on immediate troubleshooting, resulting in the discovery of various issues and disparate opinions on their root causes. This shows that defining the boundaries of an instrument's intended function is complex, raising certain philosophical and practical questions. This chapter applies the philosophy of maintenance, in particular the notion of function and malfunction, to reveal epistemic issues surrounding faulty scientific instruments. The case of the HST highlights some of the epistemic problems concerning the maintenance of scientific instruments, and it also suggests that the relation between function and malfunction should not be seen as binary. Maintenance strategies hinge on implicit or explicit function definitions that are essential for understanding how artifacts operate. However, to delineate an artifact's intended function, it is also necessary to understand when it is malfunctioning. The conventional view that an artifact is malfunctioning if it fails to fulfill its intended function is therefore insufficient. The case of the HST represents a partial malfunction somewhere between function and malfunction, demonstrating the intricate relation between science, technology, and maintenance. This approach sheds light on epistemic problems arising from flawed scientific instruments.

**Keywords** Philosophy of maintenance · Scientific instruments · Hubble Space Telescope · Malfunction

A. Karakas (✉)
Department of Philosophy and History of Science, Faculty of Economics and Social Sciences, Budapest University of Technology and Economics, Budapest, Hungary

MTA Lendület Values and Science Research Group, HUN-REN, Research Centre for the Humanities, Budapest, Hungary
e-mail: karakas.alexandra@filozofia.bme.hu



C. Didier et al. (eds.), *Engineering and Value Change*, Philosophy of Engineering and Technology 48, https://doi.org/10.1007/978-3-031-83549-0_17

## 17.1 Introduction

This chapter applies the philosophy of maintenance, in particular the notion of function and malfunction, to reveal epistemic issues surrounding scientific instruments.[1] I argue that scrutinizing the complexity of the maintenance of scientific instruments can highlight the various ways in which they affect scientific knowledge production. Discussing maintenance by examining the blurry boundaries of functionality helps us to understand how instruments enable, distort, and guide knowledge production. Focusing on function when discussing artifacts is also in keeping with the traditional way of looking at material things.

In technology and engineering, maintenance means many things: repairing, conserving, preserving, reconstructing, and replenishing, to name but a few. These different activities generally aim to keep an artifact functional and prevent or restore malfunction. As such, maintenance centers around the relation between actual or possible failure and an expectation and description of function. Therefore, establishing a proper theoretical framework for interpreting malfunction seems crucial to any maintenance strategy. Even though failure is a central issue in engineering and there is a significant amount of research about it (Del Frate, 2013; Del Frate et al., 2011; Petroski, 1992), there is no agreement on the exact meaning of failure or malfunction. I argue that function and malfunction are not binary notions, and that the relation between them is complicated and constantly in flux.

This chapter focuses mainly on the epistemic problems around the malfunction and maintenance of the technical artifacts used in science. Highlighting the heuristic value of the study of the maintenance of scientific artifacts, I expose the complicated epistemic role of scientific instruments, and more specifically, I focus on a specific type of engineering that is a means for reaching scientific goals. Scientific instruments are a particular branch of technology, with unique epistemic problems and goals, and the maintenance of these objects requires sophisticated practices. On the one hand, these special artifacts embody scientific and epistemic goals, and on the other hand, they are also influenced by engineering values, which can conflict with scientific values. The case study of the Hubble Space Telescope (HST) malfunction shows that specific issues of scientific inquiry are rooted in the problem of malfunction. The case of the HST is particularly relevant, not only because its operations started with a major problem, but also because its lifespan has been increased thanks to many maintenance, repair, and replace missions (Gainor, 2021). By pointing out the blurriness of the boundaries between functional and malfunctional performances, I highlight the epistemic relevance of the maintenance of scientific instruments.

---

[1] During the research, I was supported by the MTA Lendület Values and Science Research Group.

## 17.2 Maintaining Technology

Previously neglected, maintenance and repair studies have recently been getting more attention in science and technology studies and the philosophy of technology.[2] In engineering, the various practical aspects of maintenance have also become central issues, in contrast to previous design-focused approaches. Lately, the epistemic (Young, 2021b), ethical (McLaren, 2018; Pateraki, 2023), economic (Edgerton, 2006), cultural (Denis et al., 2015), and social aspects (Graham & Thrift, 2007) of maintenance have been recognized as philosophical problems. This is not surprising, if we look at the diversity of meanings of maintenance, care, and repair, such as cleaning, storing, conserving, preserving, and replenishing. These different activities capture the diverse notions of practice,[3] and they all highlight the general aim of keeping an artifact functional, minimizing downtime, and preventing or resolving malfunction. Jackson[4] (2014, p. 223) defines repair as something that "occupies and constitutes an aftermath, growing at the margins, breakpoints, and interstices of complex sociotechnical systems as they creak, flex, and bend their way through time"; he also emphasizes the importance of maintenance and repair for stability and change in culture (Jackson, 2016). Young (2021a, p. 356) describes maintenance as "an activity through which we provide stability to something over time." Edgerton (2006, p. 80) claims "maintenance and repair to be the most widespread forms of technical expertise."

Maintenance strategies hinge on implicit or explicit function definitions. They are essential for understanding how artifacts operate, since technological devices are usually described by their particular function. Thus, human-made objects all have specific descriptions of what they do and how they do it. There are numerous well-developed function theories of artifact performance, and many of them emphasize the role of human interaction, which is crucial in maintenance as well. Among others, Kroes and Meijers, in their introduction to a special issue on *the dual nature of technical artefacts*, have defined human-made things as designed physical structures having intentionality-related functions. "This makes them 'hybrid' objects that can only be described adequately in a way that somehow combines the physical and intentional conceptualisations of the world" (Kroes & Meijers, 2006, p. 2). Required function means the function or combination of functions of an item that are considered necessary to provide a given service connected to it. Thus, required functions specify what activities the object must be able to fulfill to be called functional. While performing the required function sometimes results in the need for maintenance (e.g., routine car maintenance), in other cases, the material of the object will degrade even without use (e.g., certain materials corrode), and thus necessitate maintenance.

---

[2] See, for instance, the volume *Maintenance and Philosophy of Technology: Keeping Things Going*, edited by Mark Thomas Young and Mark Coeckelbergh (2024).

[3] For a sociological analysis of skill, maintenance, repair, and craft, see Harper (1987).

[4] For more relevant work on repair, see Jackson (2016, 2019).

One of the core features of every maintenance strategy is a function definition: What function do we expect the artifact to fulfill, how, and to what extent? What is the description of use and the intended function? A definition of the function usually also constitutes an understanding of malfunction: If the artifact does not fulfill its intended function, then it is malfunctioning. However, in many cases, this derivation is not satisfactory for explaining failure because function and malfunction have a much more complicated philosophical and practical relationship than this somewhat simplistic dichotomy would suggest. The problem is even more pressing when it comes to scientific instruments with epistemic goals, such as telescopes, which are used for observation, exploration, and the study of various objects, with the goal of getting a better view of our universe and producing reliable knowledge about the past, the present, and the future. Telescopes have many different functions at the same time, requiring great engineering knowledge and expertise, based not just on know-how, but also on prior hypotheses. As the HST case will show, many scientific instruments cannot be said to work either properly or not at all; instead, they operate at different levels of accuracy, which are adjusted and evaluated based on the circumstances.

When maintaining scientific instruments, the practical goals of maintenance and engineering are also intertwined with more abstract epistemic (scientific) goals and values. Scientific goals, such as producing precise measurements and sufficient observational data, are crucial for determining maintenance practices, since the degree of precision or accuracy of a particular instrument is deeply interwoven with and depends on the circumstances of use. As I will show later, establishing the boundaries of function and malfunction in scientific instruments is particularly complex. The HST case will highlight the complexity of functional performances, as well as the epistemological aspects of maintenance and repair strategies.

If we accept that functionality is key when understanding artifacts, then maintenance, broadly speaking, either refers to retaining an artifact in a preferred and accepted functional state by various means, or to restoring it to a somewhat functional state after a malfunction has occurred. And the goals of different maintenance activities are as varied as their means: to reduce the probability of failure; to provide information about the object's ideal use and current life-cycle stage; to minimize downtime; to ensure cost-effectiveness; and, in general, to maximize performance. Establishing the boundaries of functionality is thus one of the key aspects of the practice and philosophy of maintenance,[5] and it is also crucial for analyzing epistemic problems concerning scientific inquiry.

There are many different categorizations of maintenance strategies, but the underlying approaches, such as preventive or reactive (or run-to-failure) maintenance, are driven by the notion of malfunction, albeit in different ways. On the one hand, reactive maintenance means that maintenance only happens once failure has occurred; thus, this strategy "waits for machine or equipment failure before any

[5] This is not to say that function is *the only* crucial feature of maintenance. However, in this chapter, I specifically focus on the connection between function and maintenance.

maintenance action is taken. It is, in truth, a no-maintenance approach of management" (Mobley, 2004, p. 2). On the other hand, preventive maintenance is time-driven, and its actual practice varies significantly, from minor adjustments to more complex scheduled operations. Preventive and reactive strategies differ in terms of the stage in the life cycle when maintenance is performed and in the methods used to prevent malfunction and possible degradation. A correct maintenance routine can prevent an artifact from breaking to keep it functional; it can repair a failed technological device, or even improve its overall quality. As a result, maintenance is an umbrella term that also centers around the notion of failure and function on a practical and philosophical level. Broadly speaking, these strategies aim to keep a technical artifact operating for as long as possible, and in this sense, they are led by both engineering values, such as cost-benefit analysis, and by other values of a more epistemic nature. Although these goals can occasionally conflict, they are always densely interwoven, especially in the case of sophisticated devices and more extensive technological systems. Consequently, at its core, the purpose of any maintenance strategy is twofold: keeping artifacts functional and preventing failure. The problem is that the difference between functional and malfunctional artifact performance is not always clear-cut.

In general, malfunction means a sort of deviation from function. Consequently, a common understanding of artifact performance is that if we want to know when something is functioning, then we also have to know when it is malfunctioning. However, I claim that this binary notion is not satisfactory for explaining the full spectrum of artifact performance, let alone the performance of scientific instruments. The reason for this is that malfunction can be partial, it can be subtle, and it can remain unnoticed for an extended time while the object is seemingly functional in other ways. Therefore, knowing the function of an artifact does not necessarily explain the malfunction of the artifact, as was the case with the HST, in which the instrument functioned in some way, but malfunctioned at the same time.

With scientific instruments, it is especially hard to disentangle different types of performances, since they can have multiple technical as well as practical and symbolic functions, which may also overlap. These instruments can be faulty in many ways, and they can cause distortions, exaggerations, and misinterpretations of phenomena. These errors are occasionally detected at some point in the production of scientific knowledge, but sometimes they remain unnoticed for extended periods. At other times, malfunctioning instruments may even lead to erroneous conclusions, faulty analysis, or the corroboration of flawed hypotheses.

Scientific artifacts are a unique cluster within the great group of things that require maintenance. On the one hand, they are technical objects, and so their design is led by one or multiple values that are prominent in engineering.[6] Proponents of value monism argue that ultimately, there is always one overarching value in engineering to which every other value could potentially be reduced. However, how to

[6] The literature on values in engineering is vast, and it would be beyond the scope of this chapter to endeavor to evaluate it. For a comprehensive volume on the topic, see van den Hover et al. (2015).

define this particular value is already subject to debate. Van de Poel (2015, p. 668) emphasizes that there are many candidates, such as "social utility, profit, customer satisfaction, or efficiency." He therefore argues for value pluralism, claiming that since it is very hard to define the *ultimate* value to which engineers should subscribe, it is evident that value pluralism should be accepted. This is even more so the case if we consider engineering as a heterogeneous practice with different fields: Some values may be highly regarded in some fields, while being irrelevant in others. Broadly speaking, professional codes of ethics for engineers agree on the importance of safety (Hansson, 2009), adherence to standards of competence, care (Pritchard, 2009), ethics (Mitcham & Briggle, 2009), and a socially responsible attitude overall (Durbin, 1987; van den Hover et al., 2015). However, as with many things, these values become more complicated as we zoom in and evaluate their exact meanings and the ways in which they guide engineering decisions.

In the case of scientific instruments, the set of engineering values is also intertwined with other values, since they are designed for a specific scientific goal or goals (such as measurement, observation, data collection, etc.); thus, they are also led by scientific values and principles. Traditionally, the goals of science are defined as gaining a better understanding of the universe and the causal relations between things, making better observations, etc. However, the engineering and scientific values do not always match; they can conflict, especially in the case of large, complex projects like the HST. For instance, safety, reliability, efficiency, human welfare, and staying on budget are important engineering values, but they do not necessarily play a crucial role in reaching a particular scientific goal. In a similar vein, if we accept that science aims at producing reliable knowledge and discovering truth, then these values do not necessarily match the engineering values described above. However, this does not mean that scientific values and engineering values are two distinct and isolated categories; nonetheless, they may have different underlying principles, priorities, and end goals that sometimes need to be ranked according to the objectives of the project at hand. An artifact can be many things and have different functions, depending on its context, design, and purpose, but the conflict between different aims is most prevalent and striking with scientific instruments.

When it comes to the scientific instruments used for producing knowledge, there are many cases in which artifacts function to some extent, but fail to meet expectations in other ways, for example, in terms of quality. Frequently, a scientific artifact may thus satisfy the engineering values while failing to comply with other values. It follows that scientific instruments represent value pluralism at work. However, defining what sort of value framework would be most beneficial for analyzing scientific instruments is tricky. They can be discussed using the dichotomy of epistemic and non-epistemic values, if we accept the common stance that "science is directed at the epistemic tasks of identifying true or reliable information about the world" (Elliott, 2022, p. 4). Epistemic values are usually about producing truth and knowledge, while non-epistemic values cover a wide array of other values, be they social, ethical, political, or economic.[7] This framework is most prevalent in the philosophy of science, particularly in the science and values literature. Another framework relevant for material instruments is the distinction between intrinsic and instrumental value, which is mostly used in the philosophy of technology and engineering. Van de Poel (2015, p. 669) describes intrinsic values as those "that are strived for their own sake," while instrumental values are "strived for the sake of other values." As scientific instruments are objects designed for an epistemic goal, their maintenance requires specific descriptions of the engineering and other values at play, and their malfunction description only makes sense within this framework of intertwined values.

## 17.3 Maintaining Scientific Instruments

Scientific knowledge is embedded in the instruments of science: finding, using, making, maintaining, and redesigning things for different scientific purposes are crucial aspects of scientific inquiry. The process of producing scientific knowledge with these instruments can be divided into different phases, from design through use to recycling. At each phase, scientists heavily depend on the skill sets of various other agents,[8] and are bound to their particular epistemic cultures. For example, maintenance and repair work is frequently delegated to technicians and lab workers, mainly because they have the required technical knowledge for operating and troubleshooting artifacts. These skills not only keep instruments running, but they have also led to discoveries, settled debates, and established facts.

Preserving an artifact in a preferred functional state can mean avoidance of use, and it can mean moderate and appropriate use, a practice of continuous maintenance, or a change of some aspects of the object due to breakage. That is why maintenance can only be understood in relation to use, intended function, and malfunction, and how these different aspects determine the object in question. However, after an instrument is put to use, a whole range of maintenance practices starts to take shape. Werrett (2019) notes that in the seventeenth century, different users, household members, servants, technicians, and other agents played different roles in preventing breakage, maintaining consistency, and generally keeping an artifact functional. The world is usually a dangerous place for instruments: "volatile reactions, humidity, extraordinary temperatures, unreliable porters, dust and dirt, rain and damp, pests, vermin, and just plain clumsiness could lead to damage, taking away properties deemed valuable in a thing" (Werrett, 2019, p. 87). Moreover, artifacts are also

[7] The science and values literature is vast, and there are many debates about the nature of and the possible categorization of epistemic and non-epistemic values. For an introduction, see Elliott (2022).

[8] Denis and Pontille (2020) argue that maintenance and repair studies help to replace the image of the lone innovator, and the innovator-centric approach to technology in general.

used by and maintained by agents with different background knowledge, resulting in diverse practices.

Even though science and material culture studies have long recognized the teamwork inherent in knowledge production, the maintenance of scientific instruments is still a somewhat neglected aspect of the intertwined relation between material culture and epistemology—despite the fact that the philosophy of maintenance fits with earlier studies by ethnographers and sociologists of science who have focused their attention on scientific laboratories. Knorr Cetina also emphasizes the intertwined and complex roles of professionals working in labs, resulting in a social organization based on practice, experiments, expertise, materials, and devices. Lab work is not always clearly defined, however: "Laboratories employ caretaking personnel for the sole purpose of tending to the waste, the used glassware, the test animals, the apparatus, and the preparatory and maintenance tasks of the lab. Scientists are not only researchers, but also caretakers of the laboratory" (Knorr Cetina, 1999, p. 38). This leads to a collaborative process involving scientists, lab workers, technicians, and other people around the lab, resulting in knowledge shaped by different epistemic backgrounds, expertise, and expectations.

In their ethnographic study of science technicians, Barley and Bechky (1994) claim that even though there are many studies on the social settings of science labs, the role of technicians is still understudied. They argue that technicians not only provide the data used in scientific arguments, but they also "maintain materials, operate instruments, conduct experiments, and record data" (Barley & Bechky, 1994, p. 87). In short, their contribution prepares the ground for knowledge production. Shapin (1989) argues that the role of technicians has been overlooked for one simple reason: their lower social status. Shapin also highlights the importance of skilled technicians and helpers in science, mentioning, for example, the several amanuenses who wrote down Robert Boyle's thoughts after other assistants had read scientific texts to him. But this is not all there is to it: Often, "several chemical assistants work in the laboratory, tending distillations, amalgamations, and rectifications, making observations, and recording them for Boyle's use. Other assistants are carrying out experiments with the air-pump or hydrostatic instruments" (Shapin, 1989, p. 554). In addition, another group of assistants conducted experiments based on Boyle's notes and returned with results for his use.[9] A consequence of this is that Boyle, like many of his peers, performed very few of his experiments himself. In the seventeenth century, it was a common practice to delegate certain technical and operational tasks to assistants and other technicians. There are many reasons for this, but it is quite likely that Boyle did not have the necessary level of general knowledge, let alone the technical skills needed to maintain particular instruments and keep them running.

Collins and Kusch have done fieldwork in a physics laboratory to test whether they would be able to operate an air pump using only the available instructions. After trying to operate these modern-day air pumps, they remind us that the case of

[9] However, these assistants are largely absent from the history of science (Shapin, 1989).

Robert Boyle's air pump[10] is undoubtedly also tied to the practice of maintenance. The success of Boyle's air pump partly depended on the skills of his technicians to maintain the instrument, which led to the widespread use of air pumps. More and more people

> were shown how to make a leather seal, more were introduced to the esoterica of "sallad oil" and water, more were shown how and when to pump, how to avoid damage and distortion, how to seal cracks with a stinking mixture of cheese parings and lead oxide, what to watch for, how to know if the pump was leaking, how long it took to produce a satisfactory vacuum, and when to stop and search for problems. (Collins & Kusch, 1998, p. 185)

Collins and Kusch (1998, p. 166) quote Bob Draper, the lab technician who helped them, as saying that a big problem with operating the air-pump was knowing how to spot anomalies and cure leaks properly. The problem lies in the many possible types of anomalies, where even the smallest deviation might cause practical and epistemic problems. For instance, "a hair caught between the rubber seal and the seating of the vacuum chamber would be sufficient to cause a leakage; if the oil in the diffusion lamp was old, it would take a long time to heat up to the point where it could be used efficiently" (Collins & Kusch, 1998, p. 166). Consequently, an instrument's anomalies and malfunctions have often guided the direction of practice and theory as well. The heuristic value of faulty experiments, malfunctional instruments, and other flawed investigations is also emphasized by Shapin (1989, p. 558), who notices that the otherwise invisible technicians mainly appear in descriptions of failed experiments. In many cases, however, it was difficult to decide whether the experiment was not done correctly, the instrument was malfunctioning, or the hypothesis was just plain wrong.[11]

Disentangling the effects of instruments and revealing the causal relationships between the results of an experiment and the underlying hypotheses can be tricky—especially if we also factor in the repair and maintenance work usually delegated to various agents. The problem is rooted in the difficulties of giving a proper description of function while also accounting for malfunction, particularly with scientific instruments. To determine when and what type of maintenance is needed, there has to be a description of function and malfunction, as well as an understanding of the (possibly clashing) engineering and scientific values of the artifact that determine its performance.

---

[10] For a detailed discussion, see Shapin and Schaffer (1985).

[11] Schickore discusses how microscopy manuals, a new genre of texts that focused on flaws, appeared in the eighteenth century. These manuals "determined the limits of microscopy as exactly as possible and characterized and estimated the range of potential sources of error" (Schickore, 2001, p. 127).

## 17.4 Maintaining HST

The HST case is steeped in scientific and engineering values and, at the same time, highlights the complicated relation between function and malfunction. Scientific and engineering endeavors all share the need to avoid malfunction as much as possible, since it is costly and disruptive, and everybody just wants to keep things functional. How should we approach the maintenance of scientific artifacts, then? First, we need to take into account that judging whether an instrument is performing adequately under certain circumstances can be both a technical and a theoretical issue at the same time. There are also conflicting scientific and engineering goals and values, such as cost and benefit and observational success, for example. We face a multidimensional approach to artifact performance, embedded in scientific and engineering values and philosophical problems.

Apart from illustrating some of the epistemic problems concerning the maintenance of scientific instruments, the case of the HST also highlights that it is beneficial to avoid looking at the relation between function and malfunction as binary. The HST is one of the best-known and most iconic scientific instruments, and it was designed to be maintained in space. The expectations were high before and after its launch in April 1990, but when the first images arrived a couple of months later, they were not as great as expected. The first image received showed blurred starlight, lacked fine detail, and failed to capture faint sources of light. The public was rather underwhelmed by the images.[12] The evaluation of their quality was based on scientific values—they were supposed to represent distant objects clearly—as well as on engineering values.

The issue aligned with the results of previous checkups in orbit, which revealed that the telescope could not focus properly for some unknown reason. NASA convened the Hubble Space Telescope Optical Systems Board of Investigation to find out what happened with its optics. Troubleshooting began immediately to determine why the telescope was not focusing properly or what other factors might be causing the poor image quality.[13] The investigation consisted of reviewing reports, documentation, and other data related to the design, construction, and testing of the telescope, carried out by the different agencies involved. Reports by NASA and optical technicians pointed out different problems and reached different conclusions regarding the exact cause of the optical problem, which affected both the Wide Field Camera and the Faint Object Camera.

The board discovered that the HST's mirror had the wrong curvature, ten times larger than what would have been acceptable, which caused a specific optical flaw. As a result, the primary mirror was also "too flat near its outer edge" (NASA). The

[12] The public reputation of HST changed after July 1994. Initially, the press even called it the *repaired Hubble*, pinpointing the telescope's malfunction, and this somewhat sarcastic name was only changed to Hubble when new, improved images were released (Gainor, 2021, p. 118).

[13] For a detailed discussion of the maintenance strategies of extremely large telescopes, see Ortega et al. (2022).

reason for the optical flaw was a phenomenon called spherical aberration: half a wavelength distortion produced significant optical problems. Spherical aberration refers to a non-convergence of light rays, whereby the outer parts of the lens do not bring light into the same focus as the central part of the lens. This caused a reduction in quality, leading to blurry images.

At the beginning of the investigation, there were different opinions regarding the exact cause of spherical aberration. Two principal questions needed to be answered: "By what sequence of mishaps did the Hubble Space Telescope acquire its unfortunate spherical aberration? And how did this severe optical flaw escape notice until after the HST was launched into orbit last spring?" (Schwarzschild, 1990, p. 19). In other words, the task was to determine what caused the problem, and how it remained unnoticed during several maintenance checkups.

At first, the investigation pointed to multiple possible explanations for the spherical aberration. Lallo (2012) mentions that the metering rods used to set the distance between the telescope's lenses and mirrors could have caused a malfunction. They were coated with a non-reflective paint, which at some point chipped off. "When the beam hit the cap it was reflected as if it were the rod end. The distance between cap and rod end was 1.3 millimetres—exactly enough to cause the aberration now seen in the Hubble primary mirror" (Joyce, 1990). Once discovered, the issue was fixed, but the repair work may have caused even more problems in this case. Because the lens had been distorted, it no longer fit into its housing, which was spotted during a checkup. The technicians then "fixed the problem with some domestic washers" (Schaffer, 2011, p. 706), potentially causing further issues. Another problem was that the wide field mirror, the HST's main mirror, was polished in the wrong way, with edges that were slightly too flat, causing distortion. NASA's Optical System Failure Report describes the possible causes by analyzing documents and reverse-engineering scenarios that may have resulted in spherical aberration. However, the report claims that a major issue was that the data indicating these problems were available during the telescope's construction, but were not adequately investigated (Allen et al., 1990).

Even though there was a reason for the telescope's spherical aberration, its malfunctioning is not clear-cut in other ways, which shows that a binary notion of function and malfunction is not sufficient in explaining the performance of scientific instruments. Claiming that an image is poor or not good enough only makes sense if we also consider the circumstances and the actual instrument. In his book on the HST, Chris Gainor (2021, p. 62) discusses the circumstances under which the first images were released, emphasizing that telescopes—especially ones launched into orbit—first need to be adjusted before producing pictures of good quality. At the same time, first images do not only play a scientific role, but also a representational one, as they contribute to the public's view on the matter. While even the HST's first images were significantly better than those produced by previous telescopes, they were seen as underwhelming by the public and the scientific community alike.

By this standard, HST was functioning to some extent, as it was producing images and was also performing other required functions. Thus, it would not make much sense to argue that it was not functioning at all. However, claiming that it was not malfunctioning would not make much sense either. Its performance was evaluated on the basis of expectations, a certain understanding of what it should do and how it should fulfill its intended function. These theoretical assumptions guided the maintenance and repair work, which was delegated accordingly in order to correct the malfunction of the telescope.

The malfunction of the HST connects to many wider issues. First, the maintenance aspect: How do we determine when maintenance starts in the case of complex artifacts? Does it start even before the object is put into operation? And if so, when does engineering end and maintenance begin? This raises the question of the role of engineers, users, technicians, and other maintenance staff. As the function of an object is maintained over time through different processes, its functionality depends on corrections made by the user, meaning that malfunction also depends on the user. Different values and frameworks also determine roles and the allocation of maintenance and engineering tasks. Some failures can be attributed to a lack of maintenance, which could even lead to the normalization and relativization of certain mistakes.

## 17.5 Summary

The case of the HST highlights the complicated relation between malfunction and function: The telescope only partially malfunctioned and was otherwise fully functional. Looking at function and malfunction as non-binary concepts enables a more nuanced understanding of the epistemic problems affecting complex scientific instruments. Emphasizing the maintenance aspect of scientific instruments exposes many issues inherent in the production of knowledge. For instance, the technical skills involved in maintenance and repair work not only help to keep an artifact functional for longer, but may also contribute to its widespread use. This increases the availability of the object for knowledge production, allowing more people to be involved in scientific work. The HST also highlights the complexity of disentangling the effects of instruments, observational data, hypotheses, and theories. A focus on agency and maintenance can also be a helpful way to analyze epistemic problems concerning artifacts, especially in the philosophy of science, which has so far neglected to pay much attention to maintenance. The next step of this project will be to look at the history of science and technology from a maintenance perspective, in order to discover possible epistemological consequences of maintenance or the lack thereof, and to develop a deeper understanding of the maintenance of scientific artifacts with epistemic goals.

## References


Allen, L., Angel, R., Mangus, J. D., Rodney, G. A., Shannon, R. R., & Spoelhof, C. P. (1990). *The Hubble Telescope optical systems failure report.*

Barley, S. R., & Bechky, B. A. (1994). In the backrooms of science: The work of technicians in science labs. *Work and Occupation, 21*(85), 85–126.

Collins, H., & Kusch, M. (1998). *The shape of actions: what humans and machines can do.* MIT Press.

Del Frate, L. (2013). Failure of engineering artifacts: A life cycle approach. *Science and Engineering Ethics*, 913–944. https://doi.org/10.1007/s11948-012-9360-0

Del Frate, L., Franssen, M., & Vermaas, P. E. (2011). Towards a trans-disciplinary concept of failure for integrated product development. *International Journal of Product Development, 14*(1–4), 72–95. https://doi.org/10.1504/IJPD.2011.042294

Denis, J., Mongili, A., & Pontille, D. (2015). Maintenance & repair in science and technology studies. *Tecnoscienza—Italian Journal of Science & Technology Studies, 6*(2), 5–15.

Denis, J., & Pontille, D. (2020). Why do maintenance and repair matter? In A. Blok, I. Farías, & C. Roberts (Eds.), *The Routledge companion to actor-network theory* (pp. 283–293). Routledge. https://minesparis-psl.hal.science/hal-02172939v2

Durbin, P. T. (1987). Toward a philosophy of engineering and science in R & D settings. In P. T. Durbin (Ed.), *Technology and responsibility* (pp. 309–328). Springer. https://doi.org/10.1007/978-94-015-6940-8

Edgerton, D. (2006). *The shock of the old. Technology and global history since 1900.* Profile Books.

Elliott, K. C. (2022). *Values in science.* Cambridge University Press. https://doi.org/10.1017/9781009052597

Gainor, C. (2021). *Not yet imagined: A study of hubble space telescope operations.* NASA.

Graham, S., & Thrift, N. (2007). Out of order: Understanding repair and maintenance. *Theory, Culture & Society, 24*(3), 1–25. https://doi.org/10.1177/0263276407075954

Hansson, S. O. (2009). Risk and safety in technology. In A. Meijers (Ed.), *Handbook of the philosophy of science, vol 9, Philosophy of technology and engineering sciences* (pp. 1069–1102). Elsevier.

Harper, D. A. (1987). *Working knowledge: Skill and community in a small shop.* The University of Chicago Press.

Jackson, S. J. (2014). Rethinking repair. In T. Gillespie, P. Boczkowski, & K. Foot (Eds.), *Media technologies: Essays on communication, materiality and society* (pp. 221–240). MIT Press.

Jackson, S. J. (2016). Speed, time, infrastructure: Temporalities of breakdown, maintenance and repair. In J. Wajcman & N. Dodd (Eds.), *The sociology of speed: Digital, organizational and social temporalities* (pp. 169–185). Oxford University Press.

Jackson, S. J. (2019). Repair as transition: Time, materiality and hope. In I. Strebel, A. Bovet, & P. Sormani (Eds.), *Repair work ethnographies: Revisiting breakdown, relocating materiality* (pp. 337–347). Palgrave Macmillan.

Joyce, C. (1990, December 1). A flake of film foiled Hubble. *New Scientist.* https://www.newscientist.com/article/mg12817450-600-a-flake-of-film-foiled-hubble/

Knorr Cetina, K. (1999). *Epistemic cultures. How the sciences make knowledge.* Harvard University Press.

Kroes, P., & Meijers, A. (2006). Introduction: The dual nature of technical artefacts. *Studies in the History and Philosophy of Science, 37*, 1–4.

Lallo, M. D. (2012). Experience with the Hubble Space Telescope: 20 years of an archetype. *Optical Engineering, 51*(1), 011011. https://doi.org/10.1117/1.oe.51.1.011011

McLaren, D. P. (2018). In a broken world: Towards an ethics of repair in the Anthropocene. *The Anthropocene Review, 5*(2), 136–154. https://doi.org/10.1177/2053019618767211

Mitcham, C., & Briggle, A. (2009). The interaction of ethics and technology in historical perspective. In A. Meijers (Ed.), *Handbook of the philosophy of science, vol 9, Philosophy of technology and engineering sciences* (pp. 1147–1191). Elsevier.

Mobley, K. R. (2004). *Maintenance fundamentals*. Elsevier.
NASA. Hubble's Mirror flaw. https://science.nasa.gov/mission/hubble/observatory/design/optics/hubbles-mirror-flaw/. Downloaded: 1 July 2024.
Ortega, M. A., Hardie, K., Gajadhar, S., & Goodrich, R. W. (2022). Maintenance concept for the extremely large telescopes. *Journal of Astronomical Telescopes, Instruments, and Systems, 8*(02). https://doi.org/10.1117/1.jatis.8.2.021506
Pateraki, M. (2023). Bringing care and maintenance under the same skin: Deep brain stimulation and temporalities of care. *Cahiers François Viète* [Online], III-15 | Online since 1 November 2023, connection on 1 July 2024. http://journals.openedition.org/cahierscfv/4401. https://doi.org/10.4000/cahierscfv.4401
Petroski, H. (1992). *To engineer is human: The role of failure in successful design*. Vintage Books. https://doi.org/10.1086/354258
Pritchard, M. S. (2009). Professional standards in engineering practice. In A. Meijers (Ed.), *Handbook of the philosophy of science*. Elsevier.
Schaffer, S. (2011). Easily cracked: scientific instruments in states of disrepair. *Isis, 102*(4), 706–717. https://doi.org/10.1086/663608
Schickore, J. (2001). Ever-present impediments: Exploring instruments and methods of microscopy. *Perspectives on Science, 9*(2), 126–146.
Schwarzschild, B. (1990). Hubble investigation board finds out what went wrong. *Physics Today, 43*(11), 19–21. https://doi.org/10.1063/1.2810750
Shapin, S. (1989). The invisible technician. *American Scientist, 77*(6), 554–563.
Shapin, S., & Schaffer, S. (1985). *Leviathan and the air-pump. Hobbes, boyle, and the experimental life*. Princeton University Press.
van den Hover, J., Vermaas, P. E., & van de Poel, I. (Eds.). (2015). *Handbook of ethics, values, and technological design. Sources, theory, values and application domains*. Springer.
van de Poel, I. (2015). Design for values in engineering. In J. van den Hoven, P. E. Vermaas, & I. van de Poel (Eds.), *Handbook of ethics, values, and technological design: Sources, theory, values and application domains* (pp. 667–690). Springer. https://doi.org/10.1007/978-94-007-6970-0_25
Werrett, S. (2019). *Thrifty science: Making the most of materials in the history of experiment*. The University of Chicago Press. https://doi.org/10.1353/tech.2019.0123
Young, M. T. (2021a). Maintenance. In D. P. Michelfelder & N. Doorn (Eds.), *The Routledge handbook of the philosophy of engineering* (pp. 356–368). Routledge.
Young, M. T. (2021b). Now you see it (now you don't): Users, maintainers and the invisibility of infrastructure. In M. Nagenborg, T. Stone, M. González Woge, & P. E. Vermaas (Eds.), *Technology and the city* (Vol. 36, pp. 101–119). Springer Nature. https://doi.org/10.1007/978-3-030-52313-8_6
Young, M. T., & Coeckelbergh, M. (Eds.). (2024). *Maintenance and the philosophy of technology. Keep things going*. Routledge.

**Alexandra Karakas** BA in Art and Design Theory (2014, MOME), MA in Design Theory (2016, MOME), PhD Philosophy (2023, ELTE). Assistant Professor, Department of Philosophy and History of Science, Budapest University of Technology and Economics. Research Fellow at the MTA Lendület Values and Science Research Group, Institute of Philosophy, HUN-REN Research Centre for the Humanities. Her research interests include science and technology studies, and the history and philosophy of science, with a current focus on the epistemic role of artefacts in science and technology.

# Part IV
# Theoretical Approaches to Value Change in Design

# Chapter 18
# Artificial and Natural Functions: A Pragmatic Taxonomy

**Clint Edward Hurshman**

**Abstract** This chapter argues for a pragmatic reconceptualization of the relationship between the functions of artificial and organic entities. While much prior writing has relied on a distinction between artifact functions and biological functions (which I call the *ontological taxonomy*), I introduce an orthogonal distinction between *artificial* and *natural* functions. This *pragmatic taxonomy* is ontologically neutral, in the sense that artificial and natural functions are not restricted to artifacts, biological characters, or any other specific class of objects. Which concept should be applied to a particular object, then, does not depend on the type of object under consideration, but on the question or questions one means to answer in ascribing the function. I show that the pragmatic taxonomy avoids two shortcomings of the ontological taxonomy and should therefore replace it in the philosophical study of functions. I further argue that giving up the view that intentions have a special claim over the functions of artifacts can be useful for the purpose of designing for value change.

**Keywords** Functions · Artifacts · Selection · Philosophy of biology

## 18.1 Introduction

Gruba Kaśka is a water pump in Warsaw, Poland, that uses eight clams to monitor toxin levels in the city's water supply. When the water becomes too toxic, the clams close their shells. Sensors glued to their shells detect this closing and automatically cut off the water supply (Pelka, 2019). The clams'—or the toxin detectors'—closing behavior clearly has a function. First, because freshwater clams are sensitive to pollution, the closing behavior has helped them to survive. In this sense, natural selection seems to have given the behavior the function of *avoiding ingesting*

C. E. Hurshman (✉)
University of Kansas, Lawrence, KS, USA
e-mail: clinthurshman@gmail.com



C. Didier et al. (eds.), *Engineering and Value Change*, Philosophy of Engineering and Technology 48, https://doi.org/10.1007/978-3-031-83549-0_18

*pollutants*. Second, however, the pump is *designed* to have toxin detectors that close when the level of pollutants becomes too high, in order to keep the city's water supply safe for humans. The system has toxin detectors—the clams are *there*—in order to facilitate this use. In this sense, the toxin detectors seem to have the function of *keeping the water supply safe for humans* in virtue of the system's design. Which, then, is the proper function of these objects?[1] And more generally, if one wants to know what objects like these are supposed to do, should one look to the system's design, or to their complete history of selection?

To answer these questions, this chapter argues for a pragmatic reconceptualization of the functions of artifactual and organic entities. Much prior writing has relied on a distinction between "artifact functions" and "biological functions." This terminology implies that in assigning a function to an object, one should start from ontology: whether the object under consideration is artificial or natural somehow determines whether its proper function, if it has one, is determined by design or by selection. According to this *ontological taxonomy*, what grounds the function of a given object depends somehow on the type of object it is. It is often taken as obvious or even tautological that the functions of artifacts are determined by agents' intentions. On the other hand, given that biological characters seem to have genuine functions, but that they are (mostly) not designed, their functions must have a different grounding, usually identified with natural selection (Millikan, 1984).[2] Neither of these prevailing views has gone unchallenged, but evolutionary accounts of artifact functions (e.g., Elder, 2007; Hurshman, 2023; Preston, 2013a) and design-based accounts of biological functions (Plantinga, 1993) have typically occupied a marginal position.

Thus, on the mainstream view, functional ascriptions are partly *anchored* in ontology.[3] That is, the rules for what counts as a function depend on the ontological status of the object in question.[4] While not always articulated explicitly, it is implicit in the way that philosophers have often tended to draw a sharp line between biological and technical functions.

---

[1] Proper functions are distinguished from *causal-role* (Cummins, 1975) and *accidental* functions by their normative force: proper functions describe what objects are *supposed* to do, and an object that fails to perform its proper function is *malfunctioning*. Throughout this chapter, I will use "function" specifically to refer to *proper* functions.

[2] The selected-effects theory is not the only theory of proper functions in biology, but I focus on it because it is the most prominent.

[3] I borrow the distinction between grounding and anchoring from Brian Epstein (2015). Grounding describes the conditions that make some fact true; for instance, the fact *Joe Biden is the president* is made true (in a non-causal way) by the fact that *Joe Biden won the majority of electoral votes*. Epstein distinguishes grounding from anchoring, which "sets up the rules" for some category. Thus, the fact *Joe Biden is the president* is anchored in the fact that *winning the majority of electoral votes is sufficient for being president*. Put differently, anchoring describes the "boundaries" of the concept of president (Epstein, 2019).

[4] For example, Nanay (2010) writes: "The function of my corkscrew is to open bottles. The function of my heart is to pump blood. These two function-attributions are of different kinds" (p. 412).

However, this view has at least two problems. First, the relevant ontological distinctions are suspect. The toxin detectors in Gruba Kaśka, for example, are both clams and parts of a designed system. Such living artifacts are neither novel nor unusual (see Preston, 2013b; Sperber, 2007), also including domesticated plants and animals, a hedge used as a fence, and the wood that makes up my desk. As these examples show, ontology often does not provide clear anchoring conditions for functional ascriptions, because the categorizations themselves are ambiguous (Carrara & Mingardo, 2013).

Second, even if these ontological boundaries can be settled, the ontological taxonomy is inconsistent, or at least heterogeneous in a way that demands a defense. If selection is sufficient to ascribe a function to a biological character, consistency seems to demand that the same should be true of other types of objects that undergo selection. Likewise, if intentions are necessary to ascribe functions to artifacts, they should be necessary for functional ascriptions in general (see Ruse, 2002). The ontological taxonomy seems to entail that proper functional ascriptions to artifacts and to biological characters have different contents, and thus appears inconsistent. Perhaps proper functional ascriptions to different types of entities simply invoke separate concepts, as marked by the labels "biological" and "artifact." If these are simply separate notions, then there is no inconsistency, but the justification for this heterogeneity itself remains unclear. That is, if separate concepts of function are at work, it remains unclear why one would have a special claim over artifacts, and another over organic characters.

One response to these difficulties is to give a *unified* account of function talk in these various domains, typically by reducing the functions of artifacts to selection (e.g., Boorse, 2002, Perlman, 2009; cited in Artiga, 2023). By doing so, one avoids inconsistency and obviates the need to delineate the boundaries of the artifactual. The impulse to give such an account has served as a "regulative idea" in the literature (Vermaas and Houkes, 2013, p. 214). Selection is a multiply realizable notion and therefore invites adaptation to market selection (Schulz, 2020), technological innovation (Ziman, 2000), cultural evolution (Boyd & Richerson, 1988), and processes at various levels within biology (Garson, 2019). If the properties ascribed to artifacts and organic entities by means of functional ascriptions are somehow related—as they seem to be—then unification is an attractive option. However, it seems clear that function talk serves different purposes in engineering, biology, social science, and other contexts in which the notion is used. The notion used by an engineer designing a product, for example, is expected to perform different work than that of a biologist seeking to understand a trait. This suggests that a single, unified account will be either untenable or so general as to be unhelpful.

In this chapter, I suggest a third way of understanding the relationship between evolutionary and intentional function theories. I argue that there are, indeed, multiple concepts of proper function at work: there is likely a difference between the contents of functional ascriptions to a bird's wing and to an airplane's wing. However, I'll argue that the difference does not follow (directly) from the fact that one is an artifact and one is a biological character. Rather, it results from the different purposes that are salient in biological and engineering contexts. I'll argue that the typical distinction between artifact functions and biological functions is not theoretically significant on its own, and indeed a red herring in the effort to differentiate function concepts. Instead, I argue for a pragmatic pluralism, on which two notions of function—which I call *artificial* and *natural*—are motivated by the different forms of work that they are suited to perform. This pragmatic taxonomy can explain the appeal of the ontological taxonomy but it is ontologically neutral: neither type of function has a privileged claim over a particular domain or type of object. Indeed, artifacts, biological characters, and other objects can have functions of either, neither, or both kinds. This *pragmatic taxonomy* is preferable because it avoids the shortcomings noted above while also reflecting the diversity of interests with which philosophers study functions.

The structure of the chapter is as follows. In Sect. 18.1, I examine the grounding conditions for functional ascriptions on the prevailing theories of technical and biological functions. I argue that there is no clear reason that these must be applied as narrowly as they traditionally have been, and I introduce my pragmatic taxonomy. In Sect. 18.2, I examine the discursive tasks that design-based and selection-based function theories are best suited to perform. Drawing on Sally Haslanger's (2016) discussion of pluralism about sex ascriptions, I argue that functional ascriptions should be anchored not in ontology, but by reflecting on the discursive task at issue in one's inquiry. The final section concludes by reconsidering the case with which this chapter began, the clams in Gruba Kaśka. Given the topic of this volume, I focus on the implications of the pragmatic taxonomy for the functions of artifacts, though the argument has implications for functions in biology and elsewhere, as well.

## 18.2 Grounding Functional Ascriptions

### *18.2.1 Artifacts and Artificial Functions*

Artifacts are objects produced by humans. Technical artifacts are intentionally designed to serve our purposes (Vermaas et al., 2011). The most widely accepted theory of the functions of technical artifacts, the *intentionalist* theory, takes them to be determined by the intentions of designers (Dipert, 1993, 1995; Hilpinen, 1993; Houkes & Vermaas, 2010). Randall Dipert (1995), for example, distinguishes artifacts from instruments and tools. An instrument is any object that an agent uses to achieve some goal (on the basis of properties that she believes the object to have, and that she believes are a means to achieving the goal). Thus, a rock that I use as a paperweight is an instrument, even if it has not been modified in any way. A tool is an object that has been intentionally modified to have some set of properties, and that is used as an instrument on the basis of those properties. Thus, a rock that has been sharpened for use as an arrowhead is a tool (even if the user does not know that it has been sharpened for this purpose). Finally, an *artifact* is a tool that intentionally signals its status as a tool. It communicates, for example, by affordances or artifact cues (see Davis, 2020; Norman, 2013) that it has been produced to be used in a particular way. It bears the conspicuous mark of human design, and its function consists in those uses for which it was designed and for which it signals it ought to be used.

Houkes and Vermaas' (2010) ICE theory is the most widely accepted version of the intentionalist theory. They call it the "ICE" theory because it combines *intentional*, *causal-role*, and *evolutionary* elements. Roughly, for Houkes and Vermaas, an artifact has a function (or rather, a function ascription is justified) if and only if (1) it was designed to perform that function, (2) the designer believes that the artifact has the physicochemical properties necessary to be at least minimally capable of performing the function, and (3) the designer communicates this function in the form of a *use plan* to users.[5] In this way, they argue that functional ascriptions "highlight... practically relevant capacities" (Vermaas & Houkes, 2013, p. 222).

The account thus explains the rational justificatory force of functional ascriptions, via a social division of epistemic labor (Houkes, 2006; see also Houkes & Pols, 2013, Mebius, 2022). The fact that an artifact has the function of producing effect *e* seems to give agents reason to use it if they want to produce effect *e* (see Franssen, 2006). I don't understand how my microwave works, but it nevertheless seems rational for me to use it if I want to heat food, since this is its function. On the ICE theory, a justified functional ascription to an artifact requires a use plan communicated by someone with beliefs about the artifact's function-relevant properties. As long as I follow the use plan communicated to me by the designer, I can reasonably expect it to produce the desired effects because the designer presumably *does* have some understanding of the microwave's mechanism.

Note, however, that (design-based) functional ascriptions can play this role even when applied to objects other than artifacts—that is, to mere *instruments* and *tools*, in Dipert's sense. As Vermaas and Houkes (2013) point out, given the social-epistemic role of functional ascriptions, the relevant design is not exclusively—and sometimes not even primarily—the design of artifacts themselves, but the design of *plans* involving artifacts. If non-artifacts are involved in plans, they seem to have the same type of *function* relative to that plan that an artifact would have. For example, consider a plan for using a tree as a shelter. This plan can be backed by a social division of knowledge—the "designer" of the plan may know the capacities of the tree relevant to its serving this purpose—even though the tree itself is not designed. This tree, I suggest, has the *function* of providing shelter in the sense that its use for this purpose is rationally justified in the same way that the use of artifacts according to their functions is typically justified. For this reason, I will call design-based functions—ascribed to artifacts or otherwise—*artificial functions*.

---

[5] Houkes and Vermaas (2010, chapter 2) describe a *plan* as a description of a series of actions, specifying their aim, order, and objects necessary for successful execution. A recipe is a paradigmatic example. A *use plan* is one that involves artifacts.

### 18.2.2 Biological and Natural Functions

In contrast, biological functions are typically understood to be grounded in natural selection. Larry Wright (1973) argued that functions are a form of etiological explanation. Specifically, to say that the function of *X* is *Z* is to say that "(a) *X* is there because it does *Z*, [and] (b) *Z* is a consequence (or result) of *X*'s being there" (p. 161). Intended functions often fit this description: my glasses are there, on my face, because they help me to see, and helping me to see is a consequence of my glasses being there. But citing intentions is not the only way to give such an explanation. Thus, philosophers of biology argued that adaptations—traits whose existence is explained by their being selected to produce particular effects—also fit this description. Wright's view could therefore vindicate the existence of non-designed functions in the biological world.

The selected-effects theory (Garson, 2019; Millikan, 1984; Neander, 1991a) builds on this insight and is the most widely accepted account of proper functions in biology. On this view, a token of a trait has the function *f* if past tokens' doing *f* explains selection for the trait. Human hearts, for example, have many effects, only some of which are relevant to explaining why hearts have been selected in the past. Those effects that explain past selection thereby explain why *my* heart now exists. I have a heart because hearts pump blood, and pumping blood is a consequence of my heart's being there. Therefore, the function of my heart is to pump blood.

Artifacts, too, are subject to various selection processes. The most obvious is intentional selection (Neander, 1991b): in general, agents are more likely to use artifacts that satisfy their intentions than ones that don't. However, this selection need not be intentional. Griffiths (1993) considers societies with simple technologies produced by trial and error:

> In societies with low-level technologies, artifacts are often designed by trial and error over periods of many generations. The contribution that a feature makes to the performance of overall function may never be appreciated. Such features do not have an intended use but they do have functions, and they can be explained by their functions. Various shapes of a tool are tried out. One is more effective than the others because, for example, it is better balanced. This shape is copied more than the others and eventually becomes the norm in that culture. The function of the shape is to balance the tool and the occurrence of the shape can be explained by the fact that it balances it… The shape's function is, I contend, a proper function. (pp. 418–419)

Imagine a world somewhat like our own in which landline telephones have come to be exclusively used by elderly people. Suppose, further, that these telephones vary (for no intended reason) in the size of their buttons, but that larger buttons are easier to read for older users. Likely, large buttons would "spread throughout the population" of telephones. The fact that they are more legible to older users would explain why they out-compete smaller-buttoned phones, and on a selection-based theory would therefore be a proper function, even if it is unintended.

With these considerations in mind, Beth Preston (2013a) argues for an etiological theory of artifact functions, modeled after Millikan's (1984) selected-effects theory of biological functions (see also Elder, 2007). I call any functions grounded in selection—ascribed to biological traits or otherwise—*natural functions*.[6]

Note that the relationship between artifacts and the selection pressures they face is bidirectional, as artifacts themselves influence the contexts into which they are introduced (Verbeek, 2011; Winner, 1980). By analogy, as niche-construction theorists have argued in biology, natural selection is a dynamic process whereby organisms and environments shape each other (Odling-Smee et al., 2003). Similarly, while agents intentionally pose selection "problems" to artifacts, artifacts can work on individuals and their environments in ways that influence their selection pressures. If such niche-constructing effects explain the etiology of an artifact, then they are included in its natural function. Verbeek (2011), for example, argues that obstetric ultrasounds have restructured the practice of antenatal medicine—modifying the responsibilities of doctors and patients, and the moral status ascribed to fetuses—and thereby firmly entrenched their continued use in that practice. In this way, natural functions can include values embedded in technologies (see Klenk, 2021; Winner, 1980), as well as artifacts' role in changing values (see Hopster, 2022; Van de Poel, 2021).

## 18.3 Anchoring Functional Ascriptions

To take stock, functional ascriptions to artifacts are typically grounded in design, and those to biological traits in selection. However, as I have shown, it is not clear that they must be applied so narrowly. Designed systems and plans can implicate non-designed components, and selection processes occur outside the biological world. Thus, I introduced a new taxonomy—including *artificial* and *natural functions*—which runs orthogonal to the ontological taxonomy.

To defend this taxonomy, I will draw from Sally Haslanger's (2016) argument for pluralism about sex ascriptions. Haslanger points out that across various medical, scientific, and social contexts, sex ascriptions are grounded in various ways. In some cases, they are grounded in social facts, such as self-identification, while elsewhere they are grounded in natural facts, such as chromosomes or morphology. Here, Haslanger doesn't privilege any of these grounding conditions as the "right" ones, nor does she conclude that "anything goes" (p. 140). Instead, she points out

[6] For this chapter's purposes, I focus on etiological or *backward-looking* function theories. However, I intend for the term "natural function" to apply also to present-focused and forward-looking theories that are based in selection. In the social sciences, Pettit (2007) points out that persistence or resilience, rather than etiology, is often the central explanandum. Pettit (1996) and Schulz (2021) thus develop theories of social functions based on *present* selection pressures acting on a social entity: an entity's function consists in those of its effects that promote its resilience or persistence into the future. I have argued for a parallel, presentist theory of artifact functions on which the function of an artifact consists in the effects of tokens that explain persistence of the artifact kind (Hurshman, 2023). In my view, these are all natural function theories, since they are rooted in selection.

that the concept is embedded in a variety of practices and disciplines where it performs different forms of work, and that the frameworks ideal for serving one purpose may not be adequate for serving others:

> It does not matter whether those who employ different frameworks in different contexts agree unless they need to communicate; and when they need to communicate, there are mechanisms available to disambiguate their terms (though not always successfully employed!). What matters is whether their conclusions about what sex is give good answers to their questions, where good answers involve tracking the parts of reality that do the needed descriptive, explanatory, normative work. (p. 134)

Because there is no "value-free starting point" (p. 139) for thinking about sex, one must start by specifying the question being asked, and reflecting on the values guiding one's inquiry. One must "theorize with a purpose" about sex.

The history of biological thinking about *species* reflects a similar shift away from fundamental ontology and toward adequacy for purpose (Reydon, 2014). Species were long treated as paradigmatic natural kinds—collections of objects sharing some set of essential properties. However, it is unclear what essential properties could be shared by all and only members of a particular species; while organisms are often grouped according to conditions like genetic and morphological similarities and the ability to produce fertile offspring, these features often leave the boundaries between species fuzzy. That is, they fall short of providing necessary and sufficient conditions for membership. Nevertheless, species play an ineliminable inductive role in biology. Thus, starting in the 1960s, theories of species deemphasized the metaphysical status of group members while prioritizing the inductive utility of such groupings (e.g., Ghiselin, 1966, 1974; Hull, 1976, 1978; cited in Reydon, 2014), a trend Reydon (2014) calls "an epistemological turn in thinking about kinds" (p. 130). As this turn illustrates, an ameliorative approach—one guided by an inquiry's purpose—is a way to overcome ontological ambiguity.

Likewise, the concept of proper function is embedded in a variety of practices and disciplines, and I suggest that it must be theorized with a purpose. If the concepts of artificial and natural functions lend themselves to different purposes, then they need not be seen as conflicting. For this reason, I call the distinction between the two the *pragmatic taxonomy*.

As Houkes (2006) explains, the role of a *designer* comes with a rich suite of moral and social responsibilities. If others' use of artifacts is guided by their prescriptions, it's important that they "live up to" the epistemic authority that artificial (viz. intentionalist) theories attribute to them. Moral norms, as well as legal and professional ones, require engineers to ensure that the artifacts they design are reliable for the purposes for which they prescribe using them. Thus, when they ascribe functions to artifacts, it is particularly important that the prescribed uses are adequately justified by the social division of epistemic labor described above. Artificial function theories provide a useful framework for the rational-justificatory work that engineers' functional ascriptions ought to perform.[7]

Natural function theories lend themselves to different sorts of work. They provide forms of explanation characteristic of the theories of functions in the sciences (namely biology and social sciences) after which they are modeled, which theories are themselves the product of reflection on the explanatory work that functional ascriptions are deployed to perform. If one's interest in artifacts is comparable to that of an evolutionary biologist seeking to explain the persistence of a trait in terms of its effects (see Millikan, 1984; Preston, 2013a), or of a social scientist aiming to explain the persistence of a social entity (see Hurshman, 2023; Pettit, 1996; Schulz, 2021), then *natural function* is the relevant concept. For these purposes, an evolutionary theory of artifact functions usefully unifies philosophers' understanding of artifacts and other subjects of inquiry.

I don't claim that justification and explanation are the only discursive tasks that functional ascriptions perform. I focus on them for reasons of space, and because they are two of the most prominent. More importantly, focusing on these two tasks helps to explain why the ontological taxonomy seems so appealing. In key contexts (engineering and evolutionary biology, respectively), artificial functions are indispensable for studying artifacts, as natural functions are for studying organisms. This fact about our interests explains why ontology seems, if imperfectly, to track cases where we deploy artificial and natural functional ascriptions. But this fact is true only approximately and contingently. Philosophers may want to *explain* the use of an artifact, independently of what one is justified in using it for. Similarly, one might want to justify the use of an organic entity in some plan in the same way that one justifies the use of an artifact; one may use a tree as a shelter, or a clam as a toxin detector, and prescribe a use plan to others for doing so. On an ontology-first approach, function ascriptions like these are merely metaphorical. On a pragmatic approach, there is nothing derivative or metaphorical about them, since the type of object is irrelevant to justifying the ascription of either an artificial or natural function to it. The pragmatic taxonomy reflects the diversity and flexibility of the values motivating the study of functions.

One might ask whether one function concept or the other can adequately perform both tasks (and any other tasks that functional ascriptions are used to perform); if this is the case, then the pragmatic approach would favor unification rather than pluralism. However, the fact that different concepts of function are currently in use in different contexts provides some evidence that different concepts are optimal for different purposes—in other words, that unification would be costly. However, I doubt that the division of labor between artificial and natural functions is perfectly clear. Artificial functions give a kind of etiological explanation for the presence of artifacts, viz. one in terms of intentional selection pressures (Neander, 1991b). For some purposes, this type of explanation is preferable to the broader, causal

[7] For example, Van Eck and Weber (2014) use the ICE theory to analyze the practices of *malfunction explanation* and *redesign* in engineering science. These practices depend on the ability to black-box certain parts of a system in the way that the ICE theory makes possible.

explanation that natural functions provide. Additionally, natural functions can contribute to the design of artifacts.[8]

One context in which natural functions can contribute to the design of artifacts is the topic of this volume, value change. Artificial functional ascriptions—particularly on Houkes and Vermaas' (2010) ICE theory—situate artifacts in plans for achieving desired outcomes. However, these use plans can be more or less conducive to *changing* plans that arise as a result of changing values. Some technologies persist in use not only because they serve their intended purposes, but because they impose constraints on the agents and societies that use them, thereby "locking in" particular trajectories. Here, the discrepancy between artificial and natural functions can be illuminating, because an artifact that remains in use because it imposes demands on its users is, to that extent, a poor fit for changing values.

Illustrative examples arise in energy systems (see Melnyk, 2022; Van de Poel & Taebi, 2022). Carbon-burning technologies like automobiles persist in feedback with infrastructures that make their use possible—for example, petroleum drilling equipment, pipelines, highways, and other car-centric planning. Their use demands and thereby contributes to the production of these sociotechnical systems, and they may therefore have the *natural function* of contributing to carbon lock-in. In contrast, solar power is not only more conducive to the emerging value of sustainability, but also compatible with a wide variety of sociotechnical contexts (Winner, 1980). Because solar panels do not impose additional infrastructural demands as strongly as carbon power, they are potentially conducive to a variety of values that might emerge after they are put into use. Designing for value change, then, may require avoiding artifacts that have the natural function of constructing a niche that makes it costly to change course.

In some cases, this means designing an artifact that is sub-optimal for one's immediate purposes; that is, it may mean designing an artifact that is worse with respect to its artificial function. Trade-offs may need to be made between an artifact's performance of its artificial function, and its propensity to take on particular natural functions, such as contributing to lock-in. In short, in designing for value change, it is valuable to consider the total effect of an artifact on our present and future values and capacities, rather than just its contribution to a pre-determined plan. Natural function theories, such as Preston's (2013a) and Hurshman's (2023) encourage this open-ended view of the functions of artifacts. In the context of design, then, both artificial functions (which pick out the roles artifacts play in plans) and natural functions (which pick out, among other things, the effects of artifacts on possible future planning) can complement one another. While I have argued that the ascription of artificial and natural functions should be guided by reflection on the purposes that such ascriptions serve, it does not follow that the division of labor between the two is always clear.

[8] In previous work (Hurshman, 2023), I argued for a natural theory of artifact functions on the basis of its power to justify uses and interventions in a particular range of cases.

Finally, the pragmatic taxonomy addresses the two shortcomings of the ontological taxonomy. First, I argued that the ontological taxonomy is inconsistent in the sense that it grounds functions in design/selection but fails to consistently apply functional ascriptions to the objects that undergo these processes, or at least heterogeneous in that it deploys different concepts of function without a clear justification for doing so. The pragmatic taxonomy embraces this heterogeneity, but recognizes that justifying this heterogeneity in terms of the forms of work that functional ascriptions perform does not necessarily reinforce the view that selection and design have special claims over the functions of organic traits and artifacts, respectively. The pragmatic taxonomy encourages philosophers to apply each concept promiscuously: natural functions to any object that undergoes selection, artificial functions to any objects that are designed or implicated in designed systems or use plans. It thus avoids the inconsistency engendered by restricting functional ascriptions of some kind to some class of objects. However, this heterogeneity must be justified by a clearly specified *purpose* in using one function concept or the other. The taxonomy's liberality must be counter-balanced by reflection on the purpose of ascribing a function in the first place.

By the same token, the pragmatic taxonomy sidesteps the ontological ambiguity of living artifacts. To diverse types of objects, functions are ascribed sometimes to justify use and sometimes to explain existence. The pragmatic taxonomy is ontologically neutral, so ontologically ambiguous objects like living artifacts present no special difficulty.

## 18.4 Conclusion

To conclude, let us return to the case with which we started: *what is the function of the closing behavior of the clams in Gruba Kaśka?* This question might mean to ask either (1) why the clams close when the level of toxins becomes too high, or (2) what the city of Warsaw is (justifiably) using the clams' closing behavior for. (1) is best answered by the clams' history of selection. Clams that close in the presence of toxins have successfully reproduced (in part, and most obviously) because closing helps them to *avoid ingesting pollutants*. Note that human intentions may factor into this etiological explanation: the designers would not have included (and thereby promoted the survival of) the clams if they did not exhibit the relevant behavior. However, intentions are simply one type of selection pressure among others. (2) is best answered by referring to the design of Gruba Kaśka. The pump's designers included the clams because they believed they had the capacities relevant to *ensuring the water supply would be safe for humans*; this is the artificial function of the closing behavior. Those who use the pump are justified in doing so on the basis of this social division of epistemic labor. The clams' closing behavior, then, may have either the natural function of avoiding ingesting pollutants, or the artificial function of ensuring a human-friendly water supply. One can only privilege one function over the other by clarifying whether one means to ask (1) or (2).

I have argued that the pragmatic taxonomy, which distinguishes *artificial* from *natural* functions, avoids two pitfalls of the ontological taxonomy while recognizing the flexibility of the purposes with which we study functions, particularly those of artifacts. Because natural functions make salient the effects of artifacts on the individuals and societies that use them, giving up the view that artifact functions come exclusively from design can contribute to designing for value change.

## References


Artiga, M. (2023). A dual-aspect theory of artifact function. *Erkenntnis, 88*(4), 1533–1554.

Boorse, C. (2002). A rebuttal on functions. In A. Ariew, R. C. Cummins, & M. Perlman (Eds.), *Functions: New essays in the philosophy of psychology and biology* (pp. 63–112). Oxford University Press.

Boyd, R., & Richerson, P. J. (1988). *Culture and the evolutionary process*. University of Chicago Press.

Carrara, M., & Mingardo, D. (2013). Artifact categorization. Trends and problems. *Review of Philosophy and Psychology, 4*, 351–373.

Cummins, R. (1975). Functional analysis. *Journal of Philosophy, 72*(20), 741–765. https://doi.org/10.2307/2024640

Davis, J. L. (2020). *How artifacts afford: The power and politics of everyday things*. MIT Press. https://doi.org/10.7551/mitpress/11967.001.0001

Dipert, R. (1993). *Artifacts, art works, and agency*. Temple University Press. https://doi.org/10.5840/monist199578218

Dipert, R. (1995). Some issues in the theory of artifacts: defining 'artifact' and related notions. *The Monist, 78*(2), 119–135. https://doi.org/10.5840/monist199578218

Elder, C. (2007). On the place of artifacts in ontology. In E. Margolis & S. Laurence (Eds.), *Creations of the mind: Theories of artifacts and their representation* (pp. 33–51). Oxford University Press.

Epstein, B. (2015). *The ant trap: Rebuilding the foundations of the social sciences*. Oxford University Press.

Epstein, B. (2019). Anchoring versus grounding: Reply to Schaffer. *Philosophy and Phenomenological Research, 99*(3), 768–781. https://doi.org/10.1111/phpr.12644

Franssen, M. (2006). The normativity of artefacts. *Studies in History and Philosophy of Science Part A, 37*(1), 42–57. https://doi.org/10.1016/j.shpsa.2005.12.006

Garson, J. (2019). *What biological functions are and why they matter*. Cambridge University Press. https://doi.org/10.1017/9781108560764

Ghiselin, M. T. (1966). On psychologism in the logic of taxonomic controversies. *Systematic Zoology, 15*, 207–215.

Ghiselin, M. T. (1974). A radical solution to the species problem. *Systematic Zoology, 23*, 536–544.

Griffiths, P. E. (1993). Functional analysis and proper functions. *The British Journal for the Philosophy of Science, 44*(3), 409–422.

Haslanger, S. (2016). Theorizing with a purpose: The many kinds of sex. In C. Kendig (Ed.), *Natural kinds and classification in scientific practice* (pp. 129–144). Routledge.

Hilpinen, R. (1993). Authors and artifacts. *Proceedings of the Aristotelian Society, 93*, 155–178. https://doi.org/10.1093/aristotelian/93.1.155

Hopster, J. (2022). Future value change: Identifying realistic possibilities and risks. *Prometheus, 38*(1), 113–123.

Houkes, W. (2006). Knowledge of artefact functions. *Studies in History and Philosophy of Science Part A, 37*(1), 102–113.

Houkes, W., & Pols, A. J. (2013). Plans for modeling rational acceptance of technology. In D. P. Michelfelder, N. McCarthy, & D. E. Goldberg (Eds.), *Philosophy and engineering: Reflections on practice, principles and process* (pp. 291–303). Springer.

Houkes, W., & Vermaas, P. (2010). *Technical functions: On the use and design of artefacts* (Vol. 1). Springer. https://doi.org/10.1007/978-90-481-3900-2

Hull, D. L. (1976). Are species really individuals? *Systematic Zoology, 25*, 174–191.

Hull, D. L. (1978). A matter of individuality. *Philosophy of Science, 45*, 335–360.

Hurshman, C. (2023). Artifacts and intervention: a persistence theory of artifact functions. *Synthese, 202*(5), 128.

Klenk, M. (2021). How do technological artefacts embody moral values? *Philosophy & Technology, 34*(3), 525–544.

Mebius, A. (2022). Against 'instantaneous' expertise. *Philosophy, Ethics, and Humanities in Medicine, 17*(1), 1–6.

Melnyk, A. (2022). An interpretation of value change: A philosophical disquisition of climate change and energy transition debate. *Science, Technology, & Human Values, 47*(3), 404–428.

Millikan, R. (1984). *Language, thought, and other biological categories*. MIT Press.

Nanay, B. (2010). A modal theory of function. *The Journal of Philosophy, 107*(8), 412–431.

Neander, K. (1991a). Functions as selected effects: The conceptual analyst's defense. *Philosophy of Science, 58*(2), 168–184. https://doi.org/10.1086/289610

Neander, K. (1991b). The teleological notion of 'function'. *Australasian Journal of Philosophy, 69*(4), 454–468.

Norman, D. (2013). *The design of everyday things: Revised and expanded edition*. Basic Books.

Odling-Smee, F., Laland, K., & Feldman, M. (2003). *Niche construction: The neglected process in evolution (MPB-37)*. Princeton University Press. https://doi.org/10.1515/9781400847266

Pelka, J. (2019). *Fat Kathy*. Motion Picture. Poland.

Perlman, M. (2009). Changing the mission of theories of teleology: DOs and DON'Ts for thinking about function. In U. Krohs & P. Kroes (Eds.), *Functions in biological and artificial worlds: Comparative philosophical perspectives* (pp. 17–36). MIT Press. https://doi.org/10.7551/mitpress/9780262113212.003.0002

Pettit, P. (1996). Functional explanation and virtual selection. *The British Journal for the Philosophy of Science, 47*(2), 291–302. https://doi.org/10.1093/bjps/47.2.291

Pettit, P. (2007). Resilience as the explanandum of social theory. In I. Shapiro & S. Bedi (Eds.), *Political contingency: Studying the unexpected, the accidental, and the unforeseen* (pp. 79–96). New York University Press.

Plantinga, A. (1993). *Warrant and proper function*. Oxford University Press.

Preston, B. (2013a). *A philosophy of material culture: Action, function, and mind*. Routledge. https://doi.org/10.4324/9780203069844

Preston, B. (2013b). Synthetic biology as red herring. *Studies in History and Philosophy of Science Part C: Studies in History and Philosophy of Biological and Biomedical Sciences, 44*(4), 649–659.

Reydon, T. A. C. (2014). Metaphysical and epistemological approaches to developing a theory of artifact kinds. In M. Franssen, P. Kroes, T. A. C. Reydon, & P. E. Vermaas (Eds.), *Artefact kinds: Ontology and the human-made world* (pp. 125–144). Springer.

Ruse, M. E. (2002). Evolutionary biology and teleological thinking. In A. Ariew, R. Cummins, & M. Perlman (Eds.), *Functions: New essays in the philosophy of psychology and biology* (pp. 33–62). Oxford University Press.

Schulz, A. W. (2020). *Structure, evidence, and heuristic: Evolutionary biology, economics, and the philosophy of their relationship*. Routledge. https://doi.org/10.4324/9781003030249

Schulz, A. W. (2021). What's the point? A presentist social functionalist account of institutional purpose. *Philosophy of the Social Sciences, 52*(1-2), 53–80.

Sperber, D. (2007). Seedless grapes: Nature and culture. In E. Margolis & S. Laurence (Eds.), *Creations of the mind: Theories of artifacts and their representation* (pp. 124–137). Oxford University Press.

Van de Poel, I. (2021). Design for value change. *Ethics and Information Technology, 23*(1), 27–31.
Van de Poel, I., & Taebi, B. (2022). Value change in energy systems. *Science, Technology, & Human Values, 47*(3), 371–379.
Van Eck, D., & Weber, E. (2014). Function ascription and explanation: Elaborating an explanatory utility desideratum for ascriptions of technical functions. *Erkenntnis, 79*(6), 1367–1389. https://doi.org/10.1007/s10670-014-9605-1
Verbeek, P. P. (2011). *Moralizing technology: Understanding and designing the morality of things*. University of Chicago Press.
Vermaas, P. E., & Houkes, W. (2013). Functions as epistemic highlighters: an engineering account of technical, biological and other functions. In P. Huneman (Ed.), *Functions: Selection and mechanisms* (pp. 213–231). Springer.
Vermaas, P., Kroes, P., Poel, I. V. D., Franssen, M., & Houkes, W. (2011). *A philosophy of technology: From technical artefacts to sociotechnical systems*. Springer.
Winner, L. (1980). Do artifacts have politics? *Daedalus, 109*(1), 121–136.
Wright, L. (1973). Functions. *Philosophical Review, LXXXII*, 139–168.
Ziman, J. (Ed.). (2000). *Technological innovation as an evolutionary process*. Cambridge University Press.

**Clint Edward Hurshman** BA Philosophy (University of Missouri–Kansas City), MA Philosophy (University of Kansas). 2019–present PhD student, University of Kansas. Research areas: Philosophy of technology, philosophy of science (esp. biology and social science), AI ethics, privacy.

# Chapter 19
# Value Change Sensitive Design: Elements of a Proces Ontological Framework and Method

**Alok Srivastava and Gunter Bombaerts**

**Abstract** This chapter addresses the need to monitor and respond to changes in values and the other core elements of a technology project or product during its lifespan—from its initial design phase through completion and into implementation and maintenance. The complexity we aim to address is that technology projects and products are situated in sociotechnical systems that undergo change in many of their parts over time. We are introducing the turn to Process Philosophy into Value Sensitive Design projects and literature to bring conceptual systems where all parts of the system undergo change over time. Values Sensitive Design (VSD) literature has traditionally applied conceptual systems called "structure and change" systems where most of the structure remains stable while some parts undergo change. We call our program of method development: Value Change Sensitive Design methods (VCSD). Our approach uses Process Ontology and Niklas Luhmann's Social Systems Theory to develop a minimal ontological system and a diagramming tool—a schema—for marking and tracking changes in the situation of technology projects and products. The minimal ontological system focuses on values and other core elements of sociotechnical systems that affect design and outcomes of projects using Value Change Sensitive Design. The minimal framework articulates a temporal duration (window) of a change process by centering on an "event" during which a change takes place. We combine this process ontological framework and diagramming tool with the tools of historical analysis called path dependence analysis and process tracing to propose a full method for discovery and characterizations of causal mechanisms underlying changes in sociotechnical systems. This chapter is a

A. Srivastava (✉)
Playful Dyads, San Francisco, CA, USA
e-mail: aloks@playfuldyads.org

G. Bombaerts
Philosophy & Ethics, Department of Industrial Engineering and Innovation Sciences, Technological University Eindhoven, Eindhoven, The Netherlands
e-mail: G.Bombaerts@tue.nl



C. Didier et al. (eds.), *Engineering and Value Change*, Philosophy of Engineering and Technology 48, https://doi.org/10.1007/978-3-031-83549-0_19

contribution to the philosophical literature in Value Change Sensitive Design and Values Change.

**Keywords** Values sensitive design · Sociotechnical systems · Ontological framework · Process ontology · Process tracing

## 19.1 Introduction

> Every systems theory that claims to relate to reality must begin with the fact that nothing remains as it is. There is change. (Luhmann, 1995)

In this chapter we have identified two gaps and opportunities in the literature and practices of Values Sensitive Design (VSD) in the field of philosophy of engineering and technology (PET) and we propose a method to address them. The complexity we aim to address is that technology projects and products are situated in sociotechnical systems that undergo change in many of their parts over time. We are introducing the turn to Process Philosophy into Value Sensitive Design projects and literature to bring conceptual systems where all parts of the system undergo change over time (Dupré & Nicholson, 2018; Seibt, 2012). VSD literature has traditionally applied conceptual systems called "structure and change" systems where most of the structure remains stable while some parts undergo change. We call our program of method development: Value Change Sensitive Design methods (VCSD).

VSD practices were developed to incorporate human values in the design process for introducing new products and novel utilities into sociotechnical systems to give engineers and technologists some capacities for the minimization of destructive and deleterious changes and the enhancement of constructive and beneficial changes that might follow their introduction (Nickel et al., 2022; Van de Poel, 2021; Winkler & Spiekermann, 2021). However sociotechnical systems undergo changes locally throughout the system and change temporally during the lifespan of a technical project and during its implementation. An extensive literature on the study of Value Change comprises mainly of retrospective case studies discovering how technology and products drive value change (de Wildt & Schweizer, 2022; van de Poel & Kudina, 2022). There is a lack of prospective approaches to study value change during the lifespan of projects and products. Prospective methods can aid in Values Sensitive Design adjustments during implementation and maintenance of technologies as values change become detectable and salient.

We focus on two gaps and opportunities to better incorporate impact on human values in the design of technologies and products in continuously changing sociotechnical systems. The first gap and opportunity we focus on is to study value change in sociotechnical systems with a more complete set of primary elements that impact value change. We identify a minimum of four primary elements of a dynamic and operating sociotechnical system: people (stakeholders and bystanders), technology (and products), values, and choices (and their consequences). The second gap and opportunity we focus on is the need to monitor and respond to changes in values and other core elements during the lifespan of a technology project—from its initial design phase through completion and into implementation and maintenance. We, the authors, see a path to develop a methodological framework for tracing and identifying the dominant and specific causes driving changes in the core elements of the sociotechnical systems. We combine the ontology of change processes in socio-technical systems from Niklas Luhmann's Social Systems Theory with the methods of process tracing for discovery of causal mechanisms of change processes in the fields of sociology and economics. Our proposed method aims to enable the routine practice of tracing and identification of causal mechanisms—causes driving changes in the core dynamical elements—in situations being addressed by VSD projects so that the Values Sensitive Design can be adjusted and adapted during the course of the projects. This type of prospective analysis in projects using Values Sensitive Design (VSD) methods will resource the wealth of retrospective studies of values change in the literature in philosophy of engineering and technology (Nickel et al., 2022).

In the body of this chapter, we will sequentially elaborate four themes to build up our methodological framework. In Sect. 19.2 we will develop the details of a process ontology approach to sociotechnical systems and articulate the minimal ontological system of four core dynamical elements—people (stakeholders and bystanders), technology/product, values, and choice. This minimal ontological system will be the basis of marking up a change process in particular case histories and will enable the application of the causal analysis process. We will describe a change process diagramming tool we have developed called "event and change process diagrams" and use it to enable process tracing methods in VSD projects. In Sect. 19.3 we will develop the selected elements of event, choice, and values from Luhmann's Social Systems Theory to set up the ontological foundations of our framework. We will also describe "event-centered diagrams of change processes." In Sect. 19.4 we will give a brief overview of the field and methods of process tracing and narrative science as developed and applied in the study of change process in sociopolitical systems and socioeconomic systems. We will also show how this method is readily integrated with our "event and change process diagrams." Finally in Sect. 19.5 we will outline the terrain of studies this methodological framework can enable. We will outline some of the lines of future work and the possible impacts our proposal might have.

## 19.2 Applying Process Ontology to Changes Affecting VSD Projects

We are introducing the turn to Process Philosophy into Value Sensitive Design projects and literature to bring conceptual systems where all parts of the system can undergo change over time. Philosophers make a distinction between two broad classes of conceptual systems: structure and change conceptual systems and process conceptual systems. In the first, structures with some stability undergo periods of change in some aspects of the structure. In the second, all things and aspects of the world are considered to be processes, like rivers with more or less amounts of change (flux) going through them. So, things with apparent stability are said to be undergoing change very slowly. Current practices in VSD projects use the structure and change class of conceptual systems. We believe that to address the two gaps and opportunities described above it will be helpful to use a process conceptual system for VSD projects.

We start with the observation that technology projects and products are situated in sociotechnical systems which undergo change in many of their parts over the lifespan of a technology project and products. This stance suggests that project design efforts need to be sensitive to changes in values during the lifespan of a project and during its implementation and not only at the beginning of the project. We have named our framework Values Change Sensitive Design (VCSD) to emphasize this stance. We will model sociotechnical systems as a process and less as a structure undergoing well-defined changes with long-term stability between changes. Our stance is supported by the studies of Value Change and surveys of methods of Values Sensitive Design focusing on the duration and temporal scales of stability of values and other key elements affecting technological projects (de Wildt et al., 2022; de Wildt & van de Poel, 2024; van der Weij et al., 2023).

To pursue our goals in this chapter we use several developments in Process Philosophy. Although the concepts "event" and "process" are peripheral in philosophy, Gottfried Wilhelm Leibniz, Gilles Deleuze, Alfred North Whitehead, and others have elaborated these concepts as important or even central concepts in their works (Seibt, 2012; Whitehead, 1929). Below we use some more recent work from process philosophers of biology and living system such as Daniel Nicholson and John Dupre (Nicholson & Dupré, 2018). As we want to explain the link with (sociotechnical) systems, we turn to a scholar less known for the use of the event concept, Niklas Luhmann.

We enable the process philosophy and ontological approach by selecting a minimal number of objects in the sociotechnical system that undergo change and impact technology projects and products. We focus on the four core elements of a sociotechnical system that interact with technology projects and products: (1) people (stakeholders who are users and choice-makers, and bystanders who might influence choice-making), (2) technology (and products), (3) values (the differential considerations choice-makers use to make choices between alternatives), and (4) choices (navigating and selecting possibilities and their consequences).

### 19.2.1 Process Ontological Approach

In process ontologies change is an integral aspect of each participating part of the system. Entities and activities are not separated into structure and change. Specifically, each entity is conceptualized as a "relational and temporal" object (Fabris, 2018; Nicholson & Dupré, 2018). The relational aspect means that each object is seen as operating at the intersection of a set of active relations—a nexus of relations (Coase, 1937, 1998; Epstein, 2016). And the temporal aspect means that each object is seen as extended through time for a "duration" during which its properties may remain stable or change in characteristic ways. Below we develop this process ontological framework into a minimal system with four characteristic elements relevant to the practice of VSD: stakeholders (people who are users and choice-makers), technology and products, values (the differential considerations choice-makers use to make choices between alternatives), and choices (navigating possibilities). Each of these objects is a nexus of situated relations that make it a durable participating part in the sociotechnical system, and they tend to have characteristic temporal consistencies and change patterns.

### 19.2.2 A Minimal Ontological System for Change Processes in VSD Projects

The growth over four decades of the field of Values Sensitive Design and the development of methods and studies has brought increased focus on ongoing change in the sociotechnical systems where new technology products and projects are introduced. While in the last two decades there has been a lot of work on values change, the surveys and critiques of the different methods of VSD have also characterized the changes in other key elements of these projects such as stakeholders and choices. The four elements of a minimal ontological system relevant to VSD projects are: people (stakeholders), technology (products, etc.), choices, and Values. These four elements constitute a minimal ontology of elements that are the core interacting and mutually influencing elements of any situation where new projects and products are to be introduced. Our process ontological approach considers the four elements as interacting together and mutually affecting changes in each other (see Fig. 19.1).

This minimal ontological system enables us to work with key ontological elements of Luhmann's Social Systems Theory introduced and developed in Sect. 19.3. Each ontic element of our minimal ontological system is a relational and temporal conceptual object. This is appropriate because VSD methods and practices confront the reality that during the project all these four elements undergo change (Friedman et al., 2017). Each of these four elements are composed of a nexus of relations that keep each of them as contributing and participating parts of the system; and this nexus of relations undergo change as the social system and project

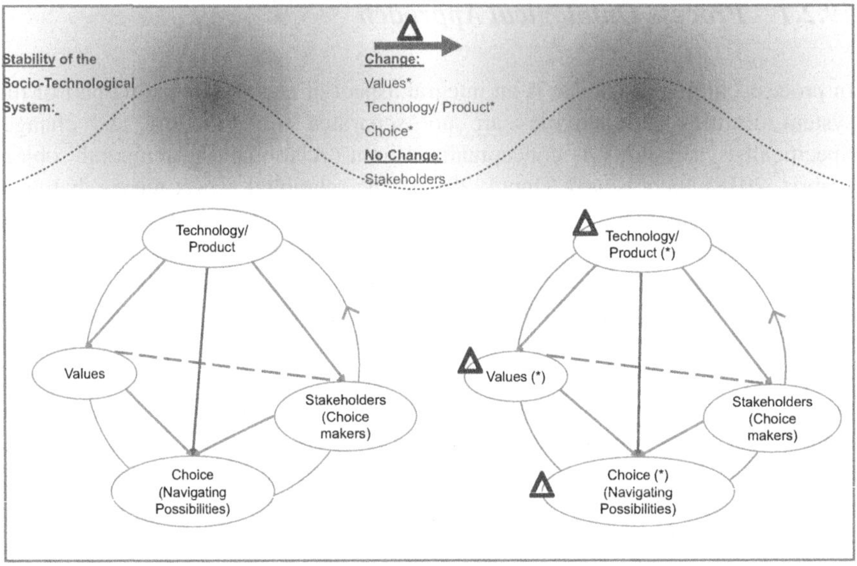


**Fig. 19.1** The Minimal Working Ontology and diagramming of changes in the arrangements in sociotechnological systems through a period of change. *The upper layer* diagrams the changing stability of the configurations of the sociotechnical system through an event of change. The sinusoidal curve emphasizes that when one part changes the other parts of the system must accommodate or change to adjust. In the middle trough we indicate the core elements that changed and did not change during this event. *The lower layer* diagrams the minimal ontological system as a four-cornered pyramid with core elements at the corners and depicts the status of the core elements before and after the "event" of change. The four-cornered pyramid diagrams are chosen to indicate the absence of a stable hierarchy between the core elements. The lines at the edges of the pyramid indicate the bidirectional dynamic between any pair of elements. For example: Values change all the time with the available choices in the bidirectional dynamics of mutual influencing. Values are constantly influencing available choices and selections during events and selections made during events are constantly influencing the strength of values in choice-making events. © Alok Srivastava and Gunter Bombaerts (2024)

proceeds. This conceptualization will let us track changes in each of these four elements during specific instances of "process change."

## 19.3 Applying Luhmann's Social Systems Theory with Diagrams

A promising theoretical and ontological framework for the development of methods for causal analysis of change processes in sociotechnical systems is Niklas Luhmann's Social Systems ((Luhmann, 1995), here onwards referred to as SS). Its ontological concepts were developed specifically to work with the highly interdependent and interactive nature of social processes. Specifically, Luhmann's concept of the "event" as the "smallest possible temporal atom, an indivisible, all-or-nothing happening" gives us the minimal change process as a starting locus and conceptual entity for our methodological framework. In addition, Luhmann's concepts of choice and values are also helpful (see below). Using the concept of "event" as "a temporal atom" and "minimal temporal change process" one of us, Gunter Bombaerts, has recently proposed "event-centered diagrams of change processes"—a diagrammatic concept to track the changes coming into an event and thus changing the elements of the system through their interdependence and interactions (see Figs. 19.2 and 19.3). These diagrams have a strong potential for marking and

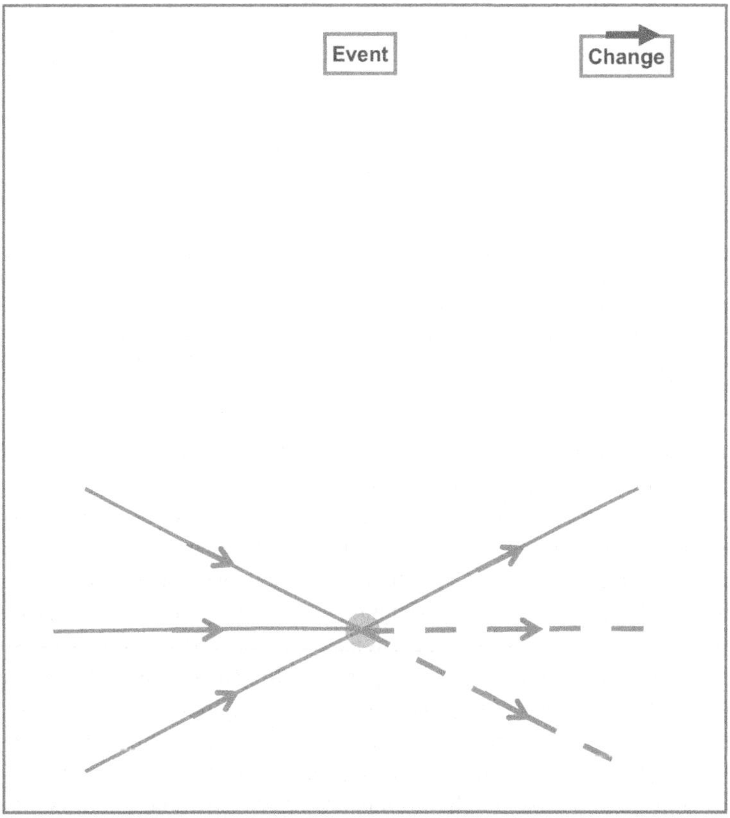


**Fig. 19.2** "Event-centered diagramming of change processes"—an event with relevant elements and a decision that is forced. The diagram shows lines of change in particular elements of a socio-technical system using Luhmann's concept of "event" as "a temporal atom" and "minimal temporal change process." The lines depict past trajectories coming from the left and possible future trajectories going to the right and intersecting at the event. The actual trajectories are indicated as solid lines and the possible trajectories that were not realized are presented as broken lines. Of the three future trajectories only one was actualized and is depicted as a solid line. © Alok Srivastava and Gunter Bombaerts (2024)

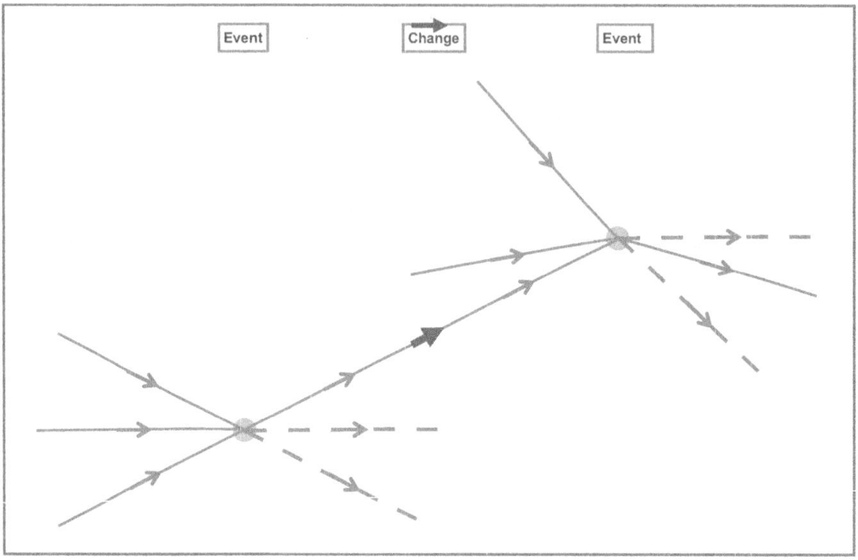


**Fig. 19.3** "Event-centered diagramming of change processes"—changes in sub-processes through two successive events showing possible causal dependency or contingency of second change. This diagram shows two successive events and indicates the influence of the change actualized during the first event on the changes actualized during the second event. This is also depiction of dependent change processes (discussed further in Sect. 19.4). © Alok Srivastava and Gunter Bombaerts (2024)

tracking possible change processes in the complex web of interdependencies and interactions between elements of a sociotechnical system.

It is important to emphasize that Luhmann's work works out the social dimensions of the processual structure of sociotechnical systems. His key entities and activities are human individuals communicating with each other, reading situations with historically accumulated "expectations"—socially contingent interpretations of available choices—and making decisions that regenerate the processual structure of the sociotechnical system they are participating in. In the sections below we detail out how Luhmann conceptualizes the interactional traffic between human individuals meeting in situations and making choices with their expectations based on incomplete information sharing move the sociotechnical system into a new changed configuration. Notice that Luhmann's work is fleshing out the interaction of the three non-technical elements of our minimal ontological system: stakeholders (people who are users and choice-makers), values (the differential considerations choice-makers use to make choices between alternatives), and choices (navigating possibilities).

### 19.3.1 Social System: Individuals with Expectations Navigating Possibilities

According to Luhmann, human individuals in their day-to-day actions experience complexity, that is the immanent constraint to connect to and participate in the sociotechnical environment, as it is not possible at any moment to connect every experience to every other. To understand how people cope with this complexity, two concepts are important: time and structure. Time is the origin of the pressure to select (SS:42). If there is an infinite amount of time available, an infinite number of possibilities can be considered. However, in a human experience, decisions must be made from moment to moment. Individuals are forced to make a selection between possibilities within a limited time and therefore cannot consider everything. This creates contingency, or the "also being possible otherwise" (SS:25). An individual will make a particular decision at a certain moment but could have made another one given more time or more information. In order to cope with this contingency, individuals develop expectations. Expectations help to narrow down the possibilities of selection (SS:96). Expectations can be thought of as a person's mental models of the available futures that are built up over time and for the particular class of situations with the incomplete information they have access to.

When two people meet, they both have expectations, including of the other. The expectation the other person has is an important element of the expectation of the first person. Person one's expectation of person two's expectation is very important for person one's selection, and vice versa. This relates to what Luhmann calls double contingency. Even more, person one's expectation of person two's expectation of person one's expectation creates feedback for person one. It might seem far-fetched, but it is exactly the expectations of the expectations of the expectations that create structure. They are structures of expectation, and there these are the only structural possibilities for social systems (SS:293). This structure represents a selection from a plurality of possibilities. As such, for Luhmann, a structure is not a concept that enables a static stability, but "an autopoietic reproduction of the system from one event to the next" (SS:286). Again, it is important to emphasize here that Luhmann is working out the social dimensions of the processual structure of socio-technical systems. By "autopoietic reproduction" he means that human individuals interacting and making choices with aid of their expectations of situations and of each other is the processual structure of social process. Through these moments (events), the choice-making by the social individuals drives the reproduction or regeneration for the social process they are participating in.

This quote above requires some extra explanation. First of all, the structure of expectations creates a system. As a famous Luhmann quote goes: "The system emerges *etsi no daretur Deus* [even if God doesn't exist]" (SS:105). As individuals from moment-to-moment create structures with their expectations, a social system emerges. In his general system theory, Luhmann mentions organisms, psychic systems, and social systems. Organisms are alive. From a systemic perspective, life is a self-organization or autopoietic mechanism. An organism uses its "life" to create "life." The same goes for psychic systems that are conscious. They use their consciousness to create consciousness. Luhmann applies this concept of autopoiesis also to explain social systems. As psychic systems interact with each other, communication emerges. In analogy to the other autopoietic systems, social systems use their communication to create communication.

### 19.3.2 Events: Temporal Atoms of Processes

Luhmann specifies how the communication and the social systems are reproduced in what he calls events. Based on Floyd Allport (1940, 1954) Luhmann defines "event" as the "smallest possible temporal atom, an indivisible, all-or-nothing happening" (SS:287). In what follows, we develop our own view on event. In Figs. 19.2 and 19.3 we develop event-centered diagramming of change processes in sociotechnical systems.

Let us give two examples. The purpose of the two cases is to illustrate Luhmann's construction of how human individuals interacting and making choices with aid of their expectations of situations and of each other is the processual structure of social process and through these moments (events), the choice-making by the social individuals drives the reproduction or regeneration for the social process they are participating in.

The first example draws from a case history in political science discussed in Sharon Crasnow's 2022 Chapter on "Process Tracing and Narrative Science" discussed in the next section, Sect. 19.4 (Crasnow, 2022). Crasnow introduces the search for causal processes in this case history of broad participation in the civil war in El Salvador this way:

(a) "Elisabeth Wood (2003) investigates the civil war in El Salvador in order to solve what she takes to be a puzzle about the case: why was there broad participation in the insurgency even though the costs of participation were so high and the rewards so minimal?" (Wood, 2003)
(b) "Participation, while broad, was not universal. Only about a third of the poor, rural class (campesinos) participated in the insurgency—a large enough rate of participation to have significant effect"
(c) "To answer her research question, Wood looks more closely at the difference between those who participated (approximately one-third of the rural poor) and those who did not. Her primary evidence is extensive interviews (more than 200) over multiple years both during and after the civil war."

To apply Luhmann's concept of complexity, expectation structure, contingency, and choice in a unit event in this case history we imagine a scenario where three people in El Salvador during the civil war in 1980 sit together in a room and decide to join an insurgency (Crasnow, 2022; Wood, 2003). Suppose one of them asks the others "Shall we join the insurgency?" In the event, some elements are present. There is fear of the vastly negative consequences of being involved, awareness of the political situation, moral commitment of being involved, knowledge of Liberation Theology, defiance to act as a reaction to beliefs that the government had overreacted to legitimate means, a feeling of authorship and pleasure in agency, and more. All these elements are present. But they are not present in the event as clearly defined elements, and not all in the same importance. As an event is a very short duration of time, all these aspects are present and exercise influence, but the influence itself is also temporal. Several outcomes are possible. As these individuals answer to the choice-making question of this situation "Shall we join the insurgency?", during this event the processual structure of the social system is regenerated.

The second example is drawn from the Values Change and Technology Design literature. Van de Poel (2021) describes a case history involving a two-level change process: "An example are mobile phones that turn out to affect traffic safety. This value was not considered relevant before in mobile phone design, and still often is not incorporated." The arrival event (first event) and ensuing widespread use of mobile phones generated a widespread occurrence (second event) of mobile phone use by drivers in traffic which increased the number of accidents. (See Fig. 19.2 for diagrammatic presentation of the first event, and Fig. 19.3 for the presentation of the second event.) We want to emphasize here that human individuals used their expectations of the utility of mobile phones to use them during activities of their lives. These choices drove a regeneration of the processual structure of the sociotechnical system they are participating in. However, the new cycle of the sociotechnical system had a new consequence—drivers in traffic suffered increased number of accidents. This new eventuality will now be incorporated into the "expectation structure" of participating individuals and will also be considered by engineers doing VSD in these situations.

As can be seen in both examples, a selection is forced in an event (SS:60) and the direction that is forced creates an indication—a directed and differential expectation of outcomes. What is possible is interpreted as the difference between different potentialities. The possibility of being actualized is then indicated in its identity as "this-and-not-something-else" (SS:66). In such an event, every complex state of affairs is based on a selection of relations among its elements, which it uses to constitute and maintain itself. The selection positions and qualifies the elements, although other relations would have been possible (SS:25) (see Fig. 19.2).

### *19.3.3 Values and Value Change*

One type of elements in events, and consequently also in communication in a social system, are values. Values according to Luhmann are "general, individually symbolized perspectives which allow one to prefer certain states or events" (SS:317). Values fall under Luhmann's notion of relations that influence the positioning.

Based on the above explanation of social system and events, we picture values as embedded in events because they are distinguishable to participants by actions of choices in situations they meet in. At this event, this smallest temporal atom, values play out in two ways. As individualized perspectives for preference, they co-determine the direction of the event. But also, when a certain direction is chosen, the values that supported this direction may become more entrenched in participants' expectations. This means that the values determine the choice or selection, but that also the choice determines the values as they will be used in the next event. As such, this provides a structure to understand how macroscopic value change gets about in a multitude of events. These individual values embedded in events thus support expectations that create structures and therefore also social systems. We can refer here to norms—shared expectations with similar preferences (SS:292).

This means that value change is not the exception that needs to be explained, rather it is the normal. Values change often with events in the bidirectional dynamics of mutual influencing. Values regularly influence event selections and event selections regularly influence the strength of values in events. As Luhmann says, "Every systems theory that claims to relate to reality must begin with the fact that nothing remains as it is. There is change" (Luhmann, 1995).

## 19.4 Process Tracing Methodologies

Our aim, in this chapter, is to develop a usable method to anticipate which changes in some elements of a sociotechnical system will drive specific changes in other elements. The term "causal mechanisms" throughout this chapter refers to the dependencies and contingencies that can be determined to operate between changes in one element and other elements, i.e., changes in one element drive changes in other elements. In this section we propose a method for carrying out discovery of causal mechanisms underlying relevant changes in specific case histories and in situations of Values Sensitive Design projects. We will be using the well-developed methods of path dependence analysis and process tracing in the fields of sociology, political science, and economics (Beach & Pedersen, 2016; Crasnow, 2017; Mahoney, 2000, 2015). These methods were developed to locate the causal mechanisms amid the complexity, and contingencies of human choices that underlie change in social processes. We will combine these methods with our minimal ontological framework developed in Sect. 19.2 and the "event centered diagramming of change processes" developed in Sect. 19.3.

### *19.4.1 Path Dependence Analysis and Process Tracing Methods*

Causality of change in social processes are studied in social science and political science by methods called path dependence analysis and process tracing (Beach & Pedersen, 2016; Crasnow, 2017; Mahoney, 2000, 2015). These methods developed over the last three decades have similar goals as our research program: to delineate mechanisms of causality and to make reliable causal claims about change in specific case histories. A critical standard of this body of work is to develop ways of making causal claims that are robust within the wealth of historical and situational data. Causal analysis in the physical sciences is studied with controlled variable experimental studies which are impossible to do for social processes. We are constrained to retrospective historical studies. However, such studies are limited in their ability to study all the possible causal processes that could influence change in a social system. Studying the path dependence of changes and then doing process tracing has allowed for more robust causal claims while focusing on a small number of changes.

The longer-term goal in the practice of process tracing is to inform design of policy which is similar to the goals of change studies in the field of Values Sensitive Design. In these fields, process tracing is used in retrospective analysis of case histories to discover causal mechanisms and drive design of policy for the future situations. We aim to integrate "event" diagrams with the methods of process tracing to study change processes in sociotechnical systems.

### *19.4.2 Discovering Causal Mechanisms in Social Processes*

The methods of path dependence analysis and process tracing study the historical sequence of events in case studies to discover causal mechanisms and specific dependencies that determine the course of the changes. James Mahoney in an early essay describes the study of contingency of the shape of a sequence of events on the multiple possible factors that could effect change (Mahoney, 2000).

> I argue that path dependence characterizes specifically those historical sequences in which contingent events set into motion institutional patterns or event chains that have deterministic properties. The identification of path dependence therefore involves both tracing a given outcome back to a particular set of historical events, and showing how these events are themselves contingent occurrences that cannot be explained based on prior historical conditions.

Mahoney's formulation above strongly resonates with the formulations by Luhmann described in Sect. 19.3. Changes of a sociotechnical system do exhibit as a "historical sequences in which contingent events set into motion institutional patterns or event chains that have deterministic properties." The methodological opportunity of path dependence analysis is this capacity to delineate the "institutional patterns or event chains" set into motion by specific changes in a specific element exhibited in a change event. These patterns or event chains in a sociotechnical system are termed "causal mechanisms." Derek Beach and Rasmus Brun Pedersen, methodologists in political science, describe the methods thus: "Process-tracing is a research method for tracing causal mechanisms using detailed, within-case empirical analysis of how a causal mechanism operated in real-world cases" (Beach & Pedersen, 2019).

Sharon Crasnow introduces the purpose and value of process tracing by quoting Mahoney (2015), as a way of "formulating and testing explanations with case studies" (Crasnow, 2017; Mahoney, 2015). She elaborates:

> The value of process tracing is thus thought to rest in how it provides evidence for causal mechanisms. Statistical and experimental methods may be able to establish a link between a dependent and independent variable, but they cannot reveal what is in the "black box"—the mechanisms through which the cause brings about the effect. The idea is that if we can trace the process, we will be able to identify the causal mechanism as it is operating in a particular case.

For sociotechnical systems, mechanisms refer to the internal arrangements of dependencies between elements that operate in patterns or event chains during a change process. She explains the core of the method and its differential benefits in her overview paper (Crasnow, 2022) in this manner:

> In order to seek traces of a process, some sense of what that process might be—for processes that are causal mechanisms, what mechanism might be operating—already must be on the table. It might be only vaguely sketched out, to be filled in later should it prove promising, or it may already be well-developed theoretically and investigated elsewhere. The hypothesis serves to guide the researcher towards the appropriate sorts of intervening factors expected to be present if that mechanism was indeed operating. These are the entities, activities and events that are elements of the causal mechanism or indicators of its operation (recognized as indicators because of other things that are known or believed).

In our case "to seek traces of a process" we have demarcated the minimal ontological system with the four core elements (entities and activities) that play significant roles in VSD projects carried out in sociotechnical systems. In Sect. 19.3 we used Luhmann's constructions of the pattern of regeneration of the arrangements sociotechnical systems set in motion by events and interactive choice-making to develop our schema for diagramming changes. This schema presents Crasnow's "entities, activities and events that are the elements of the causal mechanism or indicators of its operation." In Figs. 19.4 and 19.5 we synthesize these two tools to describe the process of tracing changes in these core elements between two successive events of noticeable change.

We hope we are successfully showing the promise of these historical methods and our ontological and diagramming tools to enable the study of change processes in VSD projects and their sociotechnical contexts. Together these methods could give us a robust capacity to discover causal mechanisms in the work of Engineering and Technology projects. Further, we believe the historical and retrospective studies that are common to the values change field can be readily modified to carry out monitoring studies of change processes taking place during the lifespan of projects.

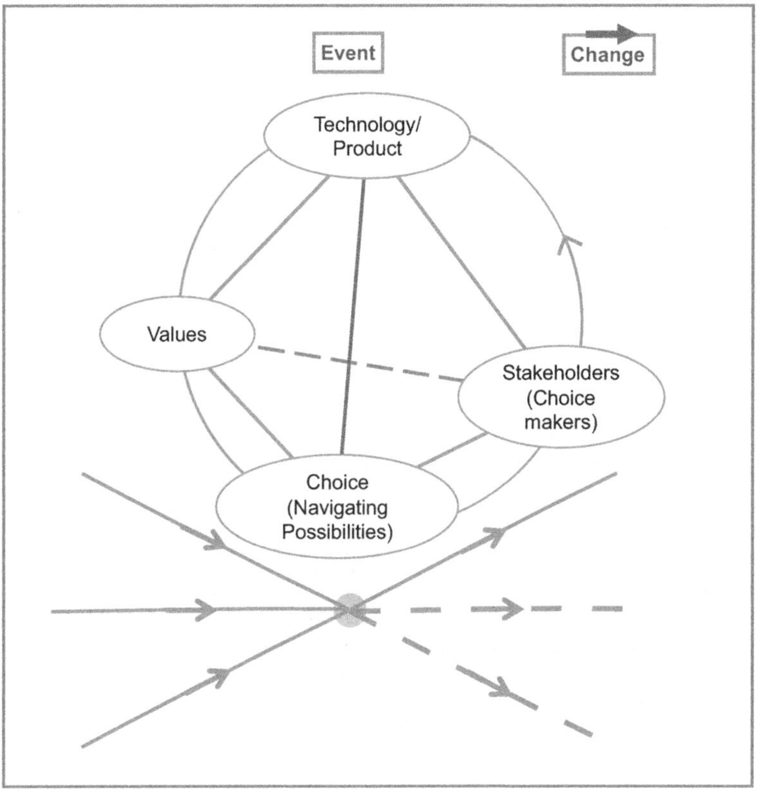


**Fig. 19.4** Tracing changes with "event-centered diagramming of change processes". This diagram shows a synthesis of Figs. 19.1 and 19.2 to enable "marking and tracking" of changes in sub-processes and core elements during an event—the minimal temporal change process. © Alok Srivastava and Gunter Bombaerts (2024)

### *19.4.3 "Event-Centered Diagrams of Change Processes"*

In this sub-section we propose to enable process tracing methods in VSD projects by combining the minimal ontology system for VSD projects developed in Sect. 19.2 and event-centered diagrams developed in Sect. 19.3. In historical sequences of changes of sociotechnical systems, the event-centered diagrams focus on a minimal temporal change event to mark and represent the incoming and outgoing sub-processes before and after specific events. The diagrams thus provide a framework to trace possible mechanistic processes—"institutional patterns or event chains" set into motion by specific changes in a specific element exhibited in a change event—that may contribute to the change in the social system. Such tracing efforts will directly help in generating hypotheses of significant causal mechanisms which can then be used to code (annotate and mark-up) specific situations and case histories. This "coding"—marking and tracking—of specific causal hypotheses will develop

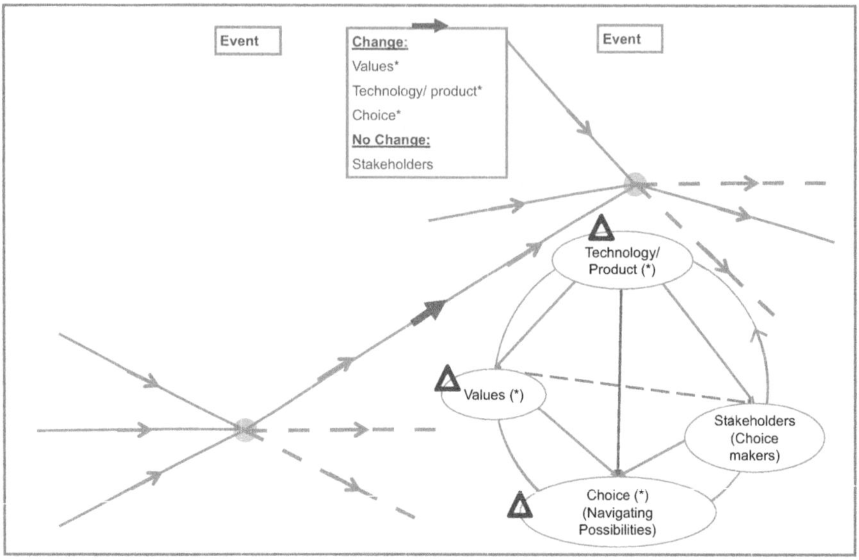


**Fig. 19.5** Tracing changes between two events with "event-centered diagramming of change processes". This diagram shows a synthesis of Figs. 19.1 and 19.3 to enable "marking and tracking" of changes in sub-processes and core elements as the sociotechnical system goes from the first event to the second event. This diagram is an illustration of "process tracing" across two events and indicates that the three core elements changed as a consequence of the first event of change. © Alok Srivastava and Gunter Bombaerts (2024)

the case studies of technology design projects and facilitate the application of process tracing methods. We illustrate the feasibility of this combination in Figs. 19.4 and 19.5.

Recently Sharon Crasnow has been working with Mary Morgan's program of narrative science to integrate process tracing methods with narrative explanation-making. Narrative science develops the methodological capacity for deriving casual explanations in complex situations where each case history is likely to be one of a kind and not readily tractable by statistical methods of study (Morgan, 2017; Morgan et al., 2022). The main methodological issue addressed by narrative science is that social systems, like geological processes (earthquakes and glaciers) and clinical patient processes (diagnosis and treatment trajectories) are populated by unique histories that do not offer large numbers of similar histories that can be analyzed by statistics of the same variables across many cases. These "N of 1" case histories are dominated by very local pattern of contingencies and dependencies and are used with some comparative analysis across loosely similar case histories to draw out causal stories. While patterns across a family of cases are helpful, the detailed longitudinal studies within each case history are differentially effective. This has been successful in geology and medical clinical science. The program and platform of narrative science developed by Morgan and colleagues makes explicit some of the methodological aspects that make such causal analysis of unique historical processes relatively successful and reliable. The methods focus on discovering how possible sequences of events in histories can be distinguished from being incidental, and unconnected happenings, versus being causally contingent and dependent on other elements of the system. These methodological advances are compatible with the complex contingencies of VSD projects in sociotechnical systems.

## 19.5 Conclusion

In this chapter, we formulated an ontological framework to enable methods for discovery of causal mechanisms underlying changes in values and other core elements of sociotechnical systems that affect design and outcomes of technology projects. Our approach addresses the need to monitor and respond to changes in values and other core elements during a technology project that are organized into the design of the project. We used Process Ontology and Niklas Luhmann's Social Systems Theory to develop a minimal ontological system and a diagramming tool—a schema—for tracing a minimal temporal change process centering on an "event" in which a change is mobilized. We propose that this ontological framework and diagramming tool of contingent changes are compatible with the tools of historical analysis called process tracing to propose a full method for discovery and characterizations of causal mechanisms underlying change in sociotechnical systems and social processes. We hope we have shown the potential of this approach and that this can be seen as a contribution to the philosophical literature in Values Sensitive Design and Values Change.

This chapter is proposing a program of research. Further research is needed to work out the details of this approach in specific case studies of specific VSD projects. As we enable this program of work, we believe more complex questions will become explorable. For example: How does the value change in an event add to a societal long-term change (de Wildt & Schweizer, 2022)? What are the consequences from the fact that values and events affect selections i.e. choice-making? This view infringes upon the universality of values, and they seem context- (or "event") dependent; and they are not relative, as every value is always used in an event embedded in a social system as a structure of expectations. This leads to a need for a relational view of values in events and social systems. Though this is compatible with the process ontologies we would be able to discover empirically the nexus of relations that justify such a conceptualization of values as a nexus of relations. In general, the more complex aspects may become addressable by this research program.

## References


Allport, F. H. (1940). An event-system theory of collective action: With illustrations from economic and political phenomena and the production of war. *The Journal of Social Psychology, 11*(2), 417–445.

Allport, F. H. (1954). The structuring of events: Outline of a general theory with applications to psychology. *Psychological Review, 61*(5), 281–303.

Beach, D., & Pedersen, R. B. (2016). *Causal case study methods: Foundations and guidelines for comparing, matching, and tracing*. University of Michigan Press.

Beach, D., & Pedersen, R. B. (2019). *Process-tracing methods: Foundations and guidelines*. University of Michigan Press.

Coase, R. H. (1937). The nature of the firm. *Economica, 4*(16), 386–405.

Coase, R. (1998). The new institutional economics. *The American Economic Review, 88*(2), 72–74.

Crasnow, S. (2017). Process tracing in political science: What's the story? *Studies in History and Philosophy of Science, 62*(April), 6–13.

Crasnow, S. L. (2022). 11—Process tracing and narrative science. In M. S. Morgan, K. M. Hajek, & D. J. Berry (Eds.), *Narrative science* (p. 229). Cambridge University Press.

de Wildt, T. D., & van de Poel, I. (2024). Modelling value change: An exploratory approach. *Journal of Artificial Societies and Social Simulation: JASSS*. https://doi.org/10.18564/jasss.5199

de Wildt, T. E., & Schweizer, V. J. (2022). Exploring value change. *Prometheus, 38*(1), 25–44.

de Wildt, T. E., van de Poel, I. R., & Chappin, E. J. L. (2022). Tracing long-term value change in (energy) technologies: Opportunities of probabilistic topic models using large data sets. *Science, Technology & Human Values, 47*(3), 429–458.

Dupré, J. A., & Nicholson, D. J. (2018). A manifesto for a processual philosophy of biology. In D. J. Nicholson & J. Dupré (Eds.), *Everything flows: Towards a processual philosophy of biology*. Oxford Academic. https://library.oapen.org/bitstream/handle/20.500.12657/29752/1/9780198779636.pdf#page=20

Epstein, B. (2016). Social ontology. In L. McIntyre & A. Rosenberg (Eds.), *The Routledge companion to philosophy of social science* (pp. 260–273). Routledge.

Fabris, F. (2018). Waddington's processual epigenetics and the debate over cryptic variability. In D. J. Nicholson & J. Dupré (Eds.), *Everything flows: Towards a processual philosophy of biology* (pp. 246–263). Oxford University Press.

Friedman, B., Hendry, D. G., & Borning, A. (2017). A survey of value sensitive design methods. *Foundations and Trends® in Human–Computer Interaction, 11*(2), 63–125.

Luhmann, N. (1995). *Social systems*. Stanford University Press.

Mahoney, J. (2000). Path dependence in historical sociology. *Theory and Society, 29*(4), 507–548.

Mahoney, J. (2015). Process tracing and historical explanation. *Security Studies, 24*(2), 200–218.

Morgan, M. S. (2017). Narrative ordering and explanation. *Studies in History and Philosophy of Science, 62*(April), 86–97.

Morgan, M. S., Hajek, K. M., & Berry, D. J. (2022). *Narrative science*. Cambridge University Press.

Nicholson, D. J., & Dupré, J. (2018). *Everything flows: Towards a processual philosophy of biology*. Oxford University Press.

Nickel, P. J., Kudina, O., & van de Poel, I. (2022). Moral uncertainty in technomoral change: Bridging the explanatory gap. *Perspectives on Science: Historical, Philosophical, Social, 30*(2), 260–283.

Seibt, J. (2012). Process philosophy. https://plato.stanford.edu/entries/process-philosophy/?trk=public_post_comment-text.

van de Poel, I. (2021). Design for value change. *Ethics and Information Technology*. https://doi.org/10.1007/s10676-018-9461-9. https://link.springer.com/article/10.1007/s10676-018-9461-9

van de Poel, I., & Kudina, O. (2022). Understanding technology-induced value change: A pragmatist proposal. *Philosophy & Technology, 35*(2), 40.

van der Weij, F., Steinert, S., van de Poel, I., Alleblas, J., Melnyk, A., & de Wildt, T. (2023). Value change and technological design. *IEEE Technology and Society Magazine, 42*(3), 25–32.
Whitehead, A. N. (1929). *Process and reality: An essay in cosmology (critical edition)*. D. R. Griffin & D. W. Sherbourne (Eds.). Macmillan.
Winkler, T., & Spiekermann, S. (2021). Twenty years of value sensitive design: A review of methodological practices in VSD projects. *Ethics and Information Technology*. https://doi.org/10.1007/s10676-018-9476-2. https://link.springer.com/article/10.1007/s10676-018-9476-2
Wood, E. J. (2003). *Insurgent collective action and Civil War in El Salvador*. Cambridge University Press.

**Alok Srivastava** MS (Integrated) Chemistry, PhD Biophysics & Biochemistry. 2001–2016 Researcher and technologist in Biotechnology and Medicine, in Cancer Biomarkers, Biosensors, Diagnostics and Personalized Medicine and Collaborative Clinical Decision-Making. Today, Independent Researcher in History Philosophy and Sociology of Science, Technology, Engineering and Medicine with focus on the interaction of experiment-making and theory-making as the basis for organizing research programs and platforms in laboratories and groups. His selective focus is on multi-disciplinary projects and teams, in which multiple styles of manipulation and measurement are brought into a coordinated process to generate "new" experimental setups. The theoretical focus is on developing process philosophy to better support theory development of studies of material and theoretical cultures in science, medical and engineering communities.

**Gunter Bombaerts** PhD Ethics (2004). Today, is Assistant Professor for Philosophy and Ethics of Technology at Eindhoven University of Technology, the Netherlands. He is coordinating the TU/e USE program and is teacher of the USE basic course on History and Ethics of Technology. His research fields include ethics in engineering education, comparative ethics and questions concerning applied ethics in the field of energy ethics, in particular on participation and innovation.

# Chapter 20
# How to Do Things with Artefacts

**Ryan Mitchell Wittingslow**

**Abstract** This chapter applies J. L. Austin's speech act theory to design function. Austin's theory, originally developed for linguistic utterances, distinguishes between locutionary content, illocutionary force, and perlocutionary outcomes. While this framework has been applied to artworks, its potential in philosophy of technology, engineering, and design remains largely unexplored. This work serves as a proof of concept, extending Austin's tripartite model to analyse design function. By drawing on Austin's original theory and its later applications in art, the chapter develops a speech act account of design function and examines how individual or conventional uses of design artefacts can diverge from their intended functions. The discussion concludes with a summary of how this framework can be used to analyse value change in design and engineering contexts.

**Keywords** Speech acts · Design · Function · Illocution · Value change

## 20.1 Introduction

In *How to Do Things with Words* (1962), J. L. Austin claims that utterances are constituted by three components: locutionary content (what an utterance literally means), illocutionary force (what the utterance conventionally achieves by virtue of those conventions governing the power of the utterance), and perlocutionary outcomes (what the utterance achieves).

Although Austin's primary concern is linguistic utterances, later theorists have applied Austin's 'speech act theory' to other domains—including, notably, to artworks. The virtue of applying a speech act approach to artworks is that Austin's tripartite model provides philosophers with a way of distinguishing between what

R. M. Wittingslow (✉)
University of Groningen, Groningen, The Netherlands
e-mail: r.m.wittingslow@rug.nl



C. Didier et al. (eds.), *Engineering and Value Change*, Philosophy of Engineering and Technology 48, https://doi.org/10.1007/978-3-031-83549-0_20

an artwork literally means (the locution) and the force of that meaning (the illocution). However, developments in this domain have, thus far and for the most part, been narrow. Prevailing accounts focus primarily upon artworks or similar expressions with offensive or explicit locutionary content and/or illocutionary force (pornography, certain examples of public statuary, and so on). Nonetheless this scholarship shows promise for application in philosophy of technology, philosophy of engineering, philosophy of design, and related domains.

Considering this conviction, this chapter is a proof of concept. It is an attempt to apply Austin's speech act theory to design function. Drawing upon both Austin's tripartite distinction and later developments of his framework as it applies to works of art, I develop a speech act account of design function and an account of the ways in which the individual or conventional uses of design artefacts can deviate from those functions. I conclude this account by offering a summary of the ways that this picture can be used to analyse value change in design and engineering.

## 20.2 Words

We begin with a summary of J. L. Austin's account of speech acts. Austin introduces the concept of a 'speech act' in *How to Do Things with Words* (1962). A speech act refers to an intentional act performed to achieve some outcome, such as promising, asking, explaining, or asserting. Austin contends that 'speech act' thus encompasses "a whole group of senses of 'doing something'" (1962, p. 94). In aid of making sense of how speech brings about states of affairs Austin begins with the posed distinction between 'constative' and 'performative' utterances.

Constatives are utterances that 'constate': they are statements that describe or otherwise state a fact about something in the world. The phrases 'the cat sat on the mat', or 'the Earth rotates around the sun' are constative in that they furnish these kinds of descriptions. Constatives are 'truth-evaluable', in that we can judge the aptness of each description true or false based upon whether they correspond to reality. 'The cat sat on the mat' can be evaluated as true if there is in fact a cat sitting on the mat.

Performatives, meanwhile, are utterances that perform an action or function, bringing about some change in the world. These strings—sentences like, 'I promise to be there', or 'I declare war on France', or 'you are under arrest'—are not truth-evaluable descriptions but constitute an attempt to bring some state of affairs into being. For this reason, we do not evaluate performatives based upon whether they are true or false. Instead, the success of a given performative is predicated on whether it meets the relevant 'felicity conditions': (a) "there must exist an accepted conventional procedure" by which a given state of affairs is brought into being, (b) the persons "must be appropriate for the invocation of the particular procedure", and (c) "the procedure must be executed […] both correctly and completely" (1962, pp. 14–15). So, for a sentence like 'I declare war on France' to be felicitous and thus successful, it must be the case that (a) there is a conventional procedure by which war can be declared, (b) the person speaking must be entrusted with the rights and responsibilities to enact such procedures, and (c) the procedure needs to be followed appropriately.

It might be tempting to think that there is some easily made grammatical distinction between the two kinds of utterances. However, Austin puts the kibosh on that idea; we cannot, he thinks, simply class utterances as either 'constative' or 'performative'. It is entirely possible for a single utterance to be simultaneously constative and performative: "there are many cases which, as language stands at present, we are not able to split into two parts in this what, even though the utterance seems to have some kind of explicit performative in it: thus 'I liken *x* to *y*', 'I analyse *x* as *y*'" (1962, p. 90). Austin's analysis of speech acts is his solution to how we make sense of performatives without trying to impose some kind of grammatical character upon them.

Austin claims that speech acts can be analysed on three distinct levels: in terms of 'locution,' 'illocution,' and 'perlocution'. Locution is the aspect of a given speech act that refers to the formal features of an utterance, including the specific sounds, words, and grammatical structure used: "to say something in the full normal sense is to do something—which includes the utterance of certain noises, the utterance of certain words in a certain grammatical construction, and the utterance of them with a certain 'meaning' in the favourite philosophical sense of that word, i.e., with a certain sense and with a certain reference. The act of 'saying something' in this full normal sense I call [...] the performance of a locutionary act" (1962, p. 94). In short, when we're talking about locution, we're talking about the contents of an utterance: all the way from the utterance's constituent phonological parts, to what it literally and constatively means.

Illocution denotes the performative aspect of speech acts. Whereas locution captures literal meaning, illocution captures what is being done by the act of speaking. To use an example, the utterance, 'Hey buddy, do you want to go outside?' is literally an invitation on the level of locution. However, the tenor of this invitation can vary wildly based upon several conventional factors. Spoken with kindness to your dog, it is an invitation to go for a walk; spoken with anger to a drunk man in a pub, it is an invitation to be punched in the face. The mere contents of the locution underdetermine what can be done with the locution. This performativity is the illocutionary 'force' that underwrites the locutionary content of a given utterance. As Austin writes:

> it might be perfectly possible, with regard to an utterance, say 'It is going to charge', to make entirely plain 'what we were saying' in issuing the utterance, in all the sense so far distinguished, and yet not at all to have cleared up whether in issuing the utterance I was performing an act of *warning* or not. It may be perfectly clear what I mean by 'It is going to charge' or 'Shut the door', but not clear whether it is meant as a statement or warning, &c. (Austin, 1962, p. 98)

The final aspect of speech acts is perlocution. Perlocution denotes the outcome resulting from the execution of the illocutionary act and encompasses the impact of the illocutionary act on the emotions, thoughts, or behaviours of others. Essentially, perlocution refers to the effects for which the speaker can be held accountable due to their illocution. The perlocution is underwritten by whether the illocution meets the relevant felicity conditions: those three conditions that must be met for the illocution to have the appropriate force.

When I say to my dog, 'Hey buddy, do you want to go outside?', the force of that offer is grounded in the fact that (a) there exists an established convention by which I take my dog for a walk, (b) I am one of the people who conventionally takes her for walks, and (c) my request conforms with her expectations and thus conforms with the felicity conditions for the utterance. This is an example of what Austin calls a perlocutionary 'object': it is an outcome that has a conventional connection to the illocution in that it is underwritten by the force of that illocution. However, were a stranger—someone who fails the second criterion—to pose my dog the same offer, it is unlikely to have the same outcome. My dog might instead cock her head and look confused, or become spooked and hide under the couch. This would be an example of a perlocutionary 'sequel': a perlocution that does not have a conventional connection to the force of the illocution (cf. 1962, pp. 101–107).

## 20.3 Artworks

We don't want to make the mistake of drawing too strong a comparison between artworks and natural languages. Unlike natural languages, artistic meaning doesn't rely on an abstract set of grammatical relationships among component parts. Art lacks the requisite lexical, morphological, syntactic, or referential features to convey meaning in a linguistic fashion. The semiotic relationships are too complex and varied for such claims to hold any real water (cf. Margolis, 1974; Mothersill, 1965).

Nonetheless, there are important similarities between languages and art. Artmaking is, like speech, a meaningful and meaning-making activity. Artworks, like speech acts, have semantic content and produce effects based, at least in part, upon that semantic content. Finally, we treat artworks as if they are a species of devolved interlocutor: we take artworks to make moral, intellectual, and emotional demands on us in a way that is like how people make demands of us, and we respond to artworks in ways typically reserved for other human beings. However, despite these similarities David Friedell and Shen-yi Liao's paper "How Statues Speak" is among the few attempts by philosophers to talk about art in the context of Austinian speech act theory, though attempts have been made using other versions of speech act theory (cf. Dixon, 2019; Goldblatt, 2011; Langton, 1993; Scarre, 2019; Shahvisi, 2021).

Inspired by Grice's distinction between natural sentence meaning and non-natural speaker meaning (cf. Grice, 1957), Friedell and Liao begin by arguing that artworks possess locutionary meaning on two levels. The first level is 'artist-meaning': that is, what the artist responsible for the artwork intended for the artwork to convey. The second is 'artwork-meaning': what the artwork means by some set of social or semiotic conventions. "Suppose you intend to write a novel with a charismatic protagonist", they write, "but mistakenly make them behave obnoxiously. Your novel's artist-meaning involves charisma, but its artwork-meaning involves obnoxiousness" (Friedell & Liao, 2022, p. 446). In this manner, artworks are like speech. What artworks are intended to mean and what they actually mean can be quite different. However, Friedell and Liao think that artworks are not only speech-like in that they *say* things. Instead, artworks are speech-like in that they *do* things.

Friedell and Liao posit that the link between the locutionary meaning of an artwork and the illocutionary force of that meaning is regulated by the various social conventions and norms that pertain to different types of artworks. They assert that the illocutionary force of an artwork is determined not only by its locutionary meaning but also by how and why that meaning is communicated to a specific artworld public. The setting in which we encounter an artwork significantly influences our response to the perceived meaning of these works, as the illocutionary forces of these environments shape our perlocutionary responses to these works. Coming face to face with a piece of esteemed modernist scribble in an art museum is a very different experience to encountering that self-same image misplaced in a kindergarten classroom. The environment circumscribes our relationship with, and reaction to, that work of art; it is what cues us to respond to the work in a certain way (also see Wittingslow, 2023a, pp. 75–95).

Friedell and Liao analyse the extent to which locutionary content and illocutionary force can deviate through the lens of political statuary. According to Friedell and Liao, statues lionising Confederate generals, slave traders, colonial masters, and similar figures, convey a clear locutionary meaning: to celebrate and honour the individuals depicted. However, whether these statues actually honour the figures they represent depends on the illocutionary force that underwrites how they are exhibited. A statue might indeed mean 'so-and-so is honourable', but for that message to be gleaned and endorsed relies upon the statue being encountered under the appropriate conventional circumstances:

> The difference between saying and doing is subtle. For example, Rossi classifies a Confederate statue as an *honorable representation* and defines this concept as "any representation of an individual in a public space that depicts that individual as an exemplar of a value or values, such as courage, integrity, or justice". But there is a difference between depicting a person as honorable and honoring that person. [...] By changing the context, we change what statues do but not what they say. For example, Elizabeth Scarbrough's proposal to leave politically controversial statues to ruin does not change their locutionary content but does change their illocutionary force: a statue that is allowed to ruin can still depict a dishonorable person as honorable, but it will no longer honor that person. (2022, p. 448)

To use an example about which I have written elsewhere (cf. Wittingslow & Lehtinen, 2024), the statue of Bristol slave-trader Edward Colston offers a good example of the difference between locution, illocution, and perlocution as Friedell and Liao conceive it. The statue was unmistakably intended to depict Colston as honourable: a local hero who, with profits earned from the slave trader, endowed schools, poorhouses, hospitals, and other public institutions. This honourable depiction constitutes the stable locutionary content of the statue. However, the illocutionary force of the statue has undergone significant changes in recent years. Until very recently, the illocutionary force of the statue was derived from its prominent position on a grand plinth in 'The Centre': a large park that is the geographic and symbolic centre of Bristol. The statue thus honoured Colston.

However, in the wake of the murder of George Floyd, the statue was pulled down by protestors on 7 June 2020. After defacing the statue with blue and red paint, the protestors pushed the statue into Bristol Harbour. Although quickly retrieved, the statue has not been returned to its plinth. It was temporarily put on public display, put into storage, and is now on permanent display—still supine and still covered in paint—at M Shed, part of Bristol Museums. While the locutionary, laudatory content of the statue has not changed throughout this process, the illocutionary force has. While the statue certainly continues to depict Colston as honourable, the transition from public statue, to museum exhibit, to stored item, and again to museum exhibit, means that the statue, having been 'allowed to ruin', no longer actually honours Colston. Due to a change in illocutionary conventions, what the statue *does* has changed. Consequently, the set of possible perlocutionary objects and sequels has also changed.

In short, Friedell and Liao's paper furnishes a compelling argument for applying some version of speech act theory to artworks. With only a little effort, Austin's distinction between locution, illocution, and perlocution offers a clear and principled way for philosophers of art to make sense of artmaking as a communicative act. Via speech acts, we can differentiate what an artwork literally means (the artwork's locutionary content) from both what the artwork does considering its context (its illocutionary force) and what people do in response to both that content and its force (the artwork's perlocutionary outcomes).

## 20.4 Things

I am convinced that there is something in both Austin's original articulation of speech act theory and Friedell and Liao's artwork reconstruction that can be used to make sense of artefact function: what function is, how it is applied, the ways in which use can deviate from that function, and so on. Moreover, I think this approach possesses the advantage of being able to accommodate how changing values in design and engineering can influence artefact function. Considering that conviction, this section is my attempt to sketch out, in general terms, how one could use Austinian speech act theory to generate a theory of function.

Before I begin in earnest, it is worth making clear what I take to be the objects of my analysis. In the case of *How to Do Things with Words*, Austin's target system is obviously language: he is interested in the ways in which meaning and circumstance intersect in such a way that words can *do* things. Friedell and Liao's "How Statues Speak" is similar, but the target system in this case is public statuary: they want to know how the meaning of an artwork relates to what we do with that work. For the purposes of this exercise, I'm interested in 'artefacts'—that is, human-made things that have some kind of function or purpose.

Artefacts, as I'll be using the term, possess teleological 'for-ness': they are *for* something. Hammers are for hammering nails; automobiles are for transporting persons and goods; rockets are for going to the moon; and so on. So, although an artefact might have many functions *simpliciter* (a hammer could be used as a doorstop; an automobile could be used as a museum piece; and so on), it is only 'for' one of those functions. We can call this teleological for-ness 'proper function'. Proper function theories commit to the notion that artefacts possess functions that are ontologically 'proper' to them. Proper functions are ontologically constitutive, in that they are what make an artefact the kind of thing that it is. According to these theories, proper functions thus play a crucial role in shaping our collective understanding of technological artefacts. Proper functions not only help us to determine what a given artefact is, but also offer normative benchmarks for whether artefacts are working 'properly'.

There are two primary approaches to understanding proper function in artefacts. The first approach (cf. Millikan, 1984, 1986, 1995; Parsons & Carlson, 2008, among others) links function to an artefact's evolutionary selection history: an artefact has a proper function if its predecessors improved their fitness by carrying out that function, leading to the continued reproduction of those artefacts with that function. The second approach associates it with the intentions of its designers. This approach argues that an artefact's proper function depends on the intentions of the designers who created it (cf. Baker, 2007; Dipert, 1993; Evnine, 2016; Hilpinen, 1993; Kroes, 2012; McLaughlin, 2001, among others). The most influential account of these is probably the ICE-theory by Vermaas and Houkes (cf. Houkes, 2006; Houkes et al., 2002; Houkes & Vermaas, 2004; Vermaas & Houkes, 2003, 2006), which states that function ascription requires: (1) the belief that the artefact has the capacity for that function, (2) justification for that belief, and (3) intentional selection and communication of that function by the artefact designers.

Both evolutionary and intentional proper function approaches excel in making clear how we can make both etiological claims about an artefact's purpose and normative claims about its performance, although they obviously differ in how those claims are grounded. What we are interested in here, however, is in how the proper function of an artefact—what an artefact is for—relates to what an artefact actually *does*: that is, how actual use coheres with and/or deviates from proper, intended use.

The account I furnish here is based upon the assumption that it is possible to analogically transpose concepts from philosophy of language—in this case, Austinian speech act theory—to philosophy of technology. This assumption is grounded in the claim that technological objects and technical systems share at least some relevant features with natural languages. While it is true that, as in the case of works of art, we do not want to oversell the similarities between technology and language (cf. Wittingslow, 2023b), we also don't want to undersell the fact that technological objects and systems do possess morphological and linguistic regularities that are relevantly language-*like*. These language-like features of technology—morphological and syntactical features, as well as perhaps a muted kind of semanticity—are what make technology an appropriate object of analysis for Austinian speech act theory.

For example, most designed things only make practical sense when properly connected and ordered as part of a larger network of other designed things. For example, using my iPhone to answer work emails requires more than just having the phone in working order: an operating chipset, a screen that turns on, a working battery, and so on. It also depends on a vast system of physical and intellectual infrastructure: energy and communication networks, mining and machining operations, questionable labour practices in the developing world, large corporate structures, education and research policies, and so on. For my phone to work as needed, there must also be an ecosystem of compatible parts—some part of the phone itself, and others in the environment in which the phone is used—all operating together correctly. As Coyne and Snodgrass write:

> Design bears certain similarities to language. There is the domain of semantics and that of syntax. There are syntactic rules (grammars) about how things go together, and there are semantic rules about the interpretation or evaluation of these assemblies. A design must meet the requirements of both the syntax and the semantics of a "design language". (1995, p. 40)

With all that in mind, let's start with locution. Following Friedell and Liao's cue, we begin with the Gricean distinction between sentence and speaker meaning—though here we will be differentiating between 'designer function' and 'design function'. Designer function, like artist meaning, refers only to the function of the artefact as intended by the designer of that artefact. Consider an automobile. Whether the vehicle in question has the assigned function of being a utility truck or a sports car or a station wagon is a matter of designer intention, however we cash that out: in terms of the individually held preferences of the designer(s), the brief or commission received by the designers, or whatever else.

Design function, however, is different. Artefacts, like artworks, can be analysed in terms of their individual elements—'line', 'volume', 'form', and so on. They can also be analysed in terms of their more complex parts. In the case of an artefact like an automobile, we identify objects like 'wheels', 'engine block', and 'chassis'. These kinds of assessments are not assessments of resemblance but are premised instead upon an understanding of what each of those parts can do. When we identify a 'wheel', for instance, we are not just identifying some circle-shaped thing. We are also identifying, at least in some sense, how the object might be used within a given system.

All of this means that, although the overall function of a technical artefact utterance is not obvious at this level of analysis, we can still glean how different parts interact with one another. We can identify, for instance, that the engine block powers an axle, which itself drives the motion of the wheels, which puts the automobile into motion. We thus get a sense of both the individual functional features of each part, and how the parts hang together in an orderly, conventional, 'grammatical' fashion. We are also able to see how the designer function and the design function of an artefact can be at odds. If a designer has intended to produce a sports car but has instead designed something with features more fitting of a utility vehicle, then we can identify a disconnect between designer and design function.

Proper function refers to what artefacts are for—that is, the locutionary content of artefacts—but does not capture what an artefact actually does. For that, we need to move to illocution. Given what I've covered so far, I doubt that my moves here will provoke any serious surprises. We already know, in the case of Friedell and Liao's Confederate statuary, that artwork illocution is a consequence not only of the locutionary content of the artwork, but also the context or environment in which the work is encountered. This is why a Confederate statue can have the locutionary meaning that 'so-and-so is honourable', but certain contextual factors—such as whether the statue has been left to ruin—mean that the artwork can fail to honour that person. Artefact doing is similarly sensitive to changes in illocutionary context. An automobile parked inside the glass and steel walls of an auto dealer is doing something quite different to that same automobile driving on the road. Moreover, this difference—window display versus mode of transport—is only an illocutionary difference. The objects are identical in a locutionary sense.

Finally, this leads us to perlocution. As already noted, whether a perlocutionary outcome is an object or a sequel has to do with whether that outcome meets the felicity conditions for the illocution. If the felicity conditions for the illocution are met, we have a perlocutionary object: an outcome that has a conventional connection to the illocution in that it is underwritten by the force of that illocution. If the felicity conditions are not met, we have a perlocutionary sequel. Seeing an automobile in the window of an auto dealer and choosing to buy it is an example of the former; seeing an automobile in the window of an auto dealer and then splattering it with paint to protest excess use of fossil fuels is an example of the latter.

## 20.5 Values

Obviously, what I've outlined here is only a brief proof of concept outlined the ways in which Austinian speech act theory can be used to make sense of design function. Nonetheless, I think that using this approach shows promise. When we talk about artefact function, we are talking about artefacts on the level of locution: their functional features, and the intentions that underwrite their creation. Illocutionary factors, meanwhile, can influence the extent to which that function is realised. This illocutionary context then influences—though does not determine—how the artefact is used; whether users produce a conventional perlocutionary object or a non-conventional perlocutionary sequel.

If nothing else, even in this very preliminary form, it meets at least three of the four desiderata (as outlined by Pieter Vermaas and Wybo Houkes) that a theory of function needs to fulfil to be properly considered fit for purpose. These desiderata are as follows:

**D1.** *Proper versus accidental*: A theory of functions should distinguish between the proper and accidental function(s) of an artefact. Proper functions can typically be understood as functions ascribed standardly to the artefact, whereas accidental functions are ascribed only occasionally.
**D2.** *Malfunction*: A theory of functions should admit an ascription of proper functions to malfunctioning artefacts, i.e., artefacts that are unable to perform their functions.
**D3.** *Physical structure*: A theory of functions should entail that for every function there exist structural conditions sufficient for its performance. Moreover, if a theory ascribes a function to an artefact, it should provide partial justification for the belief that the physical structure of the artefact satisfies such conditions.
**D4.** *Novelty*: A theory of functions should admit an ascription of proper functions to innovative or atypical artefacts. (Vermaas & Houkes, 2003, pp. 265–266)

In the case of D1, the story I've outlined here differentiates proper versus accidental function on the grounds that proper function is synonymous with locution. Accidental function is what takes place when some perlocutionary outcome does not cohere with the locutionary function of the artefact. Furthermore, one of the virtues of my Austinian account is that it outlines how different illocutionary circumstances contribute to the development of accidental functions. This account also meets D2. Indeed, not only does this picture facilitate ascriptions of malfunction, it makes clear that artefacts can malfunction for all different kinds of reasons: whether due to some disconnect between an artefact's design function and its designer function (i.e., malfunction on the level of locution), or whether the illocutionary circumstances prevent the artefact from functioning properly. This account also meets the D3 desideratum, in that the structural conditions sufficient for a given artefact's performance are captured within the design function of that artefact.

The only desideratum that this account doesn't directly address is D4. However, I think that this too can be overcome with a little bit of work. Novel artefacts are, I suspect, examples of non-conventional perlocutionary outcomes (viz., perlocutionary sequels) that are themselves artefacts. Being non-conventional, these artefacts are innovative and/or atypical. Being artefacts, these perlocutionary sequels also possess locutionary functions of both the designer and design varieties. It is on these grounds that we can ascribe function to innovative or atypical artefacts: innovative or atypical artefacts are artefactual perlocutionary sequels.

However, this is not the only virtue of this account. While preliminary, it also can be used to analyse value change in design and engineering. Under this picture, value change in design in engineering—that is, changes in attitudes to the ways in which technical artefacts are designed and used across time—is motivated not by changes in locutionary function, but by changes in illocutionary forces influencing the tenor of perlocutionary outcomes. This change in illocutionary context itself produces changes in the ways in which people interact with those objects: both conventional perlocutionary objects and non-conventional perlocutionary sequels. It is, I think, potentially a powerful way to analyse value change in design and engineering.

I use the term 'value' as defined by van de Poel and Kudina: as "*evaluative devices that carry over from earlier experiences and are (to some extent) shared in society*" (2022, p. 3; italics in the original). Under van de Poel and Kudina's conception, values should be properly understood as tools that help us to evaluate states of affairs. Values thus look and act a lot like the static and inviolable moral norms that constitute the objects of analysis in more traditional approaches to ethics. However, values differ from those kinds of moral norms in that they are subject to change under the appropriate conditions—and particularly the appropriate socio-technical conditions. That is, values can change because the design and uptake of new technologies might cause us to revisit and reappraise the foundations upon which our value judgements rest, or because they pose new and hitherto unencountered moral problems, or because they change our habits and ways of life—and thus change our evaluative expectations—in some non-trivial way. As they write:

> Technology has for example, created new moral problems: the proliferation of genetically modified food has raised concerns about global justice and responsibility, while the norms of food safety and equitable access to this technology have subsequently undergone review. Assisted reproduction technologies have defied previously existing biological horizons, enabling aged, infertile, or same-sex couples to have children. This has not only redefined what it means to be human but also fostered novel normative expectations regarding procreative rights and liberties. Or to give a more mundane example, consider how glass office doors enable the expectation of transparency in the professional setting, while simultaneously reducing the value of privacy and internalizing a surveillance gaze within employees. Technologies, thus, while being the fruits of human creativity, manifest not merely as neutral tools but also as productive elements in co-shaping how people perceive the world, each other, and themselves. (2022, pp. 1–2)

What I think is nice about the Austinian account I outline here is that it gives us an alternative way to analyse how and why technology can prompt value change, as identified by van de Poel and Kudina. Value change, I think, describes what happens when a perlocutionary outcome (whether object or sequel) of some designed thing itself triggers changes in illocutionary force—as in the case, for example, of the Colston statue that previously stood in Bristol's Centre. This change in illocutionary force influences the quality and character of subsequent perlocutionary objects and sequels, even if the locutionary function of the artefact does not change.

I end this chapter with a concrete example of this phenomenon in action. Consider shopping malls. Unlike covered shopping arcades—which are long covered pedestrian streets lined with small boutiques and cafes—shopping malls are large, enclosed buildings containing various stores, restaurants, and entertainment venues. In shopping malls, retail shops, food courts, amenities, and what are called 'anchor stores' (usually large department stores) are arranged along interior walkways, allowing visitors to browse and purchase items or services without leaving the building. Facilitating this sort of commercial activity is their locutionary function.

From Victor Gruen's invention of the modern mall in 1956 (cf. Hardwick, 2015, p. 144) to the early 2000s, there was a massive increase in mall developments all over the Western bloc. At the peak of construction in the United States, 140 new malls were being built every year (cf. Millar, 2015). This explosive growth was a consequence of at least two illocutionary factors: (a) middle- and upper-class earners moving out of urban centres and into suburbs and exurbs (a phenomenon sometimes called 'white flight', though the extent to which this flight was motivated by racial versus economic drivers is up for debate; cf. Jego & Roehner, 2006); and (b) the car taking centre stage as the primary mode of personal transport thanks to investments in road infrastructure and (in some markets) subsidies and/or tax incentives.

These developments had obvious appeal. For new suburban and exurban communities, shopping malls served a similar function to town centres and high streets in traditional urban centres: places where residents could shop, engage in leisure activities, and participate in some kind of public life. These perlocutionary objects—a fulfilment of the malls' locutionary function—could only take place with the appropriate illocutionary factors (white flight and the rise of the automobile) in place.

However, by the early 2000s things had changed significantly, with half of all America's malls pegged to close by 2030 (cf. Millar, 2015). The meteoric increase in online shopping was usually fingered as the primary (though not the only) culprit. "Online shopping in America grew 14% last year", wrote Robbie Moore in 2013, "while overall retail sales were up only 3%, according to digital business analysts ComScore and the National Retail Federation. 129 million virtual shoppers stampeded the online sales on Cyber Monday last November, spending a record $1.46 billion. Tim Worstall, a Fellow at the Adam Smith Institute in London, argues that the success of online retail portends doom for the shopping mall. 'As more shopping moves online we simply need less physical retail space'" (Moore, 2013).

What's interesting about this case is that it constitutes a clear example of a new technology—in this case, the information technology, infrastructure, and logistics systems that underwrite online shopping—that produced a new perlocutionary object. This new perlocutionary object—having consumer goods delivered directly to your home—catalysed a change in the illocutionary values that govern how and where people shop. Very quickly, consumers started to prioritise convenience over the traditional virtues of going to the mall: socialising, inspecting goods before purchase, recreation, and so on. This illocutionary change then prompted other novel perlocutionary objects: in this case, a collapse in the number of people who shopped at malls.

Of course, the story doesn't end there. It certainly looks as if rumours of the death of the suburban mall have been exaggerated; according to more recent data, consumers have started returning to malls as retailers have learned how to better coordinate their online activities with in-store offerings: "The mall as a retail concept is alive and well, with retailers pivoting to fit the demands of a digital age. Rather than jostling with brick-and-mortar stores for supremacy, each shopping format is bolstering the other. Omnichannel marketing, where a brand has both a physical and online presence, has been a major factor driving mall growth" (Rothenberg, 2023). This shift, too, is evidence of a change in values: one that attempts to reconcile the value of convenience with the value of more personalised retail experiences.

In this case the shopping mall's locutionary function remains unchanged. However, what the shopping mall is *doing* certainly has. Prior to the era of online retail, the mall was simply where people in suburbs and exurbs went to do their shopping. After the explosion of online retail, suburban malls stopped being used as shopping centres. Instead, many of these venues were resold and renovated into new structures (schools and university campuses being particularly popular choices), or simply left to moulder. And now, with the development of strategies such as 'omnichannel marketing', the mall is again doing something new. All these perlocutionary changes are the consequence of values influencing the illocutionary context in which malls operate.

## 20.6 Conclusion

In conclusion, this chapter demonstrates how J. L. Austin's speech act theory can be used to make sense design function, moving beyond its original focus on language. By using Austin's model, which includes what an utterance literally means, what it aims to achieve, and what it actually achieves, we have extended its use to technical artefacts. This chapter thus serves as a proof of concept, showing how Austin's theory can explain the intended functions of artefacts and how actual uses can differ from these functions, and is thus useful for making sense of value change within the context of philosophy of design and engineering.

## References


Austin, J. L. (1962). *How to do things with words*. Oxford University Press.

Baker, L. R. (2007). *The metaphysics of everyday life: An essay in practical realism*. Cambridge University Press.

Coyne, R., & Snodgrass, A. (1995). Problem setting within prevalent metaphors of design. *Design Issues, 11*(2), 31–61.

Dipert, R. R. (1993). *Artifacts, art works, and agency*. Temple University Press.

Dixon, D. (2019). *Alterpieces: Artworks as shifting speech acts*. Thesis. University of Cambridge.

Evnine, S. (2016). *Making objects and events: A hylomorphic theory of artifacts, actions, and organisms*. Oxford University Press.

Friedell, D., & Liao, S. (2022). How statues speak. *The Journal of Aesthetics and Art Criticism, 80*(4), 444–452.

Goldblatt, D. (2011). Taking art personally: Austin, performatives and art. *Contemporary Aesthetics, 9*.

Grice, H. P. (1957). Meaning. *The Philosophical Review, 66*(3), 377–388.

Hardwick, M. J. (2015). *Mall maker: Victor Gruen, architect of an American dream*. University of Pennsylvania Press.

Hilpinen, R. (1993). Authors and artifacts. *Proceedings of the Aristotelian Society, 93*, 155–178.

Houkes, W. (2006). Knowledge of artefact functions. *Studies in History and Philosophy of Science Part A, 37*(1), 102–113.

Houkes, W., & Vermaas, P. E. (2004). Actions versus functions: A plea for an alternative metaphysics of artifacts. *The Monist, 87*(1), 52–71.

Houkes, W., Vermaas, P. E., Dorst, K., & de Vries, M. (2002). Design and use as plans: An action-theoretical account. *Design Studies, 23*(3), 303–320.

Jego, C., & Roehner, B. M. (2006). White flight or flight from poverty? *Journal of Economic Interaction and Coordination, 1*(1), 75–87.

Kroes, P. (2012). *Technical artefacts: Creations of mind and matter: A philosophy of engineering design*. Springer.

Langton, R. (1993). Speech acts and unspeakable acts. *Philosophy & Public Affairs, 22*(4), 293–330.
Margolis, J. (1974). Art as language. *The Monist, 58*(2), 175–186.
McLaughlin, P. (2001). *What functions explain: Functional explanation and self-reproducing systems*. Cambridge University Press.
Millar, L. (2015, January 28). Half of America's malls predicted to close by 2030. *ABC News*. https://www.abc.net.au/news/2015-01-28/the-decline-of-american-shopping-malls/6050956
Millikan, R. G. (1984). *Language, thought, and other biological categories: New foundations for realism*. MIT Press.
Millikan, R. G. (1986). Thoughts without laws; cognitive science with content. *The Philosophical Review, 95*(1), 47–80.
Millikan, R. G. (1995). In defence of proper functions. In *White queen psychology and other essays for Alice*. MIT Press.
Moore, R. (2013, February 26). The death of the American mall and the rebirth of public space. *The International*.
Mothersill, M. (1965). Is art a language? *The Journal of Philosophy, 62*(20), 559–572.
Parsons, G., & Carlson, A. (2008). *Functional beauty*. Oxford University Press.
Rothenberg, E. (2023, August 20). The US mall is not dying. *CNN Business*. https://www.cnn.com/2023/08/20/business/shopping-mall-retail-growth/index.html
Scarre, G. (2019). How memorials speak to us. In J. Bicknell, J. Judkins, & C. Korsmeyer (Eds.), *Philosophical perspectives on ruins, monuments, and memorials* (pp. 21–33). Routledge.
Shahvisi, A. (2021). Colonial monuments as slurring speech acts. *Journal of Philosophy of Education, 55*(3), 453–468.
Van de Poel, I., & Kudina, O. (2022). Understanding technology-induced value change: A pragmatist proposal. *Philosophy & Technology, 35*(2), 40.
Vermaas, P. E., & Houkes, W. (2003). Ascribing functions to technical artefacts: A challenge to etiological accounts of functions. *The British Journal for the Philosophy of Science, 54*(2), 261–289.
Vermaas, P. E., & Houkes, W. (2006). Technical functions: A drawbridge between the intentional and structural natures of technical artefacts. *Studies in History and Philosophy of Science Part A, 37*(1), 5–18.
Wittingslow, R. M. (2023a). *What art does: Using philosophy of technology to talk about art*. Rowman & Littlefield.
Wittingslow, R. M. (2023b). On the use of linguistic concepts in design. *Technology and Language, 11*(2), 145–156.
Wittingslow, R. M., & Lehtinen, S. (2024). The use of light installations in architectural reconstruction: New technological solutions for the public sphere. In Z. Somhegyi & L. Giombini (Eds.), *Routledge companion to the philosophy of architectural reconstruction*. Routledge.

**Ryan Mitchell Wittingslow** BA(Hons) Art History, PhD Art History and Philosophy (University of Sydney 2014). 2016–2023 Assistant Professor in Philosophy, University of Groningen, the Netherlands. 2022–2024 Humboldt Research Fellow in Philosophy, TU Darmstadt. Today, Associate Professor in Philosophy, University College Groningen, University of Groningen. Author of *What Art Does: Using Philosophy of Technology to Talk About Art* (Rowman and Littlefield 2023). Co-founder and former editor-in-chief of *The Philosophy of the City Journal*. Research areas: philosophy of technology, aesthetics and philosophy of art, philosophy of design, political philosophy.

# Chapter 21
# Conclusion

**Christelle Didier, Aurélien Béranger, Antoine Bouzin, Hugo Paris, and Jérémie Supiot**

**Abstract** When looking at how values have evolved it is hard to maintain a neutral view of cultural differences and history, without falling into relativism. Values change, it's undeniable. They are limited by time and space, as much as the societies and individuals in which they appear. If universalism exists, i.e., a universal conception of truth, beauty or morality, it is at best a beautiful but impossible ideal, at worse a dangerous illusion that leads democracies into fascism and despotism. Yet, it seems unacceptable to say that gender equity became a matter of justice, or that racism was good then and is now bad. They were important, good, or bad before they were accepted as such. By rejecting universalism, we are exposed to another danger: relativism. Are values the expression of what is valued, or what is valuable? Can we objectively know something about values?

C. Didier (✉)
CIREL, University of Lille, Lille, France
e-mail: christelle.didier@univ-lille.fr

A. Béranger
Costech, Université de Technologie de Compiègne, Compiègne, France
e-mail: aurelien.beranger@utc.fr

A. Bouzin
Centre Émile Durkheim, Bordeaux, France
e-mail: antoine.bouzin@u-bordeaux.fr

H. Paris
ADEF, Aix-Marseille University, Marseille, France

INSA Lyon, Villeurbanne, France
e-mail: hugo.paris@insa-lyon.fr

J. Supiot
Catholic University of Lyon, Lyon, France

LUMSA, Roma, Italia
e-mail: jeremie.supiot@philosophie.ucly.fr



C. Didier et al. (eds.), *Engineering and Value Change*, Philosophy of Engineering and Technology 48, https://doi.org/10.1007/978-3-031-83549-0_21

**Keywords** Value change · Ethics · Philosophy of technology · Environmental philosophy

When looking at how values have evolved it is hard to maintain a neutral view of cultural differences and history, without falling into relativism. Values change, it's undeniable. They are limited by time and space, as much as the societies and individuals in which they appear. If universalism exists, i.e., a universal conception of truth, beauty or morality, it is at best a beautiful but impossible ideal, at worse a dangerous illusion that leads democracies into fascism and despotism. Yet, it seems unacceptable to say that gender equity became a matter of justice, or that racism was good then and is now bad. They were important, good, or bad before they were accepted as such. By rejecting universalism, we are exposed to another danger: relativism. Are values the expression of what is valued, or what is valuable? Can we objectively know something about values?

Those questions may seem too abstract for engineers. After all, we do not expect them to give their philosophical opinion on morality to be able to design technologies. Yet, thanks to the development of engineering studies, it is today quite clear that engineers often encounter ethical issues that need answers. Since artifacts have politics, engineering needs ethics. At the heart of the technical transformations of our modern societies, engineers play a fundamental role in technological controversies involving value conflicts (whether they like it or not). Engineers also contribute to the understanding of what is valued by users and citizens and the evolution of societal values. That is why they need codes of ethics to help them to evaluate what to care about and what they need to take care of (Michelfelder & Jones, Chap. 2). Of course, such codes are insufficient. Engineering schools and universities must also fight the *culture of disengagement* (which is a polite way to say that engineers seem encouraged not to care about ethics). This culture appears to be strong especially in the United States (Graeff, Chap. 3) though it seems to change (slowly) in some countries like France where environmentalist engineers are raising a distinct voice (Bouzin, Chap. 5).

However, caring about ethics is not enough. Even when they care, engineers have important problems to tackle. For example, they must investigate the context-dependency of the values or conceptions of justice they mobilize (Richter, Chap. 6). While enhancing transparency in the use of algorithms and data sets seems to be a top priority for future technologies using Artificial Intelligence (Bogucki, Chap. 7), values such as transparency and privacy (Carter & Felzmann, Chap. 4) might be less transparent and more context-sensitive that they seem to be at first glance. Likewise, the data-driven qualification and quantification of fairness seems to be a good way to "introduce elements of fairness in the policy design processes with the aim of reducing inequalities" (Quaresmini et al., Chap. 11). Yet not all engineers and designers share the same conception of fairness, and are often confronted with value pluralism, for example in international negotiations on biodiversity use and equitable benefit-sharing (Kreiken & McCarthy, Chap. 8). As Van Uffelen (Chap. 10) puts it, "this normative uncertainty stresses the importance of a public debate on justice, fostering value change in the long run." Furthermore, engineering value change in new technologies can be seen as a new form of techno-solutionism (Bögner, Chap. 9). Confronted with such a great diversity of challenging views on values, that depend on multiple conceptions of ethics and justice, engineers are confronted with an epistemic challenge: how can they know that the technologies they use, design, and maintain are really valuable for everyone?

First, engineers might want to explore matters of concern in democratic debates themselves (Supiot et al., Chap. 12). By participating in such debates, they can share their own specific understanding of technological knowledge (Kant & Poznic, Chap. 13) and open their minds to "different forms of knowledge" instead of trying to engineer the planet or the society from their own perspective (Paris et al., Chap. 15). For example, engineers that are stuck in an internalist view of cognition might tend to forget that artifacts and environments can be as manipulative as humans (Spurrett, Chap. 16). The engineers' role is never neutral: even in maintenance they need to understand what it means for an artifact to be functional (Karakas, Chap. 17). In fact, engineers always need to choose what to care about and to take care of, which means prioritizing (epistemic and ethical) values.

Defining objects by what would be their intrinsic function is more ambiguous than it seems. For example, it can be hard to distinguish between natural and artificial functions (Hurshman, Chap. 18). Furthermore, functions are context-sensitive and depend on how artifacts are perceived (Wittingslow, Chap. 20). That is why most artifacts, even when they are using intelligent systems (e.g., Intelligent Compacting Systems), still need humans to be functional (Voordijk et al., Chap. 14). Furthermore, functions, like values, evolve too, that is why we might need a more embedded Value Change Sensitive Design (Srivastava & Bombaerts, Chap. 19).

These reflections plead for us to overcome a narrow functionalist perspective on technology to better encompass the association of materials, matter of concerns, values, usages, and hijackings that constitute technological objects. This broadening also seems necessary to investigate other relationships with technology, especially in the Global South, through different frameworks and perspectives. For many scholars from Anglo-American and Dutch-influenced countries, well represented within fPET, pragmatism appears as a kind of second skin. This approach appears today in the normative position promoted by self-organized professional associations (engineering societies) in their codes of ethics (see Chap. 2). In other parts of the world, including France, where the co-editors of this volume come from, this function has been fulfilled by national laws and greater state involvement. We can hope that comparative approaches such as those developed in the field of engineering studies (e.g., Downey et al., 2015) will nourish this aspect of the discussion.

## 21.1 Outro

Engineering activities lead to political transformations of such a great importance that they challenge our democracies, especially when it comes to evaluate our technologies. Today, decision-making regarding the future of technologies is centralized in only few hands, for which engineers are decisive agents. This raises many tricky questions. Since technological choices are always partly motivated by ideological choices based on changing values, should we (re)politicize engineers and engineering, or on the contrary reinforce their impartiality? How can we think about, integrate, and practice these forms of citizenship education in the training of engineers? What role should citizens play in technoscientific decision-making? How can governments and companies go beyond the application of hard and soft law, and give engineers a sense of responsibility through their training and education? What continuity is there between ethics, standards, and laws? What's more, how can we decentralize decision-making on technological developments to put citizens at the heart of the choices that concern them?

The question of value change also relates to the description of scientific activity, and we might ask whether engineers and scientists are not directly involved in the production of their own values. Should the measurement and evaluation work of scientists be seen as value-generating, or as objectifying an intrinsic value? Does objectification contribute to the recognition of the value of a being, a material or a space? This issue brings us back to important controversies, with a long-standing critique of measurement (e.g., Vatin, 2008) and economistic temptations, i.e., the reduction of social and political life to a succession of cost-benefit calculations. Closer to us, we might mention the discussion in environmental economics surrounding the (non)substitutability of natural capital (weak and strong sustainability) by artificial capital. At stake in this debate is the relevance of compensatory mechanisms, and hence of a capitalist approach to ecological transition (Buttel, 2000; Dauguet, 2021).

As we said before, we do not expect engineers to become philosophers, but the question of their relationship to philosophers and philosophy remains open. Is it necessary and desirable for companies to hire ethicists to work with their engineers? Where does ethics washing begin and how do we integrate political and value conflicts in our understanding of value change? How can we better prepare and assist engineers to handle these issues and to deal with conflicts of values they might encounter? As Carl Mitcham (2019) analyzes, the aims of engineering are not pre-existing to the practice of the profession. These aims are defined in situ, within specific political and economic frameworks, and in the interaction between engineers and the material objects they are responsible to design and maintain. Consequently, as the chapters in this book show, engineers—both as a social group and as individuals—are fully involved with transformations of society, whether they result from environmental degradation, innovations in AI systems and the underlying security issues, democratic institutions, and procedures, and so forth. This raises the question of the involvement of the engineering community in the development of new technologies and accumulation of knowledge and know-how. Engineers are therefore compelled to select desirable values, i.e., to take positions that are necessarily subject to uncertainty and opposition from other social groups.

If social controversies have received much attention in the literature, it seems important to think about how we can integrate into the conversation non-humans that cannot speak for their own. This brings us back to the distinction we proposed earlier: are values the expression of what is valued, or what is valuable? Answering this question may depend on how we define nature, humans, and artifacts and how we understand their relations, in a context where their multiple existence involves exploitation of the two first spheres. If we are to consider that nature no longer exists, as it would have been entirely transformed by the hand of humans, we need to think in the same terms about the care and maintenance of our (techno)natural environment. This implies thinking together the affordances of the earth (Charbonnier, 2021) and those of our artifacts in relation to the dynamics of valorization/valuation. But what should we do with Western (globalized) naturalism, which nonetheless continues to value a fantasized wilderness that no longer exists? The critique of naturalism voiced by Bruno Latour (1993) and Anna Tsing (on *third nature*, see Tsing, 2015) points to the continuity of distinctions between self/other, subject/object, culture/nature, and calls for reflection on other-than-human agentivities. In this respect, is *valuing* a specifically anthropological characteristic? If there can be no seamless separation between our needs and our values, what about non-humans? Is it possible to think of a non-anthropocentric value-sensitive design?

It could be argued that such an extension of the notion of value would contribute to making it lose its meaning, and that, conversely, the concept of nature would enable us to underline and highlight the heteronomy of living things, particularly in comparison with human artifacts. Here again, we need to clarify the meaning of the notion of *value*, between a relational characterization (*what is valued*) and a more absolute moral characterization (*what is valuable*). Whatever the case, investigating the values associated with hybrid artifacts of nature and culture thus appears to be a promising area for further reflection on VSD. In the field of philosophy, this calls for a closer connection between environmental philosophy and the philosophy of technology. Empirically, these value issues could be explored among the designers and "users" of these artifacts (plants, fields, gardens, landscapes). The existential crisis imposed on us by the ecological emergency, with its already dramatic consequences and catastrophic prospects, must in any case lead us to re-examine our Western-centered categories of thought forged in the course of industrial modernity.

## References

Buttel, F. H. (2000). Ecological modernization as social theory. *Geoforum, 31*(1), 57–65.

Charbonnier, P. (2021). *Affluence and Freedom*. Polity Press.

Dauguet, B. (2021). *Mesures contre nature. Mythes et rouages de la compensation écologique*. Grevis.

Downey, G. L., Lucena, J., & Mitcham, C. (2015). Engineering ethics and engineering identities: Crossing national borders. In S. Christensen, C. Didier, A. Jamison, M. Meganck, C. Mitcham,

& B. Newberry (Eds.), *Engineering identities, epistemologies and values. Philosophy of engineering and technology* (Vol. 21). Springer. https://doi.org/10.1007/978-3-319-16172-3_5
Latour, B. (1993). *We have never been modern*. Harvard University Press.
Mitcham, C. (2019). *Steps toward a philosophy of engineering. Historico-philosophical and critical essays*. Rowman & Littlefield.
Tsing, A. (2015). *The mushroom at the end of the world: On the possibility of life in capitalist ruins*. Princeton University Press.
Vatin, F. (2008). L'esprit d'ingénieur : pensée calculatoire et éthique économique. *Revue Française de Socio-Économie, 1*(1), 131–152.

**Christelle Didier** BS Electrochemistry Eng., MA in Education, PhD in Sociology. 1993–2023 Assistant Professor in Engineering Ethics, Lille Catholic University, France. Today, Education Department, Lille University. Author of Ethique industrielle (1998); Penser l'éthique des ingénieurs (2008), Les ingénieurs et l'éthique (2008); several POET Volumes with Steen Hyldgaard Christensen (2015, 2015, 2019, 2022). Co-founder of collectif EPSI (*Études Pluridisciplinaires sur l'Ingénierie*). Research areas: Engineering cultures, identities and ethics, including historical and comparative perspective.

**Aurélien Béranger** BA Humanities and technology, Computer Science Eng., PhD student in Communication sciences and Philosophy of technology at the Costech Lab of the UTC (Université de Technologie de Compiègne). The focus of his thesis is on the low-tech movement in France. Member of Collectif EPSI (*Études Pluridisciplinaires sur l'Ingénierie*—Multi-disciplinary engineering studies). His main research interests include low-tech, technology-oriented social movements, appropriate technologies, engineering, open source, makers, and degrowth studies.

**Antoine Bouzin** Hydraulics and fluid mechanics Eng., Master of Sociology, PhD Student in Sociology and Political Science (Centre Émile Durkheim, France). Co-founder of Collectif EPSI (*Études Pluridisciplinaires sur l'Ingénierie*—Multi-disciplinary engineering studies) and Collectif Biomasse & SHS (Biomass studies from human and social sciences). Research areas: Engineering studies, social movements, public action and policies, environment, expertise.

**Hugo Paris** MA Architecture (ENSA Lyon, 2017) & MSc Civil Engineering and Urbanism (INSA Lyon, 2019), PhD Educational sciences (Aix-Marseille University, 2024). Co-founder of collectif EPSI (*Études Pluridisciplinaires sur l'Ingénierie*) and Associate Editor for the European Journal of Engineering Education. 2024–2025: Post Doc researcher at INSA Lyon, UNESCO Chair Training engineers for the socio-ecological transition. Research areas: Superior education for sustainability, teacher continuous training, curriculum studies, anthropological approach to didactics (TAD).

**Jérémie Supiot** PhD student in philosophy at the Confluence: Sciences and Humanities Research Unit (EA 1598) of UCLy (Lyon Catholic University) and LUMSA University at Rome. His thesis focuses on ethical empowerment at the interface between society and the university. It is supervised by Mathieu Guillermin (UCLy, Lyon) and Stefano Biancu (LUMSA, ROME) and is part of the Activity 4 of the NHNAI project. His main research topics are epistemology and engineering ethics. He co-organizes the ETIUS seminars at the Université Catholique de Lyon. Co-founder of collectif EPSI (*Études Pluridisciplinaires sur l'Ingénierie*), which organizes the Cultures d'Ingénieurs et Formation seminars.

The manufacturer's authorised representative in the EU is Springer Nature Customer Service Centre GmbH, Europaplatz 3, 69115 Heidelberg, Germany. If you have any concerns regarding our products, please contact ProductSafety@springernature.com

Printed and bound by CPI Group (UK) Ltd, Croydon, CR0 4YY

30/09/2026

02225589-0014